The Practical Approach in Chemistry Series

SERIES EDITORS

L. M. Harwood
Department of Chemistry
University of Reading

C. J. Moody
Department of Chemistry
Loughborough University of Technology

The Practical Approach in Chemistry Series

Organocopper reagents
Edited by Richard J. K. Taylor

Macrocycle synthesis
Edited by David Parker

High pressure techniques in chemistry and physics
Edited by W. B. Holzapfel and N. S. Isaacs

Preparation of alkenes
Edited by Jonathan M. J. Williams

Macrocycle Synthesis
A Practical Approach

Edited by
DAVID PARKER

Department of Chemistry
University of Durham, UK

OXFORD   NEW YORK   TOKYO
OXFORD   UNIVERSITY   PRESS
1996

Oxford University Press, Walton Street, Oxford OX2 6DP
Oxford New York
Athens Auckland Bangkok Bombay
Calcutta Cape Town Dar es Salaam Delhi
Florence Hong Kong Istanbul Karachi
Kuala Lumpur Madras Madrid Melbourne
Mexico City Nairobi Paris Singapore
Taipei Tokyo Toronto
and associated companies in
Berlin Ibadan

Oxford is a trade mark of Oxford University Press

Published in the United States
by Oxford University Press Inc., New York

© Oxford University Press, 1996

All rights reserved. No part of this publication may be
reproduced, stored in a retrieval system, or transmitted, in any
form or by any means, without the prior permission in writing of Oxford
University Press. Within the UK, exceptions are allowed in respect of any
fair dealing for the purpose of research or private study, or criticism or
review, as permitted under the Copyright, Designs and Patents Act, 1988, or
in the case of reprographic reproduction in accordance with the terms of
licences issued by the Copyright Licensing Agency. Enquiries concerning
reproduction outside those terms and in other countries should be sent to
the Rights Department, Oxford University Press, at the address above.

This book is sold subject to the condition that it shall not,
by way of trade or otherwise, be lent, re-sold, hired out, or otherwise
circulated without the publisher's prior consent in any form of binding
or cover other than that in which it is published and without a similar
condition including this condition being imposed
on the subsequent purchaser.

Users of books in the Practical Approach in Chemistry Series are advised that prudent
laboratory safety procedures should be followed at all times. Oxford
University Press makes no representation, express or implied, in respect of
the accuracy of the material set forth in books in this series and cannot
accept any legal responsibility or liability for any errors or omissions
that may be made.

A catalogue record for this book is available from the British Library

Library of Congress Cataloging in Publication Data
Macrocycle synthesis : a practical approach / edited by David Parker.
(The practical approach in chemistry series)
Includes bibliographical references.
1. Macrocyclic compounds–Synthesis. I. Parker, David.
II. Series.
QD400.5.S95M33 1996 547'.5–dc20   95-38369
ISBN 0 19 855841 4 (Hbk)
ISBN 0 19 855840 6 (Pbk)

Typeset by Footnote Graphics, Warminster, Wilts
Printed in Great Britain by
Bookcraft (Bath) Ltd, Midsomer Norton, Avon

Preface

This is the second volume of The Practical Approach in Chemistry Series. This series aims to provide detailed and accessible laboratory guides which are suitable for researchers who are not necessarily familiar with the area in question. The authors are respected authorities who detail key experimental procedures in a step-by-step format. Such procedures contain information on reaction equipment and conditions, work-up procedures, and other key experimental details that are 'tricks of the trade'. Whilst a comprehensive coverage of the literature is not provided, each chapter does contain background information and leading references that will aid further reading.

This book is intended therefore to be an introductory practical guide to the synthesis of macrocyclic compounds. In a relatively short text such as this, it is not possible to cover all of the major classes of synthetic macrocycles. For details of the synthesis of porphyrins, phthalocyanins, cryptophanes, and cavitands, for example, other sources of information will have to be examined. Nevertheless, the coverage given to selected examples of nitrogen, sulfur and oxygen containing mono- and bicyclic molecules, calixarenes, spherands, hemispherands, calix-spherands, torands and catenands provides a reasonable balance between certain older 'classical' ligands and some more recently defined and more elaborate macrocycles. Several of the protocols are suitable for inclusion in advanced undergraduate practical classes, while others will appeal to the specialist complexation chemist who is keen to explore the coordination chemistry of these versatile molecules.

Creative chemistry must begin with synthesis. The chemical synthesis of macrocyclic ligands that can function as selective receptors for ions and neutral molecules is a central theme of modern supramolecular chemistry. Indeed the synthesis of these tailored receptors can be a real challenge, but it is also great fun! I hope that this book conveys some of the enthusiasm of the authors for their subject.

I am indebted to the authors of the various chapters for their excellent contributions and thank Dr Geoff Lawrance for providing details of Protocol 6 (Chapter 2).

Durham D. P.
September 1995

Dedicated to those who dare to be wise.

Contents

List of contributors	xi
Abbreviations	xiii

1. Aza crowns 1
David Parker

1. Introduction	1
2. Non-templated syntheses	1
3. Templated reactions	17
References	23

2. Aza-oxa crowns 25
David Parker

1. Introduction	25
2. Tosylamide-based cyclisations	28
3. Alkylation at primary and secondary nitrogen	39
4. Conclusion	43
References	46

3. Thia, oxa-thia and aza-thia crowns 49
David Parker

1. Introduction	49
2. Cyclisation reactions forming carbon–sulfur bonds	54
3. Cyclisation reactions forming carbon–nitrogen bonds	61
References	70

4. Crown ethers 71
David B. Amabilino, Jon A. Preece and J. Fraser Stoddart

1. Introduction	71
2. Synthesis of 1,1',4,4'-tetra-*O*-benzyl-2,2':3,3'-bis-*O*-oxydiethylene-di-L-threitol, a chiral crown ether	71
3. Bis-*p*-phenylene-34-crown-10 synthesis—a receptor for π-electron-deficient aromatics	82
References	90

5. Cryptands
Bernard Dietrich

1. Introduction — 93
2. Synthesis of the cryptand[2.2.2] — 93
3. Synthesis of bis-tren — 106
4. Concluding remarks — 118
References — 118

6. Torands
Thomas W. Bell and Julia L. Tidswell

1. Introduction — 119
2. Oxidation of octahydroacridine — 121
3. Carbon functionalization of octahydroacridine — 123
4. Oxidation of alcohol **8** to ketone **9** — 128
5. Formation of a pyridine ring fused to two octahydroacridine units — 130
6. Unmasking carbonyl groups in **12** by ozonolysis of the benzylidene groups in **11** — 133
7. Functionalization of the third octahydroacridine unit — 136
8. Torand cyclization — 140
References — 142

7. Calixarenes
Arturo Arduini and Alessandro Casnati

1. Introduction — 145
2. Synthesis of calixarenes — 146
3. Functionalization at the lower rim (phenolic OH groups) — 154
4. Functionalization at the upper rim (aromatic nuclei) — 164
References — 172

8. Spherands, hemispherands and calixspherands
Willem Verboom and David N. Reinhoudt

1. Introduction — 175
2. Spherands — 177
3. Hemispherands — 181
4. Calixspherands — 199
References — 205

9. Transition metal-templated formation of [2]-catenanes and [2]-rotaxanes   207
J.-C. Chambron, S. Chardon-Noblat, C. O. Dietrich-Buchecker, V. Heitz and J.-P. Sauvage

    1. Introduction   207
    2. Transition metal-templated synthesis of catenanes   208
    3. Transition metal-templated synthesis of rotaxanes   225
    References   246

A1 List of suppliers   247
Index   251

Contributors

DAVID B. AMABILINO
Department of Chemistry, School of Chemistry, University of Birmingham, Edgbaston, Birmingham B15 2TT, UK.

ARTURO ARDUINI
Dipartimento di Chimica Organica e Industriale, Università degli Studi di Parma, Viale delle Scienze, 1–43100 Parma, Italy.

THOMAS W. BELL
Department of Chemistry, University of Nevada, Reno, NV 89557–0020, USA.

ALESSANDRO CASNATI
Dipartimento di Chimica Organica e Industriale, Università degli Studi di Parma, Viale delle Scienze, 1–43100 Parma, Italy.

JEAN-CLAUDE CHAMBRON
Laboratoire de Chimie Organo-Minérale, Université Louis Pasteur de Strasbourg, 4 rue Blaise Pascal, 67000 Strasbourg, France.

S. CHARDON-NOBLAT
Laboratoire de Chimie Organo-Minérale, Université Louis Pasteur de Strasbourg, 4 rue Blaise Pascal, 67000 Strasbourg, France.

BERNARD DIETRICH
Laboratoire de Chimie Supramoleculaire, Université Louis Pasteur de Strasbourg, 4 rue Blaise Pascal, 67000 Strasbourg, France.

CHRISTIANE O. DIETRICH-BUCHECKER
Laboratoire de Chimie Organo-Minérale, Université Louis Pasteur de Strasbourg, 4 rue Blaise Pascal, 67000 Strasbourg, France.

V. HEITZ
Laboratoire de Chimie Organo-Minérale, Université Louis Pasteur de Strasbourg, 4 rue Blaise Pascal, 67000 Strasbourg, France.

DAVID PARKER
Department of Chemistry, University of Durham, South Road, Durham DH1 3LE, UK.

Contributors

JON A. PREECE
Department of Chemistry, School of Chemistry, University of Birmingham, Edgbaston, Birmingham B15 2TT, UK.

DAVID N. REINHOUDT
Laboratory of Organic Chemistry, Faculty of Chemical Technology, University of Twente, PO Box 217, 7500 AE Enschede, The Netherlands.

JEAN-PIERRE SAUVAGE
Laboratoire de Chimie Organo-Minérale, Université Louis Pasteur de Strasbourg, 4 rue Blaise Pascal, 67000 Strasbourg, France.

J. FRASER STODDART
Department of Chemistry, School of Chemistry, University of Birmingham, Edgbaston, Birmingham B15 2TT, UK.

JULIA L. TIDSWELL
Department of Chemistry, University of Nevada, Reno, NV 89557–0020, USA.

WILLEM VERBOOM
Laboratory of Organic Chemistry, Faculty of Chemical Technology, University of Twente, PO Box 217, 7500 AE Enschede, The Netherlands.

Abbreviations

ACS	American Chemical Society specification
acac	acetylacetonate
BHT	2,6-di-*t*-butylphenol
cyclam	1,4,8,11-tetraazacyclotetradecane
cyclen	1,4,7,10-tetraazacyclododecane
DMAP	4-dimethylaminopyridine
DME	dimethoxyethane
DMF	dimethylformamide
FW	formula weight
GLC	gas–liquid chromatography
HPLC	high performance liquid chromatography
IR	infra-red
LDA	lithium diisopropylamide
Ms	methanesulfonyl
NMR	nuclear magnetic resonance
PVC	polyvinylchloride
R_f	retention factor (TLC)
t_R	retention time (HPLC)
THF	tetrahydrofuran
TLC	thin layer chromatography
TMEDA	N,N,N′,N′ – tetramethylethylenediamine
tren	tris(2-aminoethyl)amine
Ts	p-toluenesulfonyl

1

Aza crowns

DAVID PARKER

1. Introduction

Macrocyclic polyamines are versatile ligands that form well-defined complexes with a wide range of metal ions.[1,2] The selectivity and stability of metal complex formation is a sensitive function of the number of nitrogen binding sites, their relative disposition and the conformation of the macrocyclic ligand. The importance of this domain of coordination chemistry is reflected in the growing number of applications of functionalised aza crowns. For example, selectivity in complex formation is being harnessed in the development of polymer-supported aza crowns for the separation of metal ions,[3] while the kinetic stability of many of the metal complexes has permitted the development of radiolabelled complexes for *in vivo* use.[4] Protonated polyamines bind strongly to charge dense anions and selectivity in the binding of mono- and polyphosphate anions has been defined.[5]

The early syntheses of macrocyclic polyamines[6] were reported before Pedersen discovered the poly-oxa crown ethers. Representative examples of these early ligands include 1,4,7,10-tetraazacyclododecane, **1** (or [12]-N_4 which is an acronym used when referring to it), the [14]-N_4 analogue, **2** (commonly called 'cyclam'), and the medium-ring triazacyclononane, **3** or [9]-N_3. Synthetic methodology has been developed that now allows the preparation of macrocyclic polyamines using either templated or non-templated strategies.

2. Non-templated syntheses

In 1954, Stetter and Roos[7] reported that the condensation of terminal halides with bis-sulfonamide sodium salts proceeded under conditions of high dilution to give moderate yields of macrocyclic sulfonamides. Twenty years later, in a modification of this method, it was found that the reaction may be undertaken in a dipolar aprotic solvent, such as DMF, obviating the need for high dilution conditions.[8] The reaction is now mostly commonly effected by reaction of the dianion of a bis-toluenesulfonamide with a suitable bis-tosylate or bis-mesylate in anhydrous DMF. The tosylamide salt may be generated prior to cyclisation by reaction with NaH in DMF or NaOEt in ethanol, or may be formed *in situ* when the reaction is performed in DMF in the presence of caesium carbonate. The toluenesulfonyl group serves a dual purpose in these reactions. Not only does it render the secondary NH proton sufficiently acidic to permit salt formation under mild conditions, but also it acts as a nitrogen-protecting group allowing monoalkylation at the nitrogen. It may be removed by a variety of different methods, including hot concentrated sulfuric acid, HBr–AcOH in the presence of phenol, or reductive methods including Li/l. NH_3/THF–EtOH and sodium naphthalenide. Further examples of such reactions are given in Chapter 2, Section 2.

Protocol 1.
Synthesis of 1,4,7,10-tetraazacyclododecane (Scheme 1.1)

Caution! Carry out all procedures in a well-ventilated hood, and wear disposable vinyl or latex gloves and chemical-resistant safety goggles.

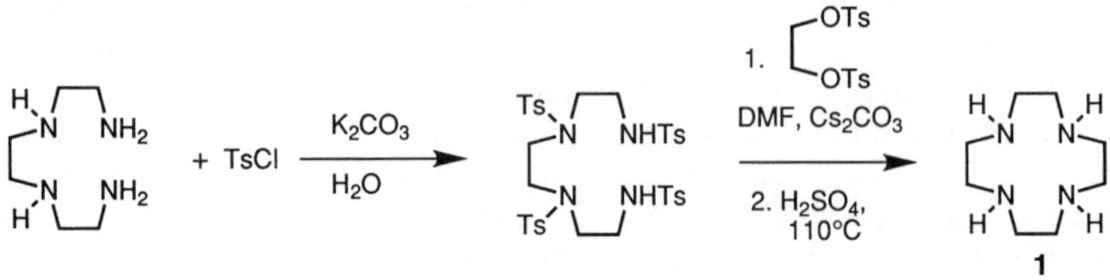

Scheme 1.1

The following procedure is representative of a co-cyclisation of a bis-toluenesulfonamide with a bis-toluenesulfonate followed by tosyl deprotection with sulfuric acid.

Equipment
- Double-necked round-bottomed flasks, 1 L and 2 L
- Water-cooled condenser
- Sintered filter funnel, porosity 3
- pH indicator strips (BDH-Merck)
- Pressure-equalising addition funnel, 250 mL
- Separating funnels, 500 mL and 250 mL
- Conical flasks, 250 mL and 500 mL
- Filter funnel
- Magnetic stirrer bar
- Gas line adaptor
- Thermostatted hot-plate stirrer
- Source of dry nitrogen or argon
- Rotary evaporator
- Oil bath
- Single-necked round-bottomed flasks, 50 mL and 500 mL
- Whatman No. 1 filter papers

1: Aza crowns

Materials

• Triethylenetetramine, 10 g, 68.4 mmol	irritant
• Fine mesh potassium carbonate, 31.3 g, 429 mmol	harmful by inhalation
• p-Toluenesulfonyl chloride, 57.5 g, 302 mmol	irritant, toxic
• Anhydrous dimethylformamide, 800 mL	irritant, harmful by inhalation
• 1,2-Bis(p-toluenesulfonato)ethane, 21.8 g, 59.2 mmol	irritant, toxic
• Dichloromethane, 300 mL	harmful by inhalation
• Methanol, 100 mL	flammable
• Diethyl ether, 100 mL	flammable, irritant
• Toluene, 540 mL	flammable
• Ethanol, 150 mL	flammable
• Chloroform, 100 mL	harmful by inhalation
• Concentrated sulfuric acid, 25 mL	corrosive, toxic
• Potassium hydroxide pellets, 50 g	corrosive, toxic
• Anhydrous potassium carbonate (drying agent), 5 g	harmful by inhalation

1. Charge a double-necked round-bottomed flask (1 L) with potassium carbonate (14.6 g, 308 mmol), triethylenetetramine (10 g, 68.4 mmol) and water (400 mL), and add a large Teflon-coated magnetic stirrer bar. Place the flask in an oil bath, attach a water-cooled condenser and heat to 80°C, while stirring the mixture vigorously.

2. Add p-toluenesulfonyl chloride (57.5 g, 302 mmol) in small portions as a solid over a period of 3 h. Heat and stir the mixture overnight, and then allow the mixture to cool when a colourless precipitate is evident. Filter the solid on a sintered filter funnel, and wash the solids in turn with water (3 × 50 mL), methanol (3 × 50 mL) and finally diethyl ether (3 × 50 mL). Dry the solid under high vacuum (<0.1 mmHg, 60°C) to yield the acyclic tetratosylamide (44.8 g, 86%), m.p. 211–212°C.

3. Dry a double-necked round-bottomed flask (2 L) in an electric oven for 1 h, and allow it to cool under a gentle flow of nitrogen gas. Prepare a solution of 1,2-bis(p-toluenesulfonato)ethane (21.8 g, 59.3 mmol) in anhydrous DMF (100 mL) in an oven-dried pressure-equalising addition funnel. Charge the double-necked flask with fine mesh anhydrous potassium carbonate (16.7 g, 121 mmol), the tetratosylamide prepared above (44.4 g, 58.2 mmol) and dry DMF (700 mL). Add a magnetic stirrer bar and stir the mixture vigorously at room temperature under nitrogen. Add the contents of the addition funnel slowly over a period of 3–4 h, and continue stirring overnight.

4. Heat the mixture to 70°C for 3 h, and then distil off the DMF under reduced pressure (<0.5 mmHg, bath temperature 70°C). Take up the residue in dichloromethane (100 mL) and water (100 mL), transfer the contents of the flask to a separating funnel (500 mL) and separate the organic layer. Wash the aqueous layer with dichloromethane (2 × 50 mL) and dry the combined organic extracts (dry K_2CO_3). Filter the mixture and remove the solvent on a rotary evaporator.

5. Take up the residual solids in hot toluene (500 mL), filter if necessary and allow the solution to cool slowly to room temperature. A white precipitate

Protocol 1. *Continued*

is deposited. Collect this solid by filtration on a sintered filter funnel (porosity no. 3), wash with cold toluene (2 × 20 mL) and dry the solid under vacuum (0.1 mmHg), to yield the cyclic tetratosylamide, m.p. >260°C, 27.8 g (61%).

6. Charge a dry single-necked round-bottomed flask (50 mL) with the cyclic tetratosylamide (15.6 g, 19.8 mmol) and add conc. sulfuric acid (25 mL). Heat the mixture under nitrogen for 40 h at 110°C (thermostatted bath temperature) to give a very dark coloured solution.

7. Pour the mixture into a conical flask (250 mL) and cool the flask in an ice bath. *Slowly* add water (20 mL) and carefully add potassium hydroxide pellets, while swirling the flask in the ice bath. Add sufficient pellets so that the pH of the viscous aqueous mixture is greater than 13 (test with pH papers). Add ethanol (150 mL) and filter the mixture under reduced pressure through a sintered filter funnel attached to a Buchner flask. Wash the residue with ethanol (3 × 10 mL), and evaporate the combined aqueous–alcoholic extracts on a rotary evaporator.

8. Take up the yellow solid residue in the minimum volume of dilute hydrochloric acid (0.1 mol dm^{-3}, *c.* 25 mL). Transfer the solution to a separating funnel (250 mL) and add dichloromethane (30 mL). Separate the aqueous layer and wash it twice more with dichloromethane (2 × 30 mL). Raise the pH of the aqueous layer to 13 by adding potassium hydroxide pellets (test with pH papers). Extract the aqueous layer with chloroform (5 × 20 mL), combine the organic extracts and dry them over anhydrous potassium carbonate. Filter the mixture and remove the solvent on a rotary evaporator to yield a colourless solid which is dried under vacuum (0.1 mmHg) at room temperature (NB do not heat, to avoid sublimation). The product (2.71 g, 79%) is sufficiently pure to be used in this form, m.p. 113–114°C.

A large number of aza crowns are made by these routes. Stepwise addition of CH$_2$CH$_2$NHTs groups occurs readily by reaction of the appropriate toluenesulfonamide with 2-bromoacetamide followed by borane reduction and *N*-tosylation. If a trimethylene spacer is desired, then reaction of the tosylamide with acrylonitrile followed by reduction and tosylation gives good yields of the desired product (Scheme 1.2). An example of such a sequence is provided by the synthesis of the symmetrical [24]-N$_6$ cycle, **4** (Scheme 1.3).[9] In this particular case, introduction of the desired C$_3$ spacer was undertaken on the electrophilic component by conjugate addition of the tosylamide with methyl acrylate followed by reduction with LiAlH$_4$ and mesylation.

It is usually straightforward to poly-*N*-acylate or *N*-alkylate an aza crown. For example *N*-methylation of ring secondary amines is conveniently achieved using the Eschweiler–Clarke reaction involving condensation of the

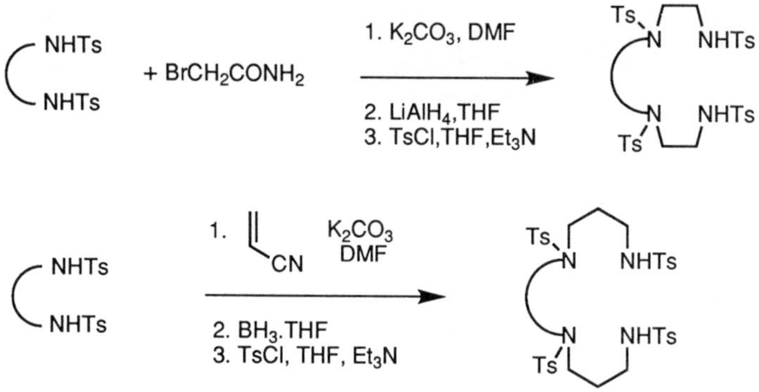

1: Aza crowns

Scheme 1.2

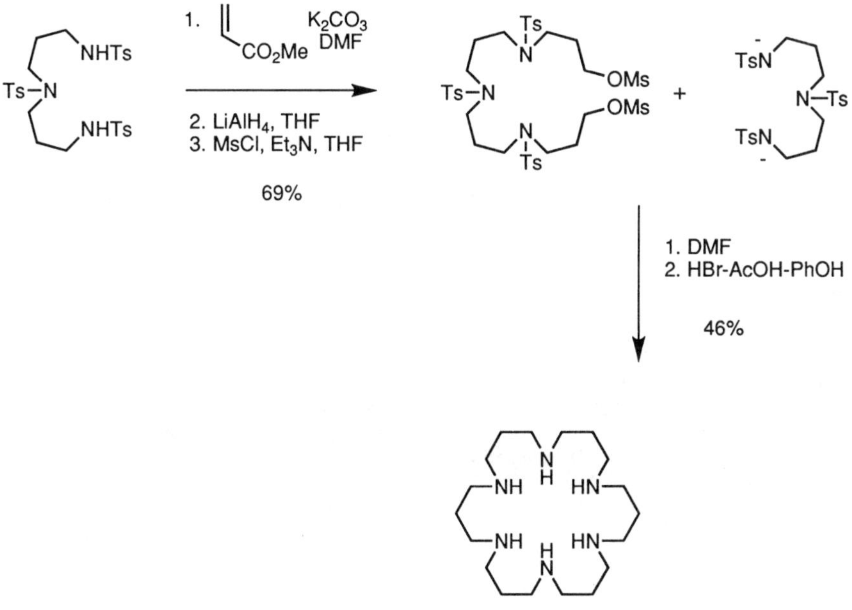

Scheme 1.3

amine with formaldehyde in the presence of formic acid. It is more difficult to selectively mono- or difunctionalise an aza crown (in the latter case with regiochemical control). Nevertheless, strategies have evolved that permit these transformations to be effected, particularly with triaza and tetraaza macrocycles. For example, the monosubstitution of [12]-N$_4$, **1**, may be effected via the intermediacy of the octahedral molybdenum tricarbonyl complex. In this organometallic complex, the fourth ring nitrogen is not metal bound and so is able to react selectively with alkylating agents. Following oxidative decomplexation, the monosubstituted ligand may be isolated in good yield.[10]

Protocol 2.
Synthesis of 1-(dimethylcarbamoylmethyl)-1,4,7,10-tetraazacyclododecane (Scheme 1.4)

Caution! Carry out all procedures in a well-ventilated hood, and wear disposable vinyl or latex gloves and chemical-resistant safety goggles.

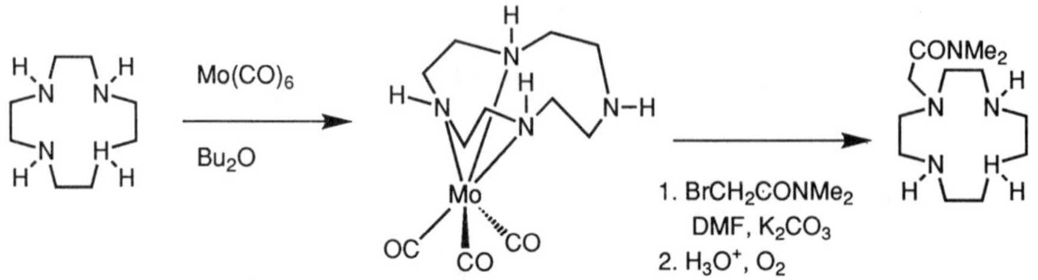

Scheme 1.4

This procedure is representative of the selective mono-N-functionalisation of aza-cycles containing more than three ring nitrogen atoms.

Equipment

- Thermostatted hot-plate stirrer
- Oil bath
- Magnetic stirrer bar
- 2 small Schlenck tubes, 250 mL with B-19 female joints
- Sintered glass filter (porosity 3) equipped with two B-19 male joints at each end
- Source of dry nitrogen or argon
- Separating funnel, 100 mL
- Rotary evaporator
- Centrifuge
- Single-necked round-bottomed flask, 500 mL
- Water-cooled condenser
- pH indicator strips (BDH-Merck)
- Filter funnel
- Whatman No. 1 filter papers
- Centrifuge tubes, 20 mL
- Septum

Materials

- 1,4,7,10-tetraazacyclododecane, 1.07 g, 6.21 mmol — irritant
- Molybdenum hexacarbonyl, 1.64 g, 6.21 mmol — toxic
- Dibutyl ether, 100 mL — irritant, flammable
- 2-Bromo-N,N-dimethylethanamide, 1.03 g, 6.21 mmol — irritant, toxic
- Anhydrous potassium carbonate (fine mesh),[a] 3 g
- Anhydrous dimethylformamide, 40 mL — irritant, harmful by inhalation
- 10% (v/v) hydrochloric acid, 20 mL — corrosive, toxic
- Potassium hydroxide pellets, 10 g — corrosive, toxic
- Anhydrous potassium carbonate (drying agent), 2 g
- Dichloromethane, 250 mL — harmful by inhalation

1. Dry a small Schlenk tube (250 mL) in an electric oven (120°C) for 2 h and then allow it to cool down to room temperature under argon gas. Charge

1: Aza crowns

the tube with 1,4,7,10-tetraazacyclotetradecane (1.07 g, 6.21 mmol) (prepared as described in Protocol 1), molybdenum hexacarbonyl (1.64 g, 6.21 mmol) and a magnetic stirrer bar, and add dibutyl ether (100 mL). Equip the Schlenk tube with a water-cooled condenser attached to the gas line via a gas inlet valve. Purge the apparatus with dry argon.

2. Cool the Schlenk tube under argon down to −78°C (dry ice–isopropanol slush bath), and then admit the tube to a vacuum (*c.* 0.1 mmHg). Pump for 60 s, then admit argon gas and allow the apparatus to warm up to room temperature. Repeat this 'freeze-thaw' process twice more. Heat the stirred mixture at 160°C under argon for 2 h. An abundant yellow precipitate of the molybdenum tricarbonyl complex develops.

3. Cool the Schlenk tube to room temperature and quickly attach a sintered glass filter surmounted by another Schlenk tube filled with argon gas (Figure 1.1). Carefully rotate the assembly of Schlenk tubes through 180° and allow the solid yellow precipitate to collect at the sinter, under argon (i.e. both taps of the Schlenk tubes are open to argon). Discard the filtrate.

4. Transfer the yellow solid to a clean, dry Schlenk tube under argon (simply replace the original reaction flask by the dry, clean one and tap the solid down into the Schlenk tube). Against a positive stream of argon gas, add fine-mesh potassium carbonate (3 g) and dry dimethylformamide (40 mL) by syringe through a septum. Quickly add 2-bromo-*N,N*-dimethylethanamide (1.03 g, 6.21 mmol) and a magnetic stirrer bar and purge the reaction mixture with inert gas by three freeze-thaw cycles (as described in point 2 above).

5. Heat to 70°C and stir the mixture for 1 h under argon. Distil off the DMF under vacuum (*c.* 0.2 mmHg, bath temperature 70°C) and take up the residue in hydrochloric acid solution (10% v/v; 20 mL). Stir the mixture at room temperature in air for 18 h. Transfer the mixture to a centrifuge tube (20 mL) and centrifuge for 5 min (≥1100 r.p.m.). Decant or carefully pipette off the supernatant liquid and transfer to a separating funnel. Wash the aqueous layer with dichloromethane (3 × 25 mL), basify the aqueous layer to pH >13 with potassium hydroxide pellets (test the pH) and extract the aqueous layer with dichloromethane (5 × 20 mL). Combine the organic extracts, dry over anhydrous potassium carbonate (*c.* 1 g) and filter the mixture. Evaporate the solvent using a rotary evaporator and dry the product under high vacuum (<0.1 mmHg) to yield a colourless oil (1.24 g, 79%) that is best characterised by carbon-13 NMR: δ_C(CDCl$_3$): 170.3 (carbonyl); 56.9 (NCH$_2$CO); 45.1, 45.7, 45.9 and 46.9 (CH$_2$N, ring); 35.1, 36.5 (NCH$_3$).

[a] Fine mesh potassium carbonate is commercially available (Aldrich, 34,782-5).

For the selective preparation of 1,7-di-*N*-substituted [12]-N$_4$ derivatives, a simpler method is available. In the diprotonated ligand, the two charged

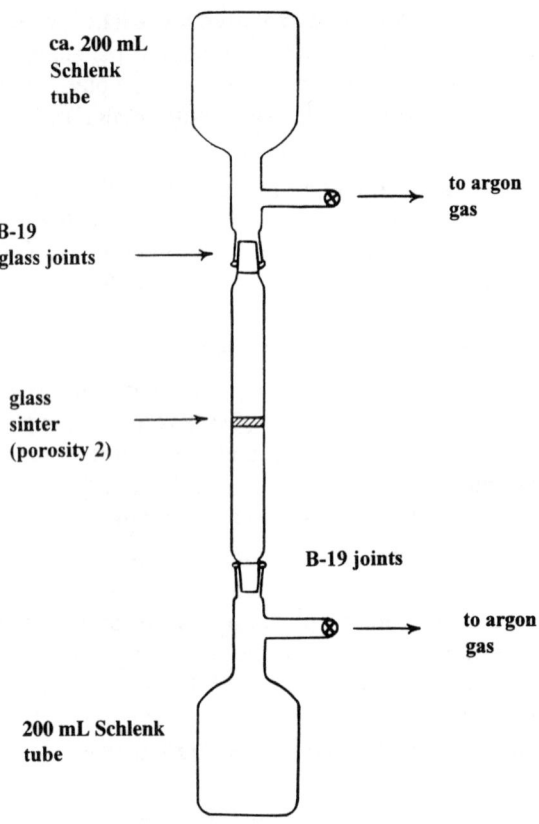

Fig. 1.1 Apparatus for filtration under inert gas.

nitrogens prefer to be 1,7-related to minimise Coulombic repulsion. Acylation at nitrogen (e.g. with chloroformate esters) under relatively acidic conditions therefore gives exclusive formation of the 1,7-derivatised carbamate product **5** (Scheme 1.5). Similarly direct reaction with toluenesulfonyl chloride in pyridine affords the di-toluenesulfonamide with the same regiochemical control.[11]

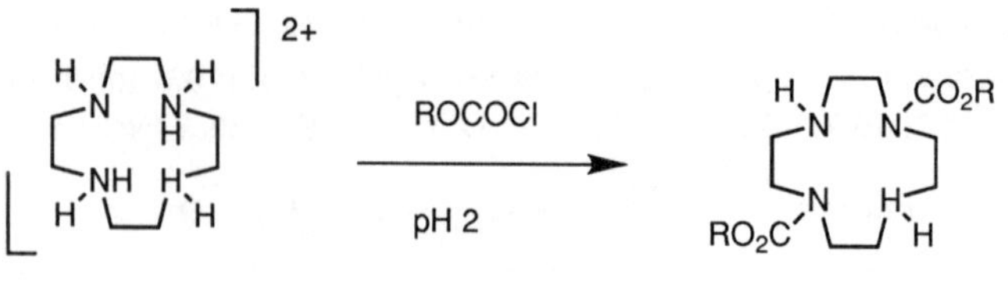

Scheme 1.5

Protocol 3.
Synthesis of 1,7-bis(toluenesulfonyl)-1,4,7,10-tetraazacyclododecane (Scheme 1.6)

Caution! Carry out all procedures in a well-ventilated hood, and wear disposable vinyl or latex gloves and chemical-resistant safety goggles.

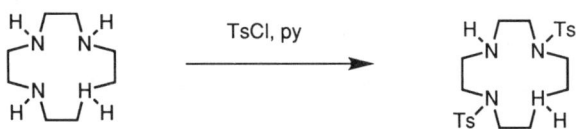

Scheme 1.6

Equipment
- Single-necked round-bottomed flask, 250 mL
- Magnetic stirrer
- Pressure-equalising addition funnel, 50 mL
- Magnetic stirrer bar
- Sintered glass filter funnel (porosity 2)
- Buchner flask, 250 mL
- Vacuum pump
- Drying tube (CaCl$_2$)

Materials
- Dry pyridine, 130 mL   **flammable, harmful by inhalation**
- Toluenesulfonyl chloride, 7.62 g, 40 mmol   **irritant, corrosive**
- 1,4,7,10-Tetraazacyclododecane, 3.44 g, 20 mmol   **irritant**
- Saturated aqueous potassium carbonate solution
- Methanol, 50 mL   **flammable**

1. Charge a dry single-necked round-bottomed flask (250 mL) equipped with a stirrer bar with recrystallised tosyl chloride[a] (7.62 g, 40 mmol) and dry pyridine[b] (100 mL). Attach a pressure-equalising addition funnel (50 mL) fitted with a nitrogen inlet adaptor and flush the apparatus with dry nitrogen gas. Add to the addition funnel a solution of 1,4,7,10-tetraazacyclododecane (3.44 g, 20 mmol) in dry pyridine (30 mL). Cool the reaction flask to 0°C in an ice bath.

2. Add the contents of the addition funnel slowly, under nitrogen, over a period of 15 min. Allow the mixture to warm up to room temperature slowly and stir for a further 3 h. Remove the pyridine on a rotary evaporator, and treat the residue with water (35 mL), vigorously stirring the mixture for 90 min.

3. Filter the product bis-toluenesulfonamide with gentle suction using a sintered glass filter funnel mounted on a Buchner flask (250 mL). Wash the product thoroughly with water (2 × 25 mL), saturated potassium carbonate solution (2 × 25 mL), water (2 × 20 mL) and methanol (2 × 25 mL). Dry the product in an electric oven (120°C) to yield a colourless solid, m.p. 234–5°C, (8.74 g, 91%).

[a] Tosyl chloride may be quickly recrystallised from CHCl$_3$–petroleum ether.
[b] Pyridine of sufficient dryness may be purchased from Aldrich (27,040-7).

Similar strategies may be adopted for the selective functionalisation of [14]-N$_4$, **2** (cyclam).[12,13] Reaction of **2** with toluenesulfonyl chloride in pyridine gives predominantly the tri-toluenesulfonamide **6**, although the use of low temperature and a deficiency of the acid chloride leads to a moderate yield of the 1,8-bis-toluenesulfonamide **7** (Scheme 1.7).

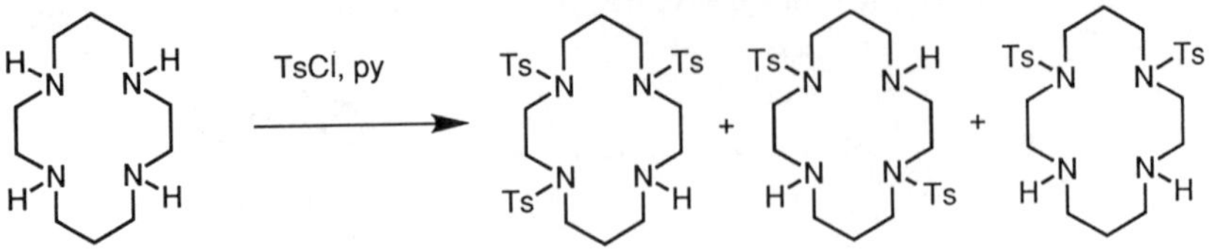

Scheme 1.7

In many cases, differential protection of ring nitrogens is a direct consequence of the cyclisation methodology used. The condensation of malonate derivatives with linear polyamines in MeOH or EtOH gives the diamide cyclisation product in moderate yield. In such a sequence two of the ring nitrogens are distinguished (Scheme 1.8),[14] allowing subsequent selective *N*-alkylation.

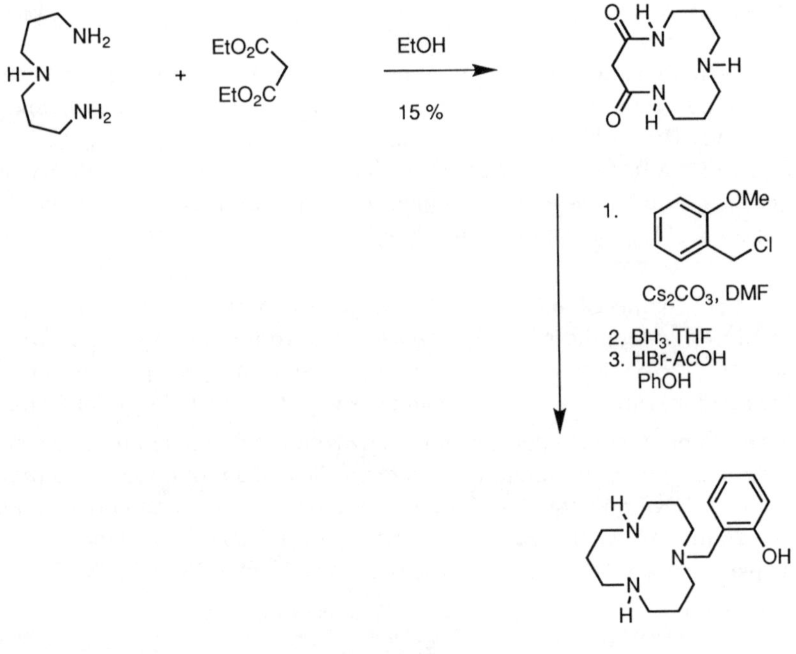

Scheme 1.8

Protocol 4.
Synthesis of 9-(p-toluenesulfonyl)-1,5,9-triazacyclododecane-2,4-dione (Scheme 1.9)

Caution! Carry out all procedures in a well-ventilated hood, and wear disposable latex or vinyl gloves and chemical-resistant safety goggles.

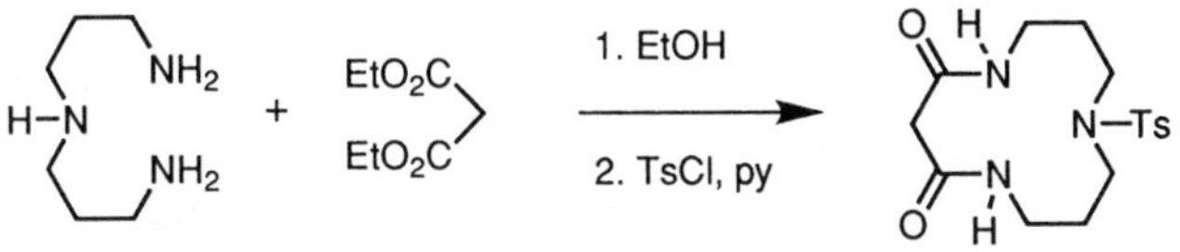

Scheme 1.9

This sequence is representative of the preparation of differentially N-protected polyazacycles involving a malonate condensation of a linear polyamine.

Equipment

- Single-necked round-bottomed flasks, 2 L and 100 mL
- Hot-plate stirrer
- Oil bath
- Magnetic stirrer bar
- Sintered glass filter funnel (porosity 3)
- Buchner flask, 100 mL
- Water-cooled condenser
- Column for 'conventional' silica gel chromatography (6 cm × 60 cm)
- Calcium chloride drying tube
- Filter funnel
- Whatman No. 1 filter papers

Materials

- 1,5,9-Triazanonane, 13.1 g, 0.1 mol   irritant
- Diethyl malonate, 16 g, 0.1 mol
- Dry ethanol, 1 L   flammable
- Dichloromethane, 3 L   harmful by inhalation
- Methanol, 3 L   flammable
- Anhydrous potassium carbonate, 2 g
- Dry pyridine, 80 mL   flammable, harmful by inhalation
- p-Toluenesulfonyl chloride, 3 g, 12 mmol   corrosive
- Silica gel for chromatography (Kieselgel 60, Merck Art. 7734)   harmful by inhalation

1. In a single-necked round-bottomed flask (2 L), prepare a solution of diethyl malonate (16 g, 0.1 mol) in ethanol (1.0 L), and add 1,5,9-triazanonane (13.1 g, 0.1 mol) and a few antibumping granules. Attach a reflux condenser surmounted by a drying tube and boil the solution under reflux for 5 d. Remove the solvent on a rotary evaporator (care: use an anti-splash device to help avoid 'bumping') and dissolve the residue in the minimum volume of dichloromethane–methanol (90:10; c. 10 mL).

2. Prepare a column for chromatography with silica gel slurried in dichloromethane, so that the column height is c. 30 cm. Load the solution

containing the crude mixture on to the column and elute the column with the following solvent gradient at the shown times:

	A = 0.88 NH$_4$OH,	B = MeOH,	C = CH$_2$Cl$_2$
t(h)	A(%)	B(%)	C(%)
0–1	1	12	87
1–3	2	23	75
3–6	4	36	60
6+	6	44	50

3. The desired product has an R_f of 0.7 (1% NH$_4$OH, 59% MeOH, 40% CH$_2$Cl$_2$) and elutes after about 7 h. Combine the eluates containing this product only and remove the solvent on a rotary evaporator to yield a colourless solid, m.p. 153–4 °C (2.99 g, 15%). This product is 1,5,9-triazacyclododecane-2,4-dione.

4. To a single-necked round-bottomed flask (100 mL), add the diamide prepared above (2.0 g, 10 mmol), dry pyridine (50 mL) and p-toluenesulfonyl chloride (3.0 g, 16 mmol). Stopper the flask and keep it in a 4 °C fridge for 48 h.

5. Pour the contents of the flask on to ice-water (c. 200 g) and filter off the resultant precipitate on a sintered glass filter funnel attached to a Buchner flask (500 mL). Wash the solid with water (3 × 20 mL). Take up the solid in dichloromethane (250 mL), dry the solution with anhydrous potassium carbonate, filter the mixture and remove the solvent under reduced pressure.

6. Recrystallise the solid residue from hot methanol (c. 20 mL) to yield a colourless solid, m.p. (dec) > 260 °C (2.53 g, 72%).

Similarly, in the so-called 'crab-like cyclisation' reaction of an α-chloroacetamide with secondary amines[15] which takes place in the presence of sodium carbonate in acetonitrile (Scheme 1.10), the product diamide possesses three distinctive nitrogen sites. Such a reaction may be carried out under medium dilution, when yields of around 50% are common. If the reactants are added slowly using syringe pumps and the reaction is undertaken at high dilution, then the yield improves to 80%.

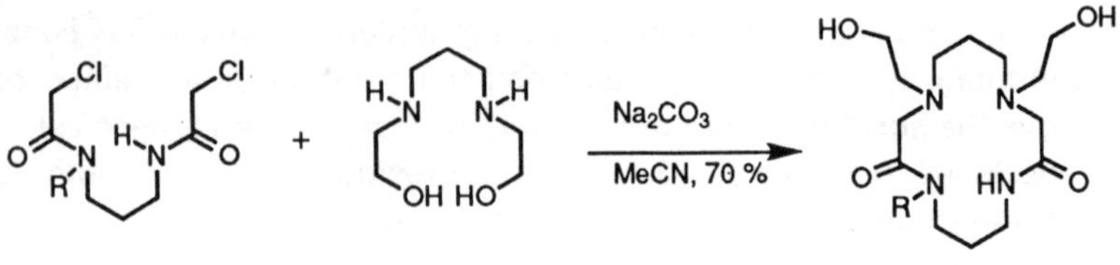

Scheme 1.10

1: Aza crowns

For the synthesis of C-functionalised aza crowns, it is necessary to introduce the required (or latent) functionality prior to cyclisation. The most common aza crowns possess a C_3 or C_2 chain between the nitrogens and a variety of methods for introducing these subunits has developed.

When a C_2 spacing group is required to be C-functionalised, the most obvious precursor is an α-amino acid derivative. Such a strategy has been developed for the synthesis of the [9]-N_3 and [12]-N_4 derivatives **8, 9** and **10**. Both **8** and **9** were synthesised from (2S)-lysine[16] while **10** has been prepared from the enantiopure p-nitrophenylalanine.[17] In the synthesis of the [9]-N_3 derivative, a key step (Scheme 1.11) involved protection of the diethylenetriamine moiety by copper(II) prior to selective acylation of the terminal primary amine and subsequent Richman–Atkins[8] cyclisation. Detosylation was effected with concentrated sulfuric acid, leaving the benzamide intact.

8   R = H, CH_2CO_2H
       CH_2PMeO_2H
   R' = H, or COPh

9

10   R = H, or CH_2CO_2H

If a C_3 unit needs to be introduced bearing a carbon chain, then there are several options. A β-amino acid could be used as a precursor in a parallel synthesis to those discussed above. Alternatively, a strategy based on malonate syntheses could be employed. Co-condensation of a C-substituted malonate with a primary diamine in boiling ethanol gives reasonable yields of cyclic diamide which may be reduced with borane in THF to yield the desired polyamine (Scheme 1.12 and Protocol 5).[18] Another possibility is to prepare a linear bis-toluenesulfonate ester from a malonate by reduction (e.g. $LiBH_4$) and tosylation, followed by a standard 'toluenesulfonamide' cyclisation reaction. Yet another variant, this time giving racemic product, is to react a coumarin derivative with a linear polyamine (Scheme 1.12).[18,19] In this case, a

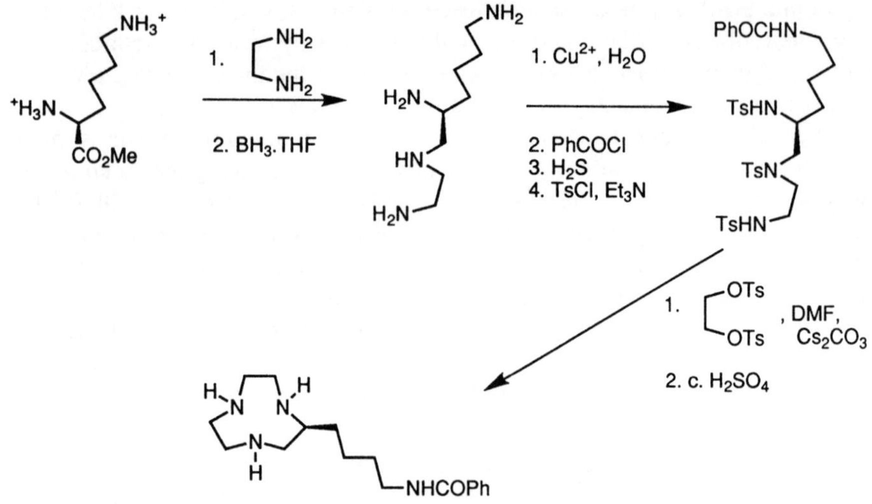

Scheme 1.11

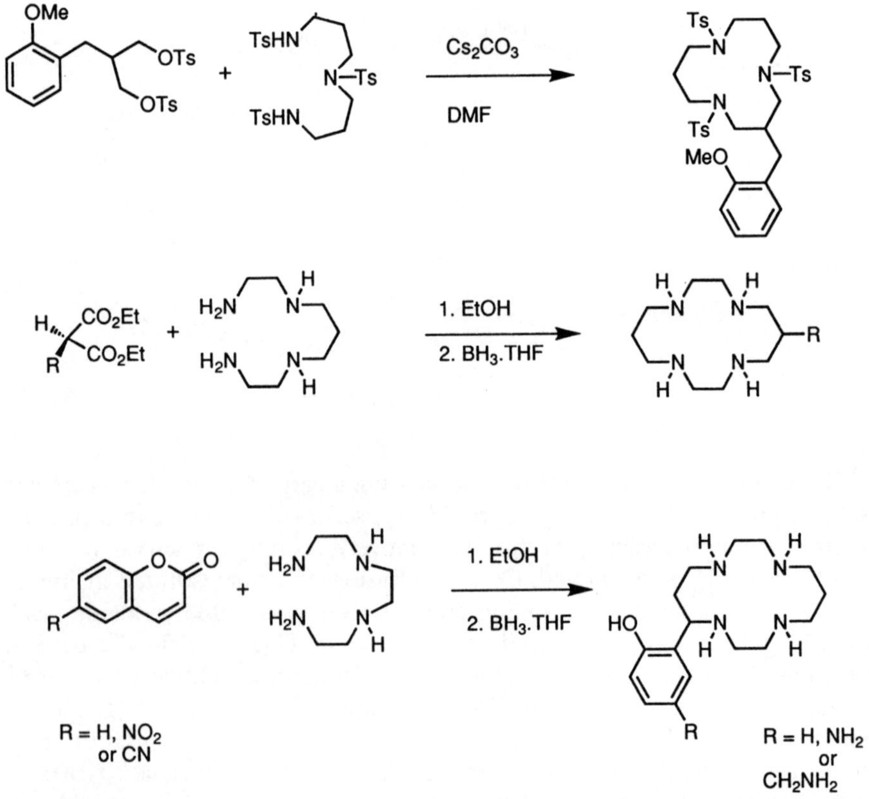

Scheme 1.12

1: Aza crowns

pendant phenolate group is released which is favourably disposed to interact with a metal bound by the four ring nitrogens.

Protocol 5.
12-(4'-Aminomethylbenzyl)-1,4,7,10-tetraazacyclotridecane (Scheme 1.13)

Caution! Carry out all procedures in a well-ventilated hood, and wear disposable latex or vinyl gloves and chemical-resistant safety goggles.

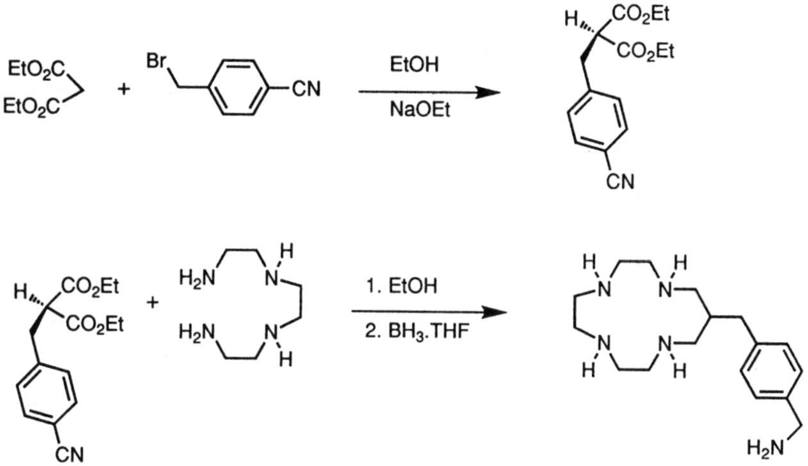

Scheme 1.13

This is an example of the synthesis of a *C*-functionalised macrocyclic polyamine bearing a suitable functional group for selective subsequent conjugation.

Equipment

- Double-necked round-bottomed flask, 500 mL and 250 mL
- Hot plate stirrer
- Magnetic stirrer bar
- Nitrogen inlet adaptor
- Sintered glass filter funnel (porosity no. 3)
- Buchner flask, 500 mL
- Kugelrohr distillation apparatus
- Source of dry nitrogen or argon
- Single-necked round-bottomed flask, 250 mL
- Oil bath

- Filter funnel
- Rotary evaporator
- Vacuum pump
- All-glass syringe with a needle lock Luer, 100 mL
- Septum
- Separating funnel, 250 mL
- Water-cooled condenser
- Calcium chloride drying tube
- Whatman No. 1 filter papers
- Pressure-equalising addition funnel, 100 mL

Materials

- Diethyl malonate, 20 g, 125 mmol
- Sodium metal, 1.4 g, 61 mmol

moisture sensitive!

Protocol 5. Continued

• α-Bromo-p-tolunitrile, 12 g, 61 mmol	irritant
• Dry dimethylformamide, 60 mL	irritant, harmful by inhalation
• Dry ethanol, 300 mL	flammable
• Diethyl ether, 500 mL	flammable, irritant
• Chloroform, 150 mL	harmful by inhalation
• 1,4,7,10-tetraazadecane, 4.18 g, 28.6 mmol	irritant
• 1 M Borane–tetrahydrofuran solution, 80 mL, 80 mmol	moisture sensitive!
• Potassium hydroxide pellets, 8 g	corrosive
• Concentrated hydrochloric acid, 10 mL	corrosive
• Dry hexane, 20 mL	flammable, irritant
• Dry methanol, 100 mL	flammable

1. Take a block of sodium metal out of the paraffin oil in which it is stored, and with a sharp knife cut approximately 1.5 g from it. Wash the block carefully in a beaker containing dry hexane to remove adhering oil and cut the block into three smaller pieces. Dry the shiny blocks with filter papers and weigh out c. 1.4 g accurately.

2. Carefully add the sodium metal (1.4 g, 61 mmol) to a double-necked round-bottomed flask, equipped with an addition funnel, a stirrer and a glass stopper, containing dry ethanol (150 mL), maintained under a gentle stream of nitrogen. Prepare a solution of diethyl malonate (20 g, 125 mmol) in dry ethanol (50 mL) and transfer this solution to the addition funnel.

3. Once the sodium has dissolved completely in the ethanol, add the contents of the addition funnel. Stir the mixture for 20 min. Meanwhile prepare a solution of α-bromo-p-tolunitrile[a] (12 g, 61 mmol) in dry DMF (60 mL) and add this solution from the addition funnel over a period of 20 min. Replace the glass stopper with a water-cooled condenser, and boil the mixture under reflux for 24 h under nitrogen.

4. Cool the mixture to room temperature and filter off the precipitate of the dialkylated malonate derivative. Extract the filtrate with diethyl ether (400 mL). Remove the ether using a rotary evaporator and distil the residue under vacuum using a Kugelrohr apparatus. The monoalkylated product is a colourless oil (b.p. 150°C, 0.1 mmHg), R_f = 0.26 (SiO$_2$, 20% ethyl acetate/ petroleum ether), 7.8 g (47%).

5. Prepare a solution of the monoalkylated malonate (7.84 g, 28.5 mmol) in dry ethanol[b] (120 mL) in a single-necked round-bottomed flask (250 mL) with a stirrer bar equipped with a condenser. Add 1,4,7,10-tetraazadecane (4.18 g, 28.6 mmol) and boil the mixture under reflux, stirring for 10 d.

6. Cool the reaction mixture to 0°C and filter off a colourless solid on a sintered glass filter funnel. Recrystallise this solid from the minimum volume of dry ethanol (c. 20 mL), and dry the solid under high vacuum, 1.9 g (20%), m.p. 246–8°C (dec.).

7. Dry a double-necked round-bottomed flask (250 mL) and a stirrer bar in an

electric oven (c. 120°C) for 1 h. Equip the flask with a septum and add the macrocyclic diamide (1.3 g, 3.95 mmol) prepared above. Attach a reflux condenser surmounted by a nitrogen inlet adaptor and flush the apparatus with dry nitrogen (or argon) gas.

8. Add slowly by syringe, via the septum inlet, a solution of borane–tetrahydrofuran solution (1 M, 80 mL). Replace the septum with a glass stopper and heat and stir the mixture to reflux under nitrogen for 72 h.

9. Cool the solution in an ice bath and very slowly add dry methanol (5 mL) by pipette down the condenser, to quench the excess borane. After the effervescence has ceased, add hydrochloric acid (1 M, 5 mL) and stir the mixture for 10 min. Evaporate the solvents using a rotary evaporator and to the residue add methanol (10 mL) and evaporate again. Repeat this procedure three times to aid removal of trimethyl borate, which forms a low-boiling azeotrope with methanol, b.p. 55°C.

10. Treat the residue with hydrochloric acid (6 M, 25 mL) and boil the solution under reflux for 3 h. Remove the water on a rotary evaporator and dissolve the residue in distilled water (30 mL). Transfer the solution to a separating funnel and wash with ether (3 × 20 mL).

11. Basify the aqueous layer by adding sufficient potassium hydroxide pellets to raise the pH to 13. Extract the aqueous layer with chloroform (4 × 30 mL). Combine the organic extracts, dry over potassium carbonate (c. 1 g), filter and evaporate the solvent to yield a colourless oil (1.2 g, 99%). The carbon-13 NMR spectrum is diagnostic: δ_C(CDCl$_3$): 140.7 (quat.); 138.8 (quat.); 128.9, 126.8 (aryl C–H); 54.6 (ArCH$_2$N); 48.8, 47.4, 47.2, 46.0 (CH$_2$N); 40.2, 38.3 (CH$_2$Ar + CHC). If necessary the product may be recrystallised as the pentahydrochloride (tetrahydrate) by dissolving the oil in dry ethanol (the minimum of c. 10 mL) and adding 0.5 mL of concentrated hydrochloric acid. The hydrochloride salt crystallises out slowly on standing.

[a] Commercially available from Aldrich (14,406–1).
[b] Distil ethanol and methanol from the corresponding magnesium alkoxide.

---

3. Templated reactions

The cyclisation reactions that are used to generate medium-ring and large-ring cyclic polyamines suffer from an unfavourable entropy term to the overall free energy change. Indeed this effect often limits the temperature at which the reaction can be carried out, because heating gives rise to an even more unfavourable $T\Delta S$ term and competing reactions (e.g. oligomerisation) may be favoured at the expense of the desired cyclisation. Such effects may be obviated if an ion can be used to act as a template for the cyclisation step. Such an effect was first reported (for poly-aza 'crown' syntheses) by Curtis[20] in the reaction of [Ni(1,2-diaminoethane)]$^{2+}$ with dry acetone (Scheme 1.14), to generate a

[14]-N$_4$ ring system in which the carbon–nitrogen double bonds may be subsequently reduced by borohydride. Many such metal templated syntheses have now been reported (a further example is in Chapter 3, Protocol 7), not only involving condensation reactions but also alkylation or acylation at metal-bound nitrogens.[21] Subsequent demetallation may be effected either by adding acid, by a ligand exchange process (e.g. adding CN$^-$, sulfide or EDTA) or following reduction of the metal if it has a suitable redox couple. A classic example is the synthesis of [14]-N$_4$ or cyclam, **2**, where the nickel ion is removed following cyanide addition (Scheme 1.15).[22]

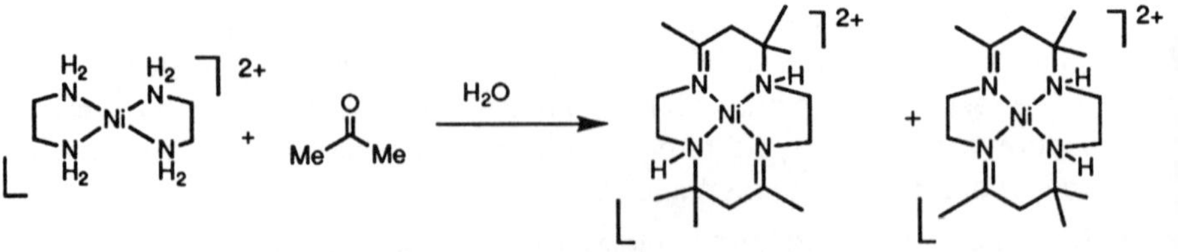

Scheme 1.14

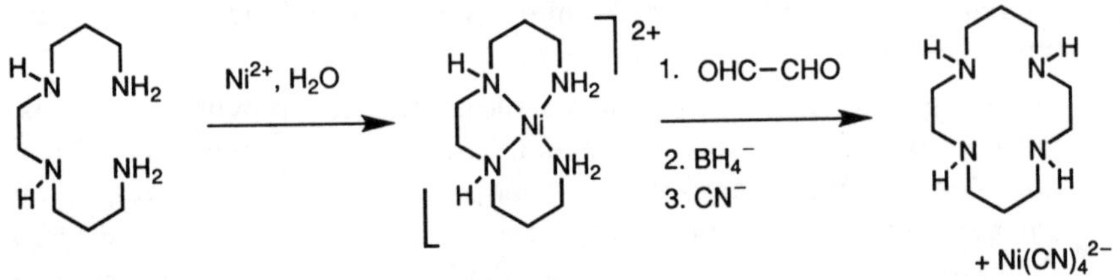

Scheme 1.15

A limited number of syntheses are templated by one or more protons wherein a hydrogen-bonding network may function as a template, correctly predisposing the electrophilic and nucleophilic components for the cyclisation step.

Protocol 6.
Synthesis of 5,12-dimethyl-1,8-diaza-4,11-diazotetradeca-4,11-diene bis(perchlorate) (Scheme 1.16)

Caution! Amine perchlorate salts are potentially explosive and should not be heated to dryness under vacuum. Manipulations should be carried out in a well-ventilated hood, transfers of perchlorate salts carried out with ceramic or glass spatulas, and latex or vinyl gloves and chemical-resistant safety goggles should be worn.

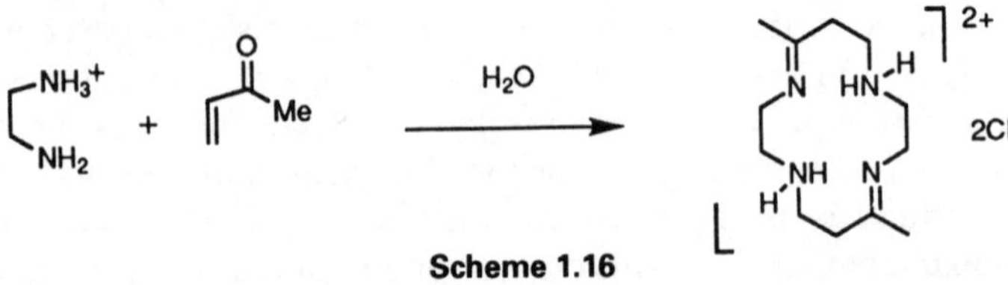

Scheme 1.16

1: Aza crowns

The following procedure is representative of a proton-templated cyclisation reaction.

Equipment

- Erlenmeyer flasks, 100 mL and 500 mL
- Sintered filter funnel (porosity 2)
- Buchner flask (500 mL)
- Pressure-equalising addition funnel (50 mL)
- Thermometer
- Vacuum desiccator containing P_4O_{10}

- Double-necked round-bottomed flask, 500 mL
- Nitrogen inlet adaptor
- Magnetic stirrer
- Stirrer bar
- Thermometer inlet adaptor

Materials

- Aqueous perchloric acid (70%), 17.2 mL   corrosive
- Ethylenediamine (12 g, 200 mmol)   irritant
- Methyl vinyl ketone (14 g, 200 mmol)   harmful by inhalation
- Dry methanol, 200 mL   flammable
- Diethyl ether, 50 mL   flammable, irritant

1. Prepare a solution of ethylenediamine (12 g, 0.2 mol) in distilled water (10 mL) in a ice-cooled conical flask (100 mL), and slowly add aqueous perchloric acid[a] (70%, 17.2 mL) with gentle swirling. Remove the water on a rotary evaporator, periodically adding distilled methanol (5 × 10 mL) to aid azeotropic removal of the solvent. Once a white solid forms, carefully transfer[b] the contents of the flask on to a crystallising dish, and place in a vacuum desiccator overnight.

2. Transfer the dried solid to a double-necked round-bottomed flask (500 mL) equipped with a stirrer bar and a thermometer and add dry methanol (150 mL), breaking up the solid gently with a glass rod. Purge the flask with nitrogen and attach a pressure-equalising addition funnel (50 mL) containing a solution of methyl vinyl ketone (14 g, 0.2 mol) in dry methanol (25 mL). Add the contents of the addition funnel slowly, over a period of 2 h, to the vigorously stirred cooled reaction flask, maintaining the temperature of the solution below 0 °C during the addition period.

3. Allow the stirred mixture to warm up slowly to room temperature and stir for an additional 3 h. Filter off the abundant white precipitate which forms on a sintered glass filter funnel attached to a Buchner flask. Wash the solid product on the filter funnel with cold methanol (3 × 5 mL), diethyl ether (2 × 25 mL) and dry the product under vacuum (0.1 mmHg) at room temperature. The product (13.4 g, 63%), m.p. 109 °C, is sufficiently pure to be used directly, but may be recrystallised from hot methanol if appropriate.

[a] 70% aqueous perchloric acid is commercially available (Aldrich 24,425–2).
[b] Use all glass or ceramic spatulas for solid transfers: do not use nickel spatulas.

A more rational approach to the synthesis of cyclic polyamines involves the cleavage of a common bond between two or more normal rings (i.e. five-, six- or seven-membered). Such an approach is well established for the synthesis of medium-ring and some large-ring carbocycles. These reactions could be regarded also as templated reactions wherein a quaternary or tertiary carbon centre serves as the 'template'. For example, tricyclic orthoamides may be readily prepared by alkylation of certain bicycloguanidines (Scheme 1.17).[23-25] These versatile intermediates may be hydrolysed to give the free triaza macrocycle or may be monoalkylated to give a mono-*N*-substituted ligand selectively. An example of such a 'carbon'-templated synthesis is provided by the synthesis of 1,5,9-triazacyclodecane (Protocol 7).

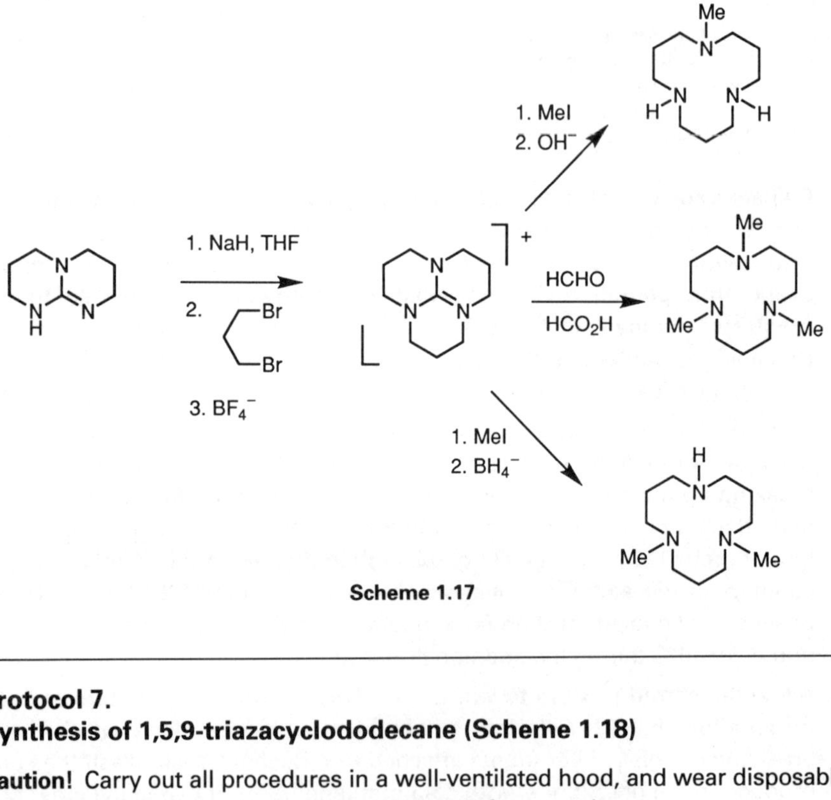

Scheme 1.17

Protocol 7.
Synthesis of 1,5,9-triazacyclododecane (Scheme 1.18)

Caution! Carry out all procedures in a well-ventilated hood, and wear disposable vinyl or latex gloves and chemical-resistant safety goggles.

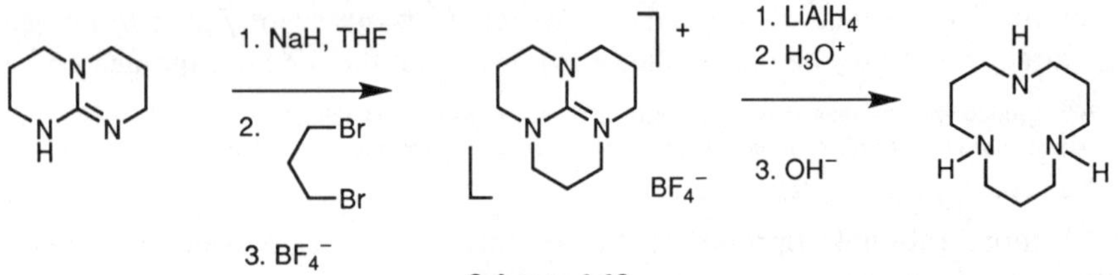

Scheme 1.18

1: Aza crowns

The following procedure is representative of the synthesis of a large-ring triazacycloalkane involving the cleavage of common bonds in normal rings.

Equipment

- Hot plate stirrer
- Oil bath
- Magnetic stirrer bar
- Separating funnels, 100 mL and 250 mL
- Kugelrohr distillation apparatus
- pH Indicator strips (BDH-Merck)
- Sintered glass filter funnel (porosity 3)
- All glass syringe with a needle-lock Luer, 5 mL and 50 mL
- Septum (Aldrich Z10, 075–7)

- Single necked round-bottomed flasks, 50 mL and 100 mL
- Double-necked round-bottomed flask, 100 mL and 500 mL
- Pressure-equalising addition funnel, 50 mL
- Source of dry nitrogen or argon
- Filter funnel
- Water-cooled condenser
- Whatman No.1 filter papers
- Gas inlet adaptor

Materials

- 1,5,7-Triazabicyclo[4.4.0]-dec-5-ene, 13.9 g, 100 mmol   irritant
- Sodium hydride, 4.5 g (60% oil suspension)   irritant, moisture sensitive
- 1,3-Dibromopropane, 20.2 g, 100 mmol   toxic, irritant
- Sodium tetrafluoroborate, 11 g, 100 mmol
- Dry ethanol, 55 mL   flammable
- Dry ether, 150 mL   irritant, flammable
- Dry tetrahydrofuran, 270 mL   flammable, irritant
- Dichloromethane, 200 mL   harmful by inhalation
- Magnesium sulfate, 10 g (drying agent)
- Potassium carbonate, 2 g (drying agent)
- Lithium aluminium hydride, 1.52 g, 40 mmol   irritant, moisture sensitive
- 15% aqueous sodium hydroxide solution, 25 mL   toxic, corrosive
- 3 M Hydrochloric acid, 50 mL   corrosive, toxic
- Distilled water, 500 mL
- Light petroleum, b.p. 40–60°C, 20 mL   flammable

1. Dry a 500 mL double-necked round-bottomed flask (with a teflon coated magnetic stirrer bar and a nitrogen inlet adaptor) in an electric oven at 125°C for 1 h. Flush the flask with dry nitrogen (or argon), and add the sodium hydride suspension[a] (4.5 g). Flush the flask again with inert gas and add light petroleum, 10 mL, by syringe. Swirl the flask, allow the sodium hydride to settle and carefully remove the supernatant liquid by syringe. Repeat this process, and then add to the flask 1,5,7-triazabicyclo[4.4.0]-dec-5-ene (13.9 g, 100 mmol) and finally freshly distilled dry THF,[b] 200 mL, by syringe. Cool the flask in an ice–salt bath (to −5°C).

2. Add by syringe, 1,3-dibromopropane (20.2 g, 100 mmol) while stirring the mixture under a gentle flow of nitrogen gas. Stir for 90 min at 0°C and allow the mixture to warm up to room temperature overnight while maintaining stirring. Add dry ethanol (5 mL) by syringe, and filter off the resultant white solid on a sintered filter funnel (NB this product is hygroscopic, so work quickly!) and quickly wash the solid product with dry THF (20 mL).[c]

3. Dissolve the solid in distilled water (100 mL) (this can be done on the filter

Protocol 7. *Continued*

funnel), and add sodium tetrafluoroborate (11 g, 100 mmol). Transfer the solution to a separating funnel (250 mL) and extract it with dichloromethane (3 × 50 mL). Dry the combined organic extracts over magnesium sulfate, filter and remove the solvent using a rotary evaporator to yield a colourless solid. Recrystallise this solid from ethanol–diethyl ether (c. 1:2, dissolve in c. 30 mL of ethanol). The product is a colourless solid, m.p. 196°C, 18.7 g (70%).

4. Dry a 100 mL double-necked round-bottomed flask and a nitrogen inlet adaptor in an electric oven at 125°C for 1 h. Charge the flask with the guanidinium salt prepared above (5.34 g, 20 mmol) and flush the apparatus with dry nitrogen (or argon). Fit a septum, add a magnetic stirrer bar and then add by syringe freshly distilled dry THF[b] (50 mL). Stir the suspension vigorously under nitrogen and cool to −5°C in an ice-salt bath. Add quickly in three portions, lithium aluminium hydride (1.52 g, 40 mmol)[d] over a period of 15 min. Stir vigorously and allow the mixture to warm up slowly to room temperature overnight.

5. Add slowly by syringe, via the septum, water (1.5 mL), followed by 15% aqueous sodium hydroxide (1.5 mL) and water (1.5 mL). Stir the mixture for 90 min and filter the mixture through a sintered filter funnel (porosity 3). Wash the filter cake with dichloromethane (2 × 15 mL). Combine the filtrate and the washings and remove the solvents using a rotary evaporator to yield a viscous oil. Distil the product under vacuum using a Kugelrohr apparatus (90°C, 0.1 mmHg) to yield the orthoamide as a hygroscopic crystalline solid (3.06 g, 85%), m.p. 36–38°C.

6. Transfer the solid (3.06 g, 16.9 mmol) into a single-necked round-bottomed flask (50 mL), add a magnetic stirrer bar, fit a water-cooled condenser and add hydrochloric acid (3 M, 25 mL). Heat and stir the solution under reflux overnight.

7. Cool the solution and basify to pH ≥13 using 15% aqueous sodium hydroxide solution. Extract the basic solution with dichloromethane (4 × 20 mL) using a separating funnel (100 mL). Dry the organic extracts with anhydrous potassium carbonate, filter and evaporate the solvent using a rotary evaporator, to yield a pale-yellow oil. Distil the oil under vacuum, using a Kugelrohr apparatus (85–90°C, 0.03 mmHg) to give a colourless oil which solidifies on standing (2.57 g, 89%), m.p. 36–37°C.

[a] NaH reacts violently with moisture releasing H_2 which may ignite. As a dispersion in oil it is safer to handle.
[b] Distil the THF from $LiAlH_4$ or sodium–benzophenone ketyl, under N_2 or Ar.
[c] Further product may be obtained by adding the filtrate to diethyl ether (50 mL).
[d] $LiAlH_4$ may inflame spontaneously in air and reacts violently with water. Take care with the disposal of residues!

References

1. Lindoy, L. F. *The Chemistry of Macrocyclic Ligand Complexes*; Cambridge University Press: Cambridge, **1989**.
2. Bradshaw J. S.; Krakowiak, K. E.; Izatt, R. M. In *The Chemistry of Heterocyclic Compounds*; Wiley: New York, **1993**, vol. 51.
3. Izatt, R. M.; Bruening, R. L.; Tarbet, B. J.; Griffin, L. D.; Bruening, M. L.; Krakowiak, K. E.; Bradshaw, J. S. *Pure Appl. Chem.* **1990**, *62*, 1115.
4. Parker, D. *Chem. Soc. Rev.* **1990**, *19*, 271; Parker, D. In *Comprehensive Supramolecular Chemistry*; Lehn, J. M.; Reinhoudt, D. N., eds; Pergamon Press: Oxford, **1996**; Vol. 10, Chapter 17.
5. Hosseini, M. W.; Lehn, J. M. *Helv. Chim. Acta* **1987**, *70*, 1312; Brand, G.; Hosseini, M. W.; Ruppert, R. *Helv. Chim. Acta* **1992**, *75*, 721.
6. Stetter, H.; Mayer, K.-H. *Chem. Ber.* **1961**, *94*, 1410.
7. Stetter, H.; Roos, E.-E. *Chem. Ber.* **1954**, *87*, 566.
8. Richmans, J.; Atkins, T. J. *J. Am. Chem. Soc.* **1974**, *96*, 2268.
9. Dietrich, B.; Hosseini, M. W.; Lehn, J.-M.; Sessions, R. B. *Helv. Chim. Acta* **1983**, *66*, 1262.
10. Norman, T. J.; Parker, D.; Pulukkody, K.; Royle, L.; Broan, C. J. *J. Chem. Soc., Perkin Trans. 2* **1993**, 605.
11. Desreux, J. F.; Dumont, A.; Jacques, V.; Qixiu, P. *Tetrahedron Lett.* **1994**, *35*, 3707.
12. Helps, I. M.; Morphy, J. R.; Parker, D.; Chapman, J. *Tetrahedron*, **1989**, *45*, 219.
13. Chapman, J.; Ferguson, G.; Parker, D. *J. Chem. Soc., Dalton Trans.* **1992**, 343.
14. Helps, I. M.; Jankowski, K. J.; Nicholson, P. E.; Parker, D. *J. Chem. Soc., Perkin Trans. 1* **1989**, 2079.
15. Krakowiak, K. E.; Bradshaw, J. S.; Izatt, R. M. *Synlett* **1993**, 611.
16. Cox, J. P. L.; Craig, A. S.; Jankowski, K. J.; Parker, D.; Helps, I. M.; Beeley, N. R. A.; Boyce, B. A.; Millar, K.; Millican, A. T.; Eaton, M. A. W. *J. Chem. Soc., Perkin Trans. 1* **1990**, 2567.
17. Moi, M. K.; Meares, C. F.; De Nardo, S. J. *J. Am. Chem. Soc.* **1988**, *110*, 6266.
18. Morphy, J. R.; Kataky, R.; Parker, D.; Eaton, M. A. W.; Alexander, R.; Millican, A. T.; Harrison, A.; Walker, C. A. *J. Chem. Soc., Perkin Trans. 2* **1990**, 573; Tabushi, I.; Taniguchi, Y.; Kato, H. *Tetrahedron Lett.* **1977**, 1049.
19. Kimura, E.; Koike, T.; Takahashi, M. *J. Chem. Soc., Chem. Commun.* **1985**, 385.
20. Curtis, N. F. *J. Chem. Soc.* **1960**, 4409; Curtis, N. F.; Curtis, Y. M.; Powell, H. J. K. *J. Chem. Soc. A* **1966**, 1015.
21. Lindoy, L. F. *The Chemistry of Macrocyclic Ligand Complexes*; Cambridge University Press: Cambridge, **1989**, Chapter 2.
22. Barefield, E. K.; Wagner, F.; Herlinger, A. W.; Dahl, A. R. *Inorg. Synth.* **1976**, *16*, 220.
23. Schmidtchen, F. P. *Chem. Ber.* **1980**, *113*, 2175; McKay, A. F.; Kreling, M. E. *Can. J. Chem.* **1957**, *35*, 1438.
24. Weisman, G. R. Vachan, D. J.; Johnson, V. B.; Gronbeck, D. A. *J. Chem. Soc., Chem. Commun.* **1989**, 794.
25. Alder, R. W.; Mowlam, R. W.; Vachon, D. J.; Weisman, G. R. *J. Chem. Soc., Chem. Commun.* **1992**, 507.

2

Aza-oxa crowns

DAVID PARKER

1. Introduction

The saturated aza-oxa crown ethers were first synthesised as intermediates in the synthesis of the nitrogen cryptands.[1] The reaction conditions used for the formation of these macrocycles involved the high-dilution technique. In this versatile method, a diamine and a diacid chloride are simultaneously added in the presence of triethylamine to a large volume of solvent, usually toluene, over an extended period of time. The major product from such a reaction is the [1+1] cyclised product, although the [2+2] adduct can often be isolated as well, in low yield. Whilst this method is still sometimes used,[2,3] particularly for cryptand synthesis (Chapter 5), it has been superseded by methods that are more convenient and which proceed under medium dilution.

A simple example of such a new reaction involves the formation of 1,4,10-trioxa-7,13-diazacyclopentadecane **1** or [15]-N$_2$O$_3$, the acronym commonly used when referring to it (Scheme 2.1). Formation of this diamine proceeds in two steps, the first of which involves the synthesis of a macrocyclic diamide. According to the availability of the precursor diamine, the reaction may proceed via one of two constitutionally isomeric diamides. Instead of using diacid chlorides, the methyl esters in methanol, typically at reagent concentrations of 0.1 mol dm^{-3}, react smoothly with the diamine at ambient temperatures to give the desired diamide in yields of around 60%. The reaction works well for the formation of the [12]-N$_2$O$_2$, [15]-N$_2$O$_3$ and [18]-N$_2$O$_4$ diaza crown ethers.[4] It is similar in nature to the reaction of dimethyl malonates with α,ω-diamines which has been used to introduce a C$_3$ unit into various simple and *C*-functionalised polyaza macrocycles (see Chapter 1, Protocol 5).[5,6] The product diamides may be reduced to the diamine using either LiAlH$_4$ or borane–tetrahydrofuran. The latter reagent is generally preferred, particularly for the reduction of tertiary amides, and it is often convenient to use the dimethyl sulfide adduct which is commercially available as a 10 M solution in THF.

Protocol 1.
Synthesis of 1,4,10-trioxa-7,13-diazacyclopentadecane (Structure 1, Scheme 2.1)

Caution! Carry out all procedures in a well-ventilated hood, and wear disposable vinyl or latex gloves and chemical-resistant safety goggles.

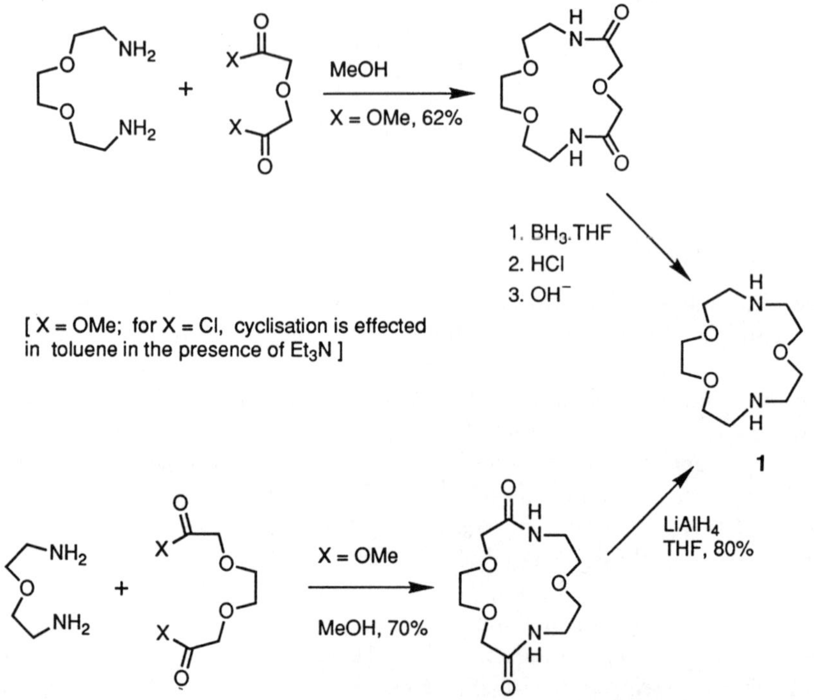

[X = OMe; for X = Cl, cyclisation is effected in toluene in the presence of Et$_3$N]

Scheme 2.1

The following procedure is representative of the direct cyclisation of a diamine with a dimethyl ester, followed by borane reduction to yield the cyclic diamine.

Equipment

- Septum (Aldrich Z10,075–7)
- Condenser
- Hot-plate stirrer
- Source of dry nitrogen or argon
- Oil bath
- Gas line adaptor
- Stirrer bar
- All-glass syringe with a needle-lock Luer, 50 ml

- Conical flasks (50 mL and 100 mL)
- pH indicator strips (BDH)
- Separating funnel (100 mL)
- Single-necked, round-bottomed flask (250 mL)
- Double-necked, round-bottomed flask (100 mL)
- Filter funnel
- Sintered filter funnel (porosity 3)
- Fluted filter paper (Whatman No. 1)

Materials

- Dry methanol, 150 mL **flammable**
- Dichloromethane, 250 mL **harmful by inhalation**

2: Aza-oxa crowns

- Hexane, 50 mL — **flammable, irritant**
- 3 M Hydrochloric acid, 10 mL — **corrosive, toxic**
- Borane–dimethyl sulfide (2 M in tetrahydrofuran), 30 mL — **flammable, toxic**
- Potassium carbonate (anhydrous), 2 g — **harmful by inhalation**
- Dimethyl glycollate, 3.80 g, 23.4 mmol
- 4,7-Dioxa-1,10-diaminodecane, 2.88 g, 19.5 mmol — **irritant**

1. Dry a single-necked round-bottomed flask, 250 mL, in an electric oven (110 °C) for 1 h, allow it to cool to room temperature under dry argon or nitrogen. Equip the flask with a magnetic stirrer bar and a water-cooled reflux condenser, with a nitrogen inlet above the condenser.

2. Add to the flask 4,7-dioxa-1,10-diaminodecane (2.88 g, 19.5 mmol) and the dimethyl ester of diglycollic acid (3.80 g, 23.4 mmol) followed by dry methanol (150 mL). Heat the solution to reflux under nitrogen, with magnetic stirring, for 24 h.

3. Remove the solvent by evaporation under reduced pressure using a rotary evaporator and dissolve the residue in a small volume of dichloromethane (c. 10 mL). Transfer the solution to a conical flask, add cyclohexane until the solution becomes turbid and maintain the solution at 5 °C.

4. Collect the colourless needles of crystalline diamide product on a sinter and dry the crystals under vacuum (0.1 mmHg, 3 h). The cyclic diamide solid (m.p. 128–9 °C, 3.0 g, 62%) may be reduced to the product directly.[a]

5. Place the diamide (4 g, 16.3 mmol) in an oven-dried (electric oven, 110 °C, 1 h) double-necked round-bottomed flask with a stirrer bar (100 mL) fitted with a reflux condenser, linked to a nitrogen line and with a septum over the other joint.

6. Add slowly by syringe, via the septum inlet, a solution of borane–dimethyl sulfide complex in tetrahydrofuran (2 M, 30 mL). Stir the mixture at room temperature, then replace the septum with a glass stopper and heat the solution under nitrogen to reflux in an oil bath.

7. To the cooled solution, slowly add[b] dry methanol (5 mL) to quench the excess borane. After the effervescence has ceased add hydrochloric acid (3 M, 5 mL) and stir the mixture for 10 min. Evaporate the organic solvents using a rotary evaporator and to the residue add methanol (10 mL) and evaporate again. Repeat this procedure three times to aid the removal of the trimethyl borate (which forms a low-boiling azeotrope with methanol, b.p. 54–8 °C).

8. To the white residue, add dichloromethane (25 mL) and dilute hydrochloric acid (20 mL). Transfer the mixture to a separating funnel and discard the lower organic layer. To the aqueous layer add sufficient solid potassium hydroxide to raise the pH above 12. Extract the aqueous layer with dichloromethane (5 × 30 mL), combine and dry (potassium carbonate) the organic extracts, filter and evaporate the solvent using a rotary evaporator, to yield a colourless solid.[c]

Protocol 1. *Continued*

9. The crystalline residue may be recrystallised from hot hexane (*c.* 90 mL, add a little dichloromethane if necessary) to yield the diamine as a colourless solid (m.p. 71°C, 2.92 g, 82%).

[a] Alternatively, the diamide may be reduced with LiAlH$_4$ in tetrahydrofuran (18 h reflux, 79%), see Chapter 5, Protocol 3 for an example of this method.
[b] A vigorous evolution of hydrogen occurs. Take care to add the first few drops particularly slowly, maintaining magnetic stirring and, if necessary, cool the reaction mixture in an ice-salt bath (−5°C).
[c] If problems occur in recrystallisation, the product may be examined by thin-layer chromatography on silica gel (R_f = 0.65, 5% ethanol in dichloromethane), and purified by column chromatography if appropriate.

---

2. Tosylamide-based cyclisations

The cyclisation method developed by Richman and Atkins[7] for the preparation of poly-aza macrocycles (discussed in Chapter 1, Section 2) is equally applicable to the synthesis of aza-oxa crown ethers. The cyclisation reaction involves co-condensation of the anion of a tosylamide (pK_a of *c.* 15) with a tosylate or mesylate in dry DMF. The deprotonated tosylamide may be generated either prior to the cyclisation step by reaction with sodium ethoxide in ethanol or *in situ* using caesium carbonate, as in the synthesis of the versatile [24]-N$_6$O$_2$ ligand **4**, which involves a [1+1] co-condensation of the *N*-tosylated diethylenetriamine **2** with the linear di-tosylate ester **3**.[8] Tosylate and mesylate esters are generally preferred over alkyl halides, with the cyclisation yield dropping in the sequence Cl>Br>I probably due to the relative ease of competitive elimination. The caesium ion does not play an active role in these cyclisation reactions but affords a means of solubilising the carbonate in the solvent. Indeed fine-mesh potassium carbonate may be used in its place without seriously compromising the reaction yield.

The toluenesulfonyl group in these reactions is also a protecting group for nitrogen, for example in **3**, but it must be removed subsequently to liberate the free secondary amine. There are various methods for achieving this deprotection step which are compatible with ether linkages. The two most useful involve either the use of hydrogen bromide in acetic acid in the

2: *Aza-oxa crowns*

presence of excess phenol, or involve a reductive method such as the use of sodium (or lithium) in liquid ammonia in THF (Scheme 2.2).

Protocol 2.
Synthesis of 1,7-dioxa-4,10-diazacyclododecane (Scheme 2.2)

Caution! Carry out all procedures in a well-ventilated hood, and wear disposable vinyl or latex gloves and chemical-resistant safety goggles.

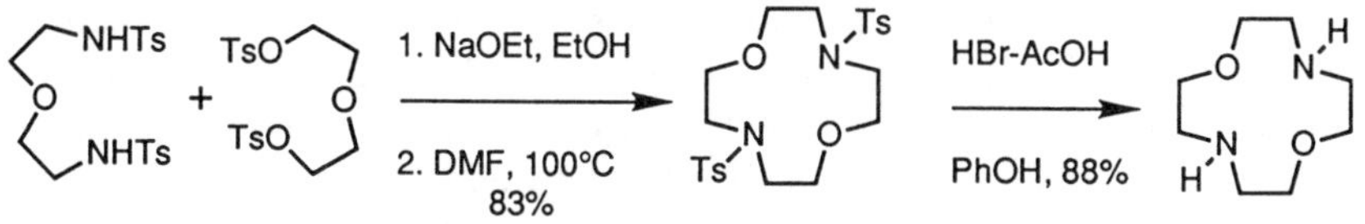

Scheme 2.2

The following procedure is representative of a direct [1+1] cyclisation reaction using a toluenesulfonamide at medium dilution.[a]

Equipment

- Hot plate stirrer (thermostatted)
- Magnetic stirrer bar
- Oil bath
- Single-necked, round-bottomed flasks (250 mL and 500 mL)
- Double-necked, round-bottomed flask (1000 mL)
- Separating funnel (500 mL)
- Vacuum pump (<0.1 mmHg)
- Water-cooled condenser
- Source of dry nitrogen
- Pressure-equalising addition funnel (250 mL)
- pH papers (BDH-Merck)
- Buchner flask (500 mL)
- Buchner filter funnel (100 mL)
- Column for ion-exchange chromatography (c. 50 × 2.5 cm)
- Conical flask (250 mL)

Materials

- Sodium metal, 0.92 g, 40 mmol   **moisture sensitive!**
- Dry methanol, 120 mL   **flammable, toxic**
- Dry dimethylformamide (HPLC grade), 320 mL   **irritant, harmful by inhalation**
- *N,N'*-bis(*p*-toluenesulfonyl)-3-oxa-1,5-diaminopentane, 8.24 g, 20 mmol
- 1,5-bis(*p*-toluenesulfonato)-3-oxapentane, 8.28 g, 20 mmol
- Ethanol (96%), 75 mL   **flammable**
- Phenol, 25 g   **corrosive, irritant**
- Hydrogen bromide in acetic acid (45%), 250 mL   **corrosive, toxic**
- Toluene for recrystallisation, 250 mL   **harmful by inhalation**
- Dichloromethane, 1000 mL   **harmful by inhalation**
- Dry hexane, 50 mL   **irritant, flammable**
- Amberlite 420 anion exchange resin (BDH-Merck) (c. 200 g)

1. Take a block of sodium metal out of the oil and with a sharp knife cut approximately 1 g from it. Wash the block carefully with dry hexane (in a small beaker) to remove adhering oil, and cut the block into three smaller pieces. Dry the shiny blocks with filter papers and weigh out 0.92 g accurately.

29

Protocol 2. *Continued*

2. Carefully add the sodium metal (0.92 g, 40 mmol) to a single-necked round-bottomed flask (250 mL) containing dry methanol (120 mL), maintained under an atmosphere of nitrogen. Attach a water-cooled condenser equipped with a nitrogen inlet, and gently heat (40°C) and stir (Teflon stirrer bar) the mixture until all of the sodium has dissolved.

3. Add N,N'-bis(p-toluenesulfonyl)-3-oxa-1,5-diaminopentane (8.24 g, 20 mmol) to the solution and stir the mixture for 1 h under reflux, when a colourless solution forms. Cool the solution and remove nearly all of the methanol solvent on a rotary evaporator. Transfer the flask to a vacuum line and remove the remaining solvent (<10 mL) under vacuum with gentle heating (0.1 mmHg, 40°C for 1 h). Take up the resultant solid in dry dimethylformamide (200 mL) and transfer the solution quickly[b] to an oven-dried double-necked round-bottomed flask (1000 mL) equipped with a Teflon stirrer bar and an air condenser attached to a nitrogen line.

4. Warm the solution to 100°C and add, via a pressure-equalising dropping funnel (250 mL), a solution of 1,5-bis(p-toluenesulfonato)-3-oxapentane (8.28 g, 20 mmol) in dry DMF (120 mL) over a period of 2 h. Heat (100°C) and stir the mixture for a further 4 h, then allow the solution to cool to room temperature and add water (200 mL) slowly over a period of 30 min. Cool the mixture in ice for 30 min and separate the white precipitate which forms by filtration on a Buchner funnel. Wash the solid repeatedly with aqueous ethanol (1:1 v/v, 4 × 50 mL) and transfer the partly dried solid to a single-necked round-bottomed flask (500 mL). Dry the flask and its contents under vacuum (0.1 mmHg, 50°C) for 2 h to yield the cyclic ditoluenesulfonamide as a colourless solid, 8.0 g, 83%, m.p. 195–6°C, R_f = 0.61 (Al_2O_3, 1% MeOH in CH_2Cl_2).

5. Add phenol (25 g) to a solution of the ditosylamide (12.5 g, 0.026 mol) in hydrogen bromide–acetic acid (45%, 250 mL) in a single-necked round-bottomed flask equipped with a magnetic stirrer bar and a reflux condenser surmounted by an air condenser.[c] Heat and stir the mixture at 80°C for 36 h.

6. After cooling the dark-red solution, remove the acetic acid on a rotary evaporator, adding four portions of toluene (each 30 mL) periodically removing solvent between each addition to drive off all the acetic acid.[d] Dissolve the residue in water (200 mL) and add dichloromethane (150 mL), transferring the mixture to a separating funnel (500 mL). Separate the layers and wash the aqueous layer with dichloromethane (4 × 100 mL). Remove the water on a rotary evaporator and dissolve the residue in the minimum volume of deionized water (c. 5 mL).

7. Prepare a column (50 cm × 2.5 cm) for anion-exchange chromatography

using Amberlite resin[e] (c. 200 g). Convert the resin to the 'hydroxide' form by successively passing down deionized water (200 mL), aqueous potassium hydroxide solution (10%, 50 mL) and then deionized water (200 mL) until the pH of the eluting solution is around 7. Add the aqueous solution of the salt (prepared in step 6) to the column and elute with deionized water (c. 200 mL), monitoring the pH of the eluant off the column. When the pH has dropped down from 11 to 7, combine the basic aqueous fractions and remove the water on a rotary evaporator, adding distilled methanol (4 × 25 mL) periodically to aid removal of the water.

8. Dry the residue under high vacuum (0.1 mmHg, 40°C, 3 h) and recrystallise the solid from hot toluene[f] (c. 100 mL) to yield colourless needles of the diamine, 4.0 g, 88%, m.p. 83–4°C.

[a] For an example of a related cyclisation reaction using caesium carbonate as a base in DMF, see Protocol 3. The starting ditoluenesulfonamide (m.p. 119–20°C) is readily prepared by reaction of the corresponding diamine with p-toluenesulfonyl chloride (aq NaOH/Et₂O).
[b] The disodium salt is hygroscopic but can be handled in air for short periods without compromising the reaction yield.
[c] Considerable volumes of HBr evolve, so the second condenser is used to help vent the evolving gas towards the top of the hood.
[d] Toluene and acetic acid form a low-boiling azeotrope (b.p. 105°C, v/v, 72:28).
[e] Before using the resin (supplied in the chloride form) it is advisable to boil the resin in dry methanol for 30 min to remove soluble impurities.
[f] A little hexane (20% v/v) aids crystallisation.

The related p-methoxybenzenesulfonyl group is a useful alternative to the toluenesulfonyl group and may be removed under milder acidic conditions, e.g. HBr–AcOH–PhOH at 50°C.[9]

By working at somewhat higher dilution, these toluenesulfonamide-based co-cyclisation reactions can provide viable yields of the [2+2] macrocyclic product, as well as the major [1+1] adduct. Reaction of the ditoluenesulfonamide of ethylenediamine **5** with the ditoluenesulfonate **6** derived from diethyleneglycol affords a 2:1 mixture of the [18]-N₄O₂ and [9]-N₂O cyclic products **7** and **8** (Scheme 2.3).[10,11] The [9]-N₂O cycle **8** is the sole product if the reaction is carried out by a different route taking advantage of the alternative disconnection. Reaction of the ditoluenesulfonate of ethylene glycol with the ditoluenesulfonamide **9** affords a high yield (79%) of the [9]-N₂O cyclic ditoluenesulfonamide **8**.

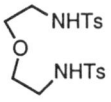

9

Protocol 3.
Synthesis of 1,10-dioxa-4,7,13,16-tetraazacyclooctadecane (Scheme 2.3)

Caution! Carry out all procedures in a well-ventilated hood, and wear disposable vinyl or latex gloves and chemical-resistant safety goggles.

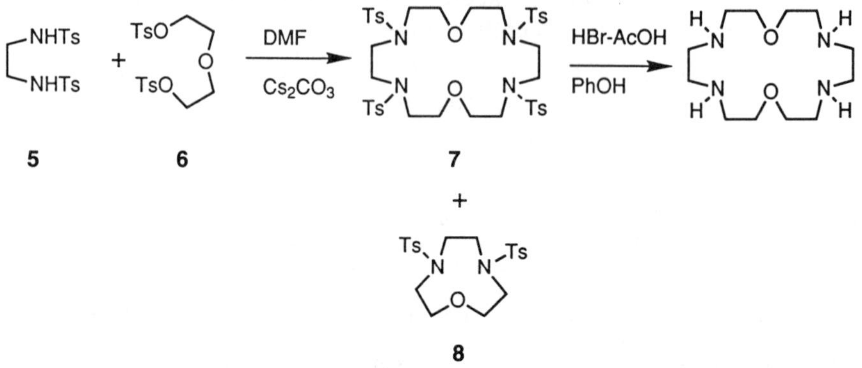

Scheme 2.3

The following procedure for the synthesis of the parent [18]-N_4O_2 macrocycle is representative of a direct [2 + 2] 'Richman–Atkins' macrocyclisation.[7]

Equipment

- Air condenser
- Hot plate stirrer (thermostatted)
- Magnetic teflon-coated stirrer bar
- Source of dry nitrogen or argon gas
- Gas line adaptor
- Vacuum pump (<0.2 mmHg)

- Water-cooled condenser
- Pressure-equalising addition funnel (10 mL)
- Two-necked, round-bottomed flask (250 mL)
- Separating funnels (500 mL and 250 mL)
- Conical flasks (250 mL and 100 mL)
- Sintered filter funnel (porosity 3)

Materials

• Dry DMF (HPLC grade), 100 mL	irritant, harmful by inhalation
• Dichloromethane, 300 mL	harmful by inhalation
• Toluene for recrystallisation, 100 mL	flammable
• Hexane for recrystallisation, 50 mL	flammable, irritant
• Diethyl ether, 50 mL	flammable, irritant
• Phenol, 2.0 g, 26.5 mmol	irritant, corrosive
• Anhydrous potassium carbonate (drying agent)	
• Anhydrous magnesium sulphate (drying agent)	
• Caesium carbonate, 8.26 g, 25.4 mmol	harmful by inhalation
• 1,5-bis(p-toluenesulfonato)-3-oxapentane, 5.00 g, 1.21 mol	cancer-suspect agent
• N,N'-bis(p-toluenesulfonyl)ethane-1,2-diamine, 4.44 g, 12.1 mmol	irritant
• Hydrogen bromide in acetic acid (45%) 100 mL	corrosive, irritant
• Aqueous potassium hydroxide solution (30%) 25 mL	corrosive
• Fluorescent silica gel TLC plates	irritant

2: Aza-oxa crowns

1. Dry all glassware and the stirrer bar in an 105°C electric oven for 2 h and allow the glassware to cool down under an atmosphere of dry nitrogen.
2. Add caesium carbonate (8.26 g, 25.4 mmol) to a solution of 1,5-bis(p-toluenesulfonato)-3-oxapentane (5.0 g, 12.1 mmol) in dry DMF (50 mL) in a two-necked round-bottomed flask (250 mL) fitted with an air condenser attached to a nitrogen line.
3. Add slowly, using a pressure-equalising addition funnel, a solution of N,N'-bis(p-toluenesulfonyl)ethane-1,2-diamine (4.44 g, 12.1 mmol) in dry DMF (50 mL) over a period of 4 h, with vigorous stirring.
4. Stir the reaction mixture at room temperature for 12 h, and then heat to 60°C for a period of 4 h, or until TLC analysis reveals that the starting materials have been consumed (silica, 2.5% methanol in dichloromethane).
5. Remove the DMF under reduced pressure (bath temperature c. 75°C at 1 mmHg pressure) and take up the residue in dichloromethane (100 mL) and water (100 mL). Transfer the mixture to a separating funnel, separate the organic layer and wash with more water (100 mL). Dry the organic layer with magnesium sulfate, (c. 2 g), filter and remove the solvent using a rotary evaporator to leave a pale-yellow solid residue.
6. Take up the residue in hot toluene (40 mL), transfer the solution to a warm conical flask (100 mL) and allow it to cool to room temperature. The 18-ring tetratoluenesulfonamide crystallises out first. Collect the product by filtration (warm sintered filter funnel) and dry it under vacuum (0.2 mmHg, 20°C) to yield a colourless solid, 1.06 g, 20%, m.p. 243–4°C.
7. The nine-membered ring ditoluenesulfonamide[a] may be obtained from the filtrate. Cool the filtrate (4°C) overnight and collect the colourless crystalline solid by filtration (sintered filter funnel) to give the [9]-ring co-product, 2.17 g, 41%, m.p. 161–2°C.
8. To a solution of phenol (2 g, 26.5 mmol) in HBr–acetic acid solution (45%, 100 mL) in a double-necked round-bottomed flask equipped with a water-cooled condenser surmounted by an air condenser attached to a water bubbler, add a magnetic stirrer bar and the tetratoluenesulfonamide as a solid (2.5 g, 2.86 mmol). Heat and stir the solution at 105°C for 1 week.[b]
9. Cool the mixture to room temperature and slowly add diethyl ether (50 mL) when a fine precipitate of the hydrobromide salt forms. Collect the salt by filtration and wash quickly with ether.[c] Take up the solid in distilled water (10 mL), basify with potassium hydroxide (pellets) until the pH is greater than 13, add dichloromethane (30 mL) and transfer the mixture to a separating funnel. Extract the aqueous layer with dichloromethane (6 × 30 mL), combine and dry the organic extracts (K_2CO_3), filter and remove the solvent on a rotary evaporator.

Protocol 3. Continued

10. Recrystallise the solid by dissolving in the minimum volume of dichloromethane (c. 2 mL) and adding hexane until the solution becomes turbid. Collect the solid by filtration (porosity 3 sinter) to give the desired tetraamine, 360 mg, 50%, m.p. 59–60°C.

[a] At higher dilution (500 mL of DMF), the yield of the 9-ring product rises to 62%.
[b] Alternatively, the tetratoluenesulfonamide may be deprotected with Li/l. NH$_3$/THF/EtOH (see Protocol 4).
[c] This salt is hygroscopic.

For the synthesis of large (>18) ring poly-aza-oxa macrocycles it is normally not possible to produce sufficient of the [2+2] cycloadduct for the reaction to be worthwhile. In these cases, a useful strategy is to add the poly-oxyethylene chain by alkylating an N-toluenesufonyl site in order to build up the desired number of heteroatoms prior to the [1+1] cyclisation reaction which again involves formation of two C–N bonds. An example of such a strategy is to be found in the synthesis of [24]-N$_4$O$_4$, **10**.[12] Attempts to prepare it directly by reaction of the ditoluenesulfonamide **5** of ethylenediamine with 'trigolditoluenesulfonate' **11** in DMF in the presence of Cs$_2$CO$_3$ resulted in the formation of the [12]-N$_2$O$_2$ cycle **12**. The preferred method involves alkylation of **5** with the chloro ether **13** to yield the crystalline dichloro intermediate **14a** (Scheme 4). This may be condensed directly with **5** to yield the cyclic tetratoluenesulfonamide in a 52% yield. Although the corresponding ditoluenesulfonate **14b** gives a higher yield (66%) in the ring-forming step, it is more difficult to purify.

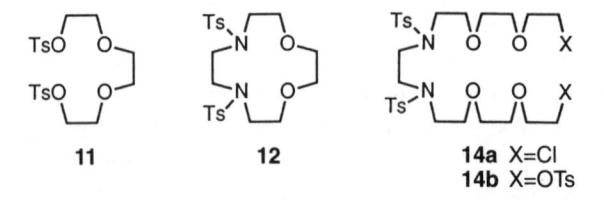

11

12

14a X=Cl
14b X=OTs

2: Aza-oxa crowns

Protocol 4.
Synthesis of 1,4,13,16-tetraoxa-7,10,19,22-tetraazacyclotetraeicosane (Structure 10, Scheme 2.4)

Caution! Carry out all procedures in a well-ventilated hood, and wear disposable vinyl or latex gloves and chemical-resistant safety goggles. The product amine should be stored under nitrogen or argon.

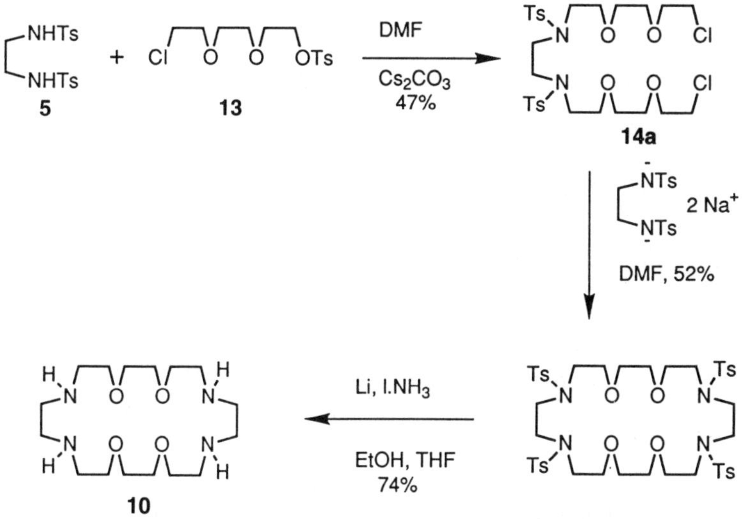

Scheme 2.4

The following procedure for the synthesis of the [24]-N_4O_4 macrocycle occurs in three steps.

Equipment

- Oil bath
- Hot-plate magnetic stirrer
- Single-necked, round-bottomed flask (250 mL)
- Single-necked, round-bottomed flask (100 mL)
- Source of dry nitrogen or argon gas
- Filter funnel
- Double-necked, round-bottomed flask (500 mL)
- Pressure-equalising addition funnel (250 mL)

- Condenser
- Magnetic stirrer bar
- Separating funnel (250 mL)
- Dry ice condenser
- Three-necked, round-bottomed flask (250 mL)
- pH indicator strips (BDH)
- Filter papers (Whatman No. 1)
- Fluorescent silica gel TLC plates

Materials

- Diethyl ether, 200 mL — flammable, irritant
- Dry tetrahydrofuran,[a] 20 mL — flammable, irritant
- Dry dimethylformamide HPLC grade, 430 mL — irritant, harmful by inhalation
- Caesium carbonate, 6.67 g, 0.02 mol — harmful by inhalation
- Sodium sulfate (anhydrous)
- Dry methanol[b] for recrystallisation, 100 mL — flammable

35

Protocol 4. Continued

• Anhydrous potassium carbonate	harmful by inhalation
• Dry ethanol[b] for recrystallisation, 50 mL	flammable
• Cylinder of ammonia gas	irritant
• Lithium metal, 1 g, 0.15 mol	flammable, moisture sensitive
• Dichloromethane, 500 mL	irritant, harmful by inhalation
• 6 M hydrochloric acid, 50 mL	corrosive, irritant
• Solid lithium hydroxide (c. 10 g)	irritant
• 1-Chloro-3,6-dioxa-8-toluenesulfonatooctane,[c] (6.38 g, 2 mmol)	toxic, irritant
• N,N'-bis(toluenesulfonyl)-1,2-diaminoethane,[c] 3.64 g, 0.01 mol	irritant
• Sodium metal, 0.35 g, 15 mmol	moisture sensitive!

1. Dry all glassware and stirrer bars in an electric oven (105°C) for 2 h.

2. Transfer the 1-chloro-3,6-dioxa-8-toluenesulfonatooctane (6.38 g, 0.02 mol) and N,N'-bis(toluenesulfonyl)-1,2-diaminoethane (3.64 g, 0.01 mol) into a single-necked 250 mL round-bottomed flask.

3. Pour into the flask 80 mL of dry dimethylformamide under a slow stream of nitrogen, add a Teflon stirring bar and add caesium carbonate (6.67 g, 0.02 mol).

4. Stir the mixture under nitrogen at room temperature for 20 h.

5. Heat for up to 3 h at 40°C, until the TLC analysis (silica, 2% methanol in dichloromethane eluant) shows that the product has formed completely (R_f=0.14).

6. Filter the mixture and remove the solvent by distillation under reduced pressure (bath temperature c. 50°C, at 0.5 mmHg using a vacuum pump).

7. Take up the solid residue in water (30 mL) and dichloromethane (50 mL), separate the organic layer in a separating funnel, wash the aqueous layer with further dichloromethane (30 mL) and combine the organic extracts. Dry the organic layer (using c. 2 g of sodium sulfate for 1 h), filter and remove the solvent using a rotary evaporator (25°C, 20 mmHg).

8. Take up the residue in the minimum volume of warm (c. 40°C) methanol and recrystallise to yield a white powder (m.p. 77–8°C, 6.02 g, 47%) which is N,N'-bis (toluenesulfonyl)-10,11-diaza-3,6,15,18-tetraoxa-1,20-dichloro-eicosane, 14a.

9. Take a small block of sodium metal out of the oil and with a sharp knife cut approximately 0.5 g from it. Wash the block with dry hexane to remove adhering oil, and cut the block into three smaller pieces. Dry the shiny small blocks with filter papers and weigh out 0.35 g accurately.

10. Carefully add the sodium metal (0.35 g, 15 mmol) to a 100 mL round-bottomed flask containing dry methanol (50 mL) and a small stirrer bar under nitrogen gas. Heat and stir the mixture to reflux for 90 min or until all of the sodium has dissolved. Cool the solution to room temperature, and add N,N'-bis(toluenesulfonyl)-1,2-diaminoethane (2.72 g, 7.4 mmol)

2: Aza-oxa crowns

under a stream of nitrogen gas. Heat the solution to reflux, with stirring (30 min), then cool and distil off the methanol. Remove the last traces of solvent by drying under high vacuum (40°C, 0.1 mmHg, 30 min).

11. Dissolve the solid in dry dimethylformamide (60 mL) and transfer the solution to an oven-dried 500 mL double-necked round-bottomed flask, equipped with a stirrer bar under a stream of nitrogen gas. Add a further volume of dimethylformamide (140 mL) and maintain under nitrogen.

12. Prepare a solution of N,N'-bis(toluenesulfonyl)-10,11-diaza-3,6,15,18-tetra-oxa-1,20-dichloroeicosane, **14a**, (4.95 g, 7.4 mmol) in dry dimethylformamide (150 mL) in a dry, pressure-equalising addition funnel and slowly add this solution, under nitrogen, over a period of 7 h, to the DMF solution of the sodium salt prepared above (step 11) which is heated to 100°C.

13. Stir the mixture for 18 h at 100°C under nitrogen, then remove the DMF by distillation under reduced pressure (bath temperature *c.* 50°C, 0.5 mmHg obtained using a vacuum pump).

14. To the solid residue add water (50 mL) and dichloromethane (100 mL) and some lithium chloride[d] (*c.* 1 g). Using a separating funnel obtain the organic layer and wash the aqueous layer with two further aliquots of dichloromethane (2 × 50 mL). Dry the combined organic extracts (*c.* 2 g of K_2CO_3), filter and evaporate the solvent using a rotary evaporator to yield a pale-yellow oil.

15. Dissolve the residue in hot ethanol (*c.* 60 mL) which, on cooling, yields a colourless solid. Collect the solid by filtration (filter papers), dry under vacuum (0.2 mmHg, 30 min) and check the purity of the product by TLC (R_f = 0.22, silica, 2% methanol in dichloromethane elutant). The colourless solid (m.p. 179°C, 3.67 g, 52%) displays the appropriate ^{13}C NMR and combustion analytical data for the tetratoluenesulfonamide intermediate.

16. Take an oven-dried three-necked round-bottomed flask (250 mL) and add the tetratoluenesulfonamide (1.96 g, 2.04 mmol), prepared as described in step 15, under a positive pressure of nitrogen. Equip the flask with a dry ice condenser, a nitrogen inlet and a stopper, and add dry tetrahydrofuran (30 mL) and dry ethanol (2 mL). Cool the suspension in a dry ice/isopropanol cold bath and add dry ice and isopropanol to the 'dry ice condenser'. Allow ammonia gas (from a cylinder) to condense into the cooled flask until about 75 mL of liquid ammonia has been added.

17. Take a lithium metal block out of its oil, and cut approximately 1 g (0.15 mol) from it. Ensure all the surfaces of the cut piece are shiny and metallic. Wash the lithium thoroughly in dry hexane (*c.* 20 mL) in a 50 mL beaker. Dry the lithium on filter papers and add it in small cut pieces to the cooled reaction flask, maintaining a strong nitrogen flow.

18. A strong blue coloration develops, and after maintaining the temperature

Protocol 4. *Continued*

at −78°C for 2 h, allow the reaction flask to warm up slowly to room temperature. During this period, the ammonia gas can be conveniently removed by allowing the reaction flask to vent (bubble through) into an anti-suck back apparatus involving a flask containing at least 500 mL of 6 M hydrochloric acid.

19. After all the ammonia has evaporated, add ethanol (2 mL) slowly drop by drop, followed by cold water (70 mL). The white solid will mostly dissolve. Remove the volatile organic solvents on a rotary evaporator and to the resultant aqueous solution add sufficient 6 M hydrochloric acid (c. 50 mL) so that the pH of the solution is less than 2. Wash the turbid solution with diethyl ether (4 × 50 mL) to leave a clear aqueous phase. Remove the water under reduced pressure (40°C, 2 mmHg), furnishing a colourless solid residue.

20. Redissolve the residue in the minimum volume of water (about 5 mL) and add sufficient solid lithium hydroxide to raise the pH above 13. Extract the aqueous layer with dichloromethane (8 times using c. 30 mL each time). Combine the organic extracts, dry with potassium carbonate, filter and evaporate the solvent using a rotary evaporator. Dry the colourless residue under high vacuum (0.1 mmHg) and from the resultant colourless oil, the product, **10**, crystallises on standing (m.p. 63–4°C, 74%). The ^{13}C NMR spectrum (CDCl$_3$) is diagnostic: δ_C 70.7 (CH$_2$O), 70.6 (CH$_2$O); 49.5 (CH$_2$N); 49.4 (CH$_2$N).

[a] THF and diethyl ether should be distilled from lithium aluminium hydride under an inert atmosphere immediately prior to use.
[b] Methanol and ethanol should be dried from the corresponding magnesium alkoxide.
[c] 1-Chloro-3,6-dioxa-8-toluenesulfonatooctane may be prepared from the commercially available 2-[2-(2-chloroethoxy)ethoxy]ethanol (Aldrich), by reaction with toluenesulfonyl chloride in THF in the presence of dry triethylamine; R_f=0.51 (silica, 1% methanol in dichloromethane). The ditoluenesulfonamide of ethylenediamine may be prepared by reaction of ethylenediamine with toluenesulfonyl chloride in pyridine at 40°C.
[d] Adding lithium salts to aqueous/dichloromethane extractions inhibits co-extraction of DMF into the organic phase.

A further example of this 'chain elongation' strategy is in the synthesis of pyridino-[24]-N$_6$O$_2$, **15**.[13] This heteroditopic ligand is most easily synthesised by linking all of the heteroatoms of one moiety of the target compound prior to cyclisation. Thus reaction of the pentatoluenesulfonamide **16** with 2,6-bis(bromomethyl)pyridine **17** affords the desired cycle in 63% yield under standard Richman–Atkins cyclisation conditions.[12] It is worth pointing out that in all of these macrocyclic aza-oxa syntheses it is imprudent to consider forming two carbon–oxygen bonds in the cyclisation step, notwithstanding the variety of methods that have been developed, based, for the most part, on cation-templated macrocyclisations (discussed in Chapter 4, page 80).

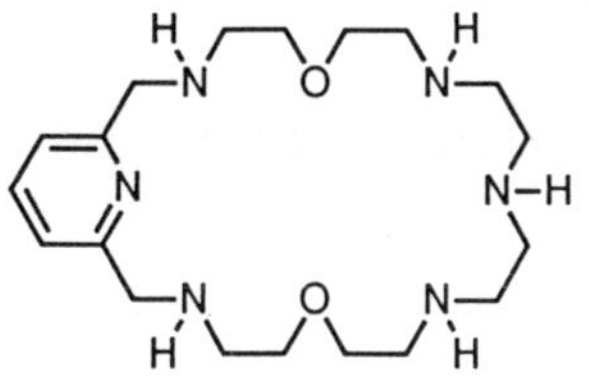

2: Aza-oxa crowns

15 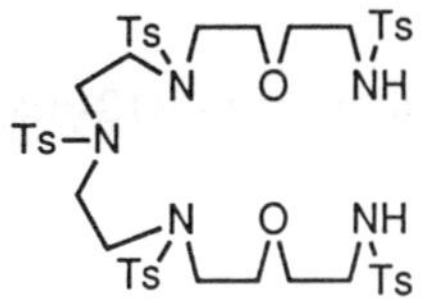 **16** 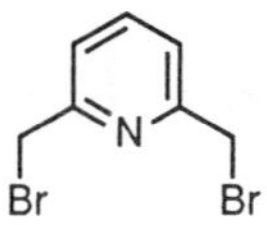 **17**

3. Alkylation at primary and secondary nitrogen

The Norwegian chemist Dale found that if a primary amine is reacted with a diiodo derivative of a short-chain polyethylene glycol, such as **18**, then reasonable yields of the cyclic mono-aza crown ether were obtained using Na$_2$CO$_3$ in acetonitrile, as in the formation of the [12]-NO$_3$ ligand **19**.[14]

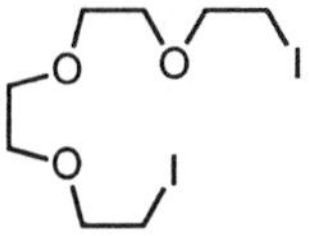

18 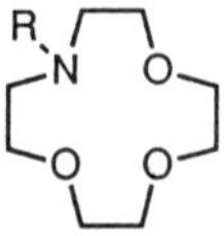 **19**

Thus for the formation of cyclic mono-aza-oxa ligands, there is no real need to use the *N*-toluenesulfonamide strategy discussed above, and the two ring C–N bonds are formed in the macrocyclisation step. This reaction is quite tolerant of other functionality in the primary amine (e.g. OH, CONR$_2$, CO$_2$Et, bpy),[14,15] and, if benzylamine is used as the starting material, the tertiary *N*-benzyl group may be removed by hydrogenolysis in aqueous acetic acid, preferably using Pearlman's catalyst, Pd(OH)$_2$ on C. The Dale reaction works well for the formation of [12] to [15]-NO$_x$ (x = 3 or 4) cycles, but if benzylamine is reacted with triethyleneglycol diiodide, then a [2 + 2] cyclisation results, leading to the formation of *N,N'*-dibenzyl-[18]-N$_2$O$_4$, rather than the expected [1 + 1] adduct.[16] In this case the sodium ion must be acting as a template for the formation of the larger ring cycle and the intermediate cyclised product is isolated as the sodium complex. This reaction, when carried out in a stepwise manner (Scheme 2.5) probably affords the best available synthesis of the [18]-N$_2$O$_4$ ring system.[17] The reaction is amenable to scale-up and the benzyl groups are readily removed by hydrogenolysis over Pd on carbon. Alternative routes are discussed earlier in this chapter and in Chapter 5 (Protocols 2 and 3)

Protocol 5.
Synthesis of 1,7,10,16-tetraoxa-4,13-diazacyclooctadecane (Scheme 2.5)

Caution! Carry out all procedures in a well-ventilated hood, and wear disposable vinyl or latex gloves and chemical-resistant safety goggles.

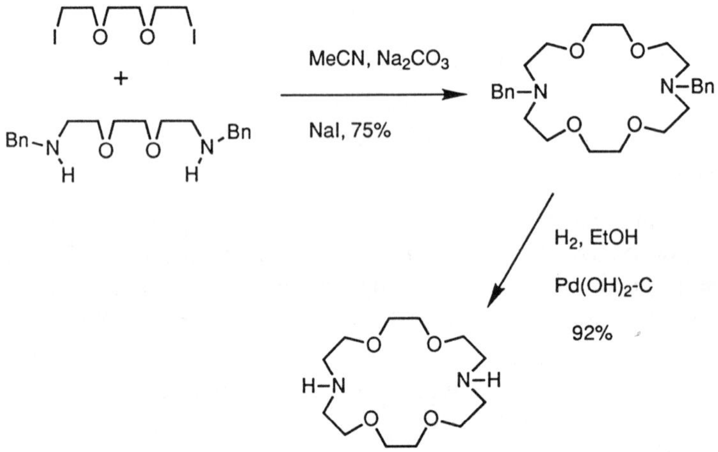

Scheme 2.5

Equipment

- Hot plate stirrer
- Oil bath
- Source of dry nitrogen
- Vacuum pump (<0.2 mmHg)
- Parr bench-top hydrogenation apparatus
- Source of hydrogen gas
- Water-cooled condenser
- Sintered filter funnel (porosity 2)
- Filter funnel
- Single-necked, round-bottomed flasks (1 L and 500 mL)
- Whatman No. 1 filter papers
- Conical flasks (250 and 500 mL)
- Stirrer bars
- Vacuum still head
- Separating funnel (500 mL)
- Buchner flask (1 L)
- Kugelrohr distillation apparatus

Materials

- Benzylamine, 192 g, 1.8 mol — irritant, toxic
- 1,2-Bis(2-chloroethoxy)ethane, 22.75 g, 0.122 mol — toxic, cancer-suspect agent, irritant
- Sodium hydroxide pellets (tech. grade), 8 g — corrosive, toxic
- Chloroform, 900 mL — flammable, harmful by inhalation
- Magnesium sulfate, 10 g (drying agent)
- 1,2-Bis(2-iodoethoxy)ethane,[a] 16.3 g, 0.44 mol — toxic, irritant, cancer-suspect agent
- Anhydrous sodium carbonate, 18.6 g, 0.176 mol — irritant
- Anhydrous sodium iodide, 2.8 g, 0.018 mol — irritant
- 96% ethanol, 100 mL — flammable
- Dry acetonitrile,[b] 600 mL — flammable
- Acetone for recrystallisation,[c] 50 mL — flammable, irritant
- Dioxane for recrystallisation, 50 mL — flammable, irritant
- Hexane for recrystallisation, 100 mL — flammable, irritant
- Palladium hydroxide on carbon, 1 g (Aldrich 21,291–1) — irritant

2: Aza-oxa crowns

1. Add benzylamine (192 g, 1.80 mol) and 1,2-bis(2-chloroethoxy)ethane (22.7 g, 0.122 mol) to a single-necked round-bottomed flask fitted with an air condenser attached to a nitrogen line. Stir the mixture for 48 h at 120°C.

2. Allow the mixture to cool to room temperature and add solid sodium hydroxide (8 g, 0.2 mol). Return the flask to 120°C and stir the mixture for 1 h. Distil off the excess benzylamine under high vacuum (0.2 mmHg at 45°C) and, when all the amine is gone, dissolve the residue in chloroform (200 mL). Wash the organic layer with water (2 × 150 mL), separate the lower organic layer, dry it over magnesium sulfate (c. 5 g, 30 min), filter and remove the solvent on a rotary evaporator.

3. Remove remaining traces of solvent and benzylamine by heating the flask under vacuum (0.05 mmHg, 60°C), to leave a pale-yellow viscous oil, 39.8 g, 99%, which is N,N'-dibenzyl-4,7-dioxa-1,10-diazadecane. Bulb to bulb distillation (Kugelrohr apparatus) at 175°C (0.2 mmHg) yields material which is sufficiently pure to be used directly in the next step.

4. Add to a single-necked 1 L round-bottomed flask, fitted with a water-cooled condenser linked to a nitrogen line, N,N'-dibenzyl-4,7-dioxa-1,10-diazadecane (12 g, 0.037 mol), 1,2-bis(2-iodoethoxy) ethane (16.25 g, 0.044 mol), anhydrous sodium carbonate (18.6 g, 0.176 mol), anhydrous sodium iodide (2.78 g, 0.018 mol) and dry acetonitrile (500 mL). Stir the mixture vigorously for 48 h and heat under reflux.

5. Allow the mixture to cool, filter off the inorganic solids on a Buchner flask and wash the residue with hot acetonitrile (2 × 50 mL). Combine the organic fractions and remove the solvent using a rotary evaporator to yield a sticky semi-solid mass. Dissolve the residue in the minimum volume of dioxane and warm acetone (1:1 v/v). This will be approximately 75 mL of solution. Leave the product to crystallise at −15°C (overnight) and filter off a colourless solid, which is the sodium complex of the macrocycle.

6. Dissolve the solid in chloroform (c. 200 mL), transfer the solution to a 500 mL separating funnel and wash with water (2 × 150 mL). Separate the organic layer and wash the aqueous layers with further chloroform (2 × 100 mL). Combine the organic extracts, dry over magnesium sulfate, remove the solvent using a rotary evaporator and recrystallise[d] the residue from boiling hexane to yield the N,N'-dibenzyl macrocycle as a colourless crystalline solid, m.p. 81–2°C, 7.26 g, 75%.

7. Transfer a solution of the N,N'-dibenzyl-1,7,10,16,-tetraoxa-4,13-diazacyclooctadecane (10 g, 0.023 mol) in ethanol (100 mL) to a 500 mL 'hydrogenation bottle' suitable for use in a low pressure Parr hydrogenation apparatus.[e] Carefully add Pearlman's catalyst (1.0 g), connect the bottle to the hydrogenation apparatus and shake for 72 h under a hydrogen pressure of 2–3 atmospheres. Filter the mixture through a bed of Celite (c. 1 cm) on a

Protocol 5. Continued

sinter funnel (porosity 2), and remove the solvent on a rotary evaporator. Recrystallise the solid residue from hot hexane to afford the desired amine as a colourless crystalline solid, m.p. 114–15°C, 5.46 g, 92%.

[a] The diiodo compound is easily prepared from the commercially available 1,2-bis(2-chloroethoxy)ethane by reaction with anhydrous sodium iodide in dry acetone.
[b] Acetonitrile may be dried by distillation from calcium hydride.
[c] Acetone is distilled from 4 Å molecular sieves. HPLC grade acetone may be used directly.
[d] Repeated recrystallisations from hot hexane may be needed to remove the yellow coloration of impurities.
[e] Alternatively, debenzylation may be effected using sodium in liquid ammonia in THF (95% yield). See Protocol 4 for a related example using lithium in liquid ammonia.

A related strategy for the formation of N-substituted aza-oxa crowns also involves alkylation at a secondary nitrogen site in the ring-forming step. Reaction of bis-α-chloroamides with secondary diamines in boiling MeCN in the presence of a carbonate base yields cyclic diamides (Scheme 2.6).[18] The reaction is also tolerant of hydroxyl functionality in the N- or C-substituent, and may in principle be used to prepare differentially N-protected poly-aza-oxa rings, although most of the examples reported so far involve N-alkyl vs N'-benzyl groups.

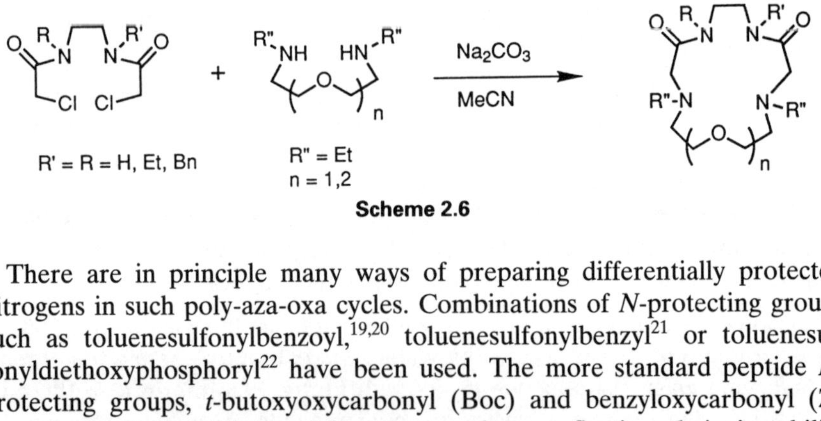

R' = R = H, Et, Bn

R" = Et
n = 1,2

Scheme 2.6

There are in principle many ways of preparing differentially protected nitrogens in such poly-aza-oxa cycles. Combinations of N-protecting groups such as toluenesulfonylbenzoyl,[19,20] toluenesulfonylbenzyl[21] or toluenesulfonyldiethoxyphosphoryl[22] have been used. The more standard peptide N-protecting groups, t-butoxyoxycarbonyl (Boc) and benzyloxycarbonyl (Z) have been used rarely in this sense, perhaps reflecting their instability towards the rather forcing macrocyclisation reaction conditions. Nevertheless, in the synthesis of the [24]-N$_4$O$_4$ precursor **21b**, obtained from the N-benzyldiamine 20 by reaction with N-benzyloxycarbonylaziridine, selective removal of the Z group in the presence of the N-benzyl group was effected by bubbling hydrogen through a methanol solution of **21a** in the presence of palladium on carbon.[23] Reaction of **21b** with the appropriate diacid chloride (Scheme 2.7) under high dilution conditions afforded the differentially N-protected diamide [24]-N$_4$O$_4$ cycle **22** which could be reduced to the diamine **23** in high yield with LiAlH$_4$.[23] Such a molecule is probably more con-

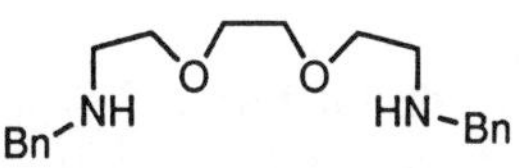

2: Aza-oxa crowns

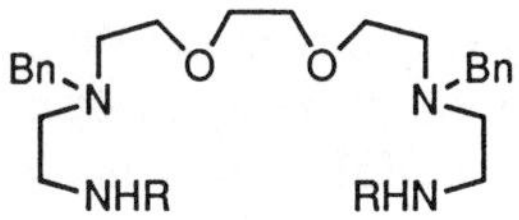

Scheme 2.7

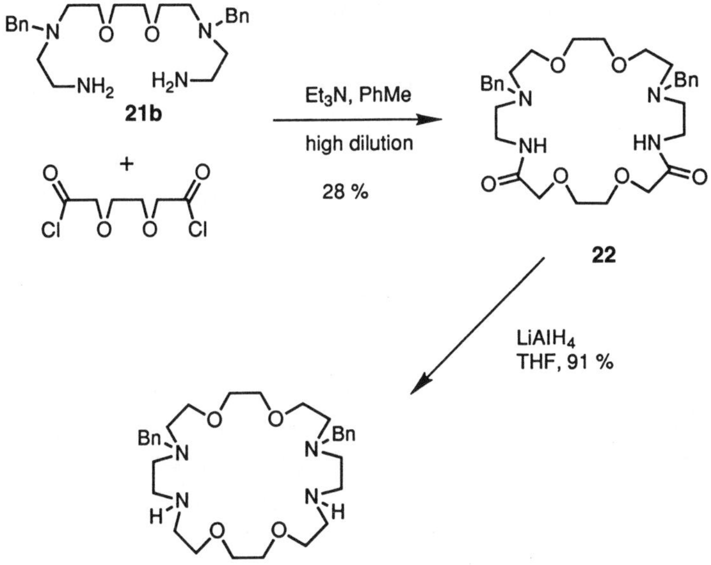

veniently prepared via the intermediacy of α-chloroacetamides,[18] as is suggested by the example in Scheme 2.6.

4. Conclusion

The synthesis of aza-oxa crown ethers is best accomplished by making carbon–nitrogen bonds in the cyclisation step. Although the original syntheses operated under conditions of high dilution and involved the co-condensation of a diamine with a diacid chloride, these methods have been supplanted by the more versatile and convenient *N*-alkylation pathways involving toluenesulfonamide or *N*-benzyl intermediates. This chapter has focused on the metal-free synthesis of saturated aza-oxa crown ethers. There are a large number of examples of the synthesis of aromatic and heterocyclic aza-oxa crown ethers that involve the co-condensation of aldehydes and amines mediated by metal ions such as Pb^{2+} and Ba^{2+}.[24,25] This *in situ* synthetic

David Parker

method is very direct, but the lack of understanding of the mechanism of the templated Schiff-base reaction has tended to inhibit rational syntheses and the selection of the amine and carbonyl components, together with the choice of metal ion, relies to a certain extent upon intuition. Nevertheless, this semi-empirical method has permitted certain [1+1] or [2+2] syntheses to proceed in high yield in single-step reactions, according to the metal ion chosen, as shown in Scheme 2.8. The macrocyclic metal complex can be reduced by sodium borohydride, which not only reduces the imine ligand but may also result in demetallation. Thus reduction of **24** (in Scheme 2.8) with methanolic borohydride yields the reduced free amine py$_2$-[30]-N$_6$O$_4$ **25**.

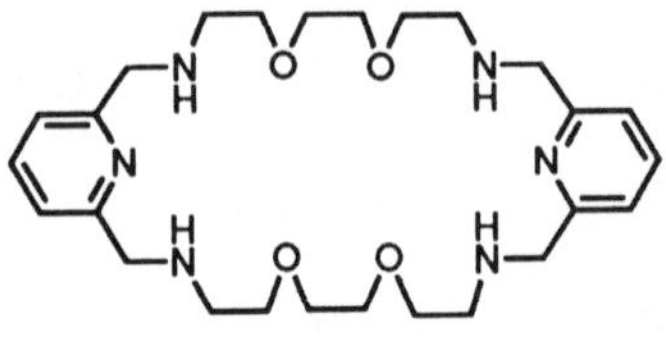

25

Protocol 6.
Synthesis of 6,9,23,26-tetraoxa-3,12,20,29,35,36-hexaazatricyclo-[29.3.1.1^{14,18}]hexatriacontane-1(35),14,16,18(36),31,35-hexaene, py$_2$-[30]-N$_6$O$_4$, (Structure 24, Scheme 2.8)

Caution! Carry out all procedures in a well-ventilated hood, and wear disposable vinyl or latex gloves and chemical-resistant safety goggles.

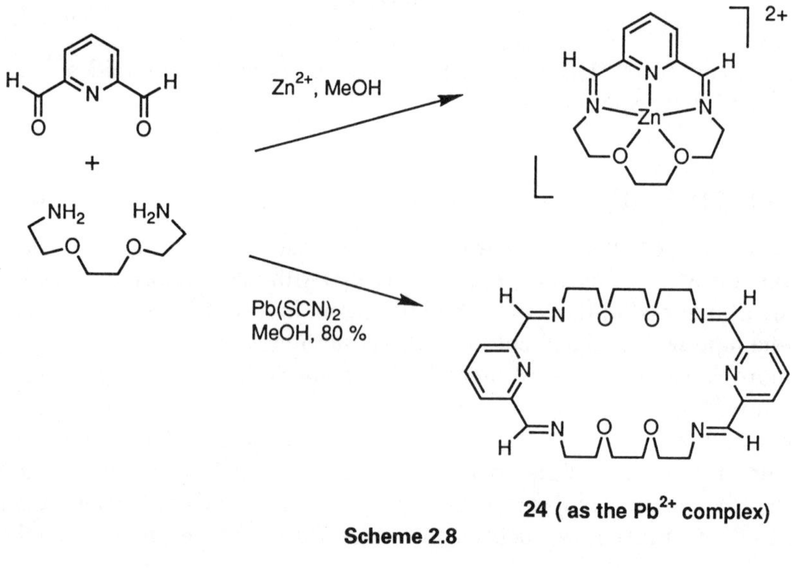

Scheme 2.8

24 (as the Pb^{2+} complex)

2: Aza-oxa crowns

The following procedure is representative of a metal-templated [2+2] co-condensation reaction of an aromatic dialdehyde with an acyclic diamine.[26]

Equipment

- Three-necked, round-bottomed flask (1 L)
- Thermostatted hot-plate stirrer
- Oil bath
- Reflux condenser
- Gas inlet tube
- Source of dry argon
- Buchner flask (100 mL)

- Separating funnel (250 mL)
- Source of dry HCl gas
- Thermometer (0–100 °C)
- Fluted filter paper (Whatman No. 1)
- Filter funnel
- Teflon magnetic stirrer bar

Materials

- Methanol, 380 mL — **flammable, toxic**
- 2,6-Diformylpyridine, 1.0 g, 7.4 mmol — **irritant**
- 4,7-Dioxa-1,10-diaminooctane, 1.1 g, 7.4 mmol — **irritant**
- Lead thiocyanate,[a] 2.5 g, 7.75 mmol — **toxic**
- Sodium borohydride, 2.2 g, 64.7 mmol — **irritant, corrosive**
- Aqueous EDTA solution (1.0 M, 20 mL, pH 9.7)
- Chloroform, 100 mL — **harmful by inhalation**
- Saturated aqueous sodium chloride solution, 50 mL
- Sodium sulfate (drying agent), 5 g
- Ethanol for recrystallisation, 50 mL — **flammable**
- Diethyl ether for recrystallisation, 50 mL — **flammable, irritant**

1. Charge a 1 L three-necked round-bottomed flask fitted with a thermometer, a Teflon magnetic stirrer bar, a gas inlet tube and a reflux condenser, with dry methanol (380 mL), and add 2,6-diformylpyridine (1 g), lead thiocyanate (2.5 g) and 4,7-dioxa-1,10-diaminooctane (1.1 g), in that order.

2. Pass dry argon gas through the solution to displace most of the dissolved oxygen and whilst maintaining a flow of argon, heat the mixture at 40 °C for 1 h. Raise the temperature to 45 °C and quickly add sodium borohydride (2.2 g) as a solid. After the addition is complete, stir the mixture for a further 30 min.

3. Cool the solution to room temperature and remove the deposit of metallic lead by filtration. Evaporate the filtrate and dissolve the residue in chloroform (100 mL) and aqueous EDTA solution (1.0 M, pH 9.7, 20 mL). Transfer the contents of the flask to a separating funnel and discard the aqueous layer.

4. Wash the chloroform solution twice with saturated aqueous brine (2 × 25 mL), dry with anhydrous sodium sulfate (c. 4 g, 2 h), filter and remove the solvent on a rotary evaporator to yield a pale-yellow oil, 1.78 g (90%), which is the free amine.

5. Treat the residue with ethanol/ether (1:1, v/v, 25 mL) and pass dry HCl gas through the solution until the solution is saturated. Allow the mixture to stand in a stoppered flask for 30 min, and isolate the colourless solid by

Protocol 6. Continued

filtration using a Buchner flask. Wash the solid with dry ether (2 × 20 mL), and recrystallise from absolute ethanol (c. 30 mL). Isolate the hydrochloride salt by filtration and dry under high vacuum to give the title compound as the tetrahydrochloride dihydrate, 1.7 g, 67%.

[a] Lead thiocyanate may be prepared directly by reaction of Pb(NO$_3$)$_2$ with KNCS.

References

1. Dietrich, B.; Lehn, J. M.; Sauvage, J. P. *Tetrahedron Lett.* **1969**, 2885; Dietrich, B.; Lehn, J. M.; Sauvage, J. P.; Blanzat, J. *Tetrahedron* **1973**, *29*, 1629.
2. Gokel, G. W.; Korzeniowski, S. H. *Macrocyclic Polyether Syntheses*; Springer-Verlag: Berlin, **1982**.
3. Krakowiak, K. E.; Bradshaw, J. S.; Zamecka-Krakowiak, D. J. *Chem. Rev.* **1989**, *89*, 156.
4. Jurczak, J.; Kaspryzk, S.; Salanski, P.; Stankiewicz, T. *J. Chem. Soc., Chem. Commun.* **1991**, 956.
5. Tabushi, I.; Okino, H.; Yuroda, Y. *Tetrahedron Lett.* **1976**, 4339.
6. Helps, I. M.; Jankowski, K. J.; Nicholson, P. E.; Parker, D. *J. Chem. Soc., Perkin Trans. 1* **1989**, 2079.
7. Atkins, T. J.; Richman, J. E.; Oettle, W. F. *Org. Synth.* **1979**, *58*, 86.
8. Comarmond, J.; Plumere, P.; Lehn, J. M.; Agnus, Y.; Louis, R.; Weiss, R.; Kahn, O.; Morgenstern-Badarau, I. *J. Am. Chem. Soc.* **1982**, *104*, 6330.
9. Fuji, T.; Sakakibara, S. *Bull. Chem. Soc. Jpn.* **1974**, *47*, 3146; Pulukkody, K. P.; Norman, T. J.; Parker, D.; Royle, L.; Broan, C. J. *J. Chem. Soc., Perkin Trans. 2* **1993**, 605.
10. Luboch, E.; Cygan, A.; Biernat, J. F. *Inorg. Chim. Acta* **1983**, *168*, 201; Biernat, J. F.; Luboch, E. *Tetrahedron* **1984**, *40*, 1927.
11. Craig, A. S.; Kataky, R.; Matthews, R. C.; Parker, D.; Ferguson, G.; Lough, A.; Adams, A.; Bailey, N.; Schneider, H. *J. Chem. Soc., Perkin Trans. 2* **1990**, 1523.
12. Parker, D.; Rosser, M.; Howard, J. A. K.; Yufit, D.; Ferguson, G.; Gallagher, J. *J. Chem. Soc., Chem. Commun.* **1993**, 1267.
13. Parker, D. *J. Chem. Soc., Chem. Commun.* **1985**, 1129.
14. Amble, E.; Dale, J. *Acad. Chem. Scand.* **1979**, *B33*, 698; Calverley, M. J.; Dale, J. *Acad. Chem. Scand.* **1982**, *B36*, 241.
15. Kataky, R.; Matthes, K. E.; Nicholson, P. E.; Parker, D.; Buschmann, H. J. *J. Chem. Soc., Perkin Trans. 2* **1990**, 1425.
16. Gatto, V. J.; Gokel, G. *J. Am. Chem. Soc.* **1984**, *106*, 8240.
17. Gatto, V. J.; Miller, S. R.; Gokel, G. W. *Org. Synth.* **1989**, *68*, 227.
18. Krakowiak, K. E.; Bradshaw, J. S.; Izatt, R. M. *Synlett* **1993**, 611.
19. Martin, A. E.; Ford, T. M.; Bulkowski, J. E. *J. Org. Chem.* **1982**, *47*, 412.
20. Cox, J. P. L.; Craig, A. S.; Helps, I. M.; Jankowski, K. J.; Parker, D.; Eaton, M. A. W.; Millican, A. T.; Millar, K.; Beeley, N. R. A.; Boyce, B. A. *J. Chem. Soc., Perkin Trans. 1* **1990**, 2567.
21. Anelli, P. L.; Montanari, F.; Qici, S. *J. Org. Chem.* **1988**, *53*, 5292.

2: Aza-oxa crowns

22. Mertes, K. B.; Mertes, M. P.; Qian, L.; Sun, Z. *J. Org. Chem.* **1991**, *56*, 4904.
23. Zinic, M.; Alihodzic, S.; Skaric, V. *J. Chem. Soc., Perkin Trans. 1* **1993**, 21.
24. Cook, D. H.; Fenton, D. E.; Rodgers, A.; McCann, M.; Nelson, S. M. *J. Chem. Soc., Dalton Trans.* **1979**, 414.
25. Nelson, S. M.; Knox, C. V.; McCann, M. *J. Chem. Soc., Dalton Trans.* **1981**, 1669.
26. Menif, R.; Chen, D.; Martell, A. E. *Inorg. Chem.* **1989**, *28*, 4633.

3

Thia, oxa-thia and aza-thia crowns

DAVID PARKER

1. Introduction

The synthesis of macrocyclic thia crown ethers dates back long before the first reports of the synthesis of Pedersen's oxa crown ethers. Back in 1920, Ray[1] reported the synthesis of 1,4,7-trithiacyclononane, **1**, which is also known by the acronym [9]-S$_3$, involving reaction of potassium hydrogen sulfide with dibromoethane. In 1934, Meadow and Reid described the isolation of 1,4,7,10,13,16-hexathiacyclooctadecane ([18]-S$_6$), **2**, albeit in a yield of 1.7%.[2] It has also been appreciated for some time that in order to avoid undesirable oligomerisations, the co-cyclisation of an acyclic dithiol with an alkyl dihalide should be carried out under conditions of high dilution. Workers at Kodak, for example, described in 1961 the synthesis of [18]-S$_2$O$_4$, **3**, in a 59% yield by working at high dilution in ethanol.[3]

1   **2**   **3**

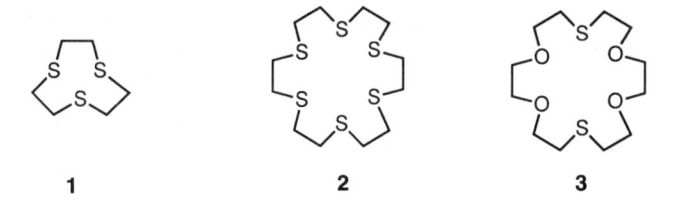

The homoleptic thia crowns, such as **1** and **2**, are not as amenable to versatile metal-templated syntheses as the all-oxygen crowns (see Chapter 4). However in 1981, Kellogg and Buter reported a high yielding and general synthesis of macrocyclic thioethers that did not require conditions of high dilution but that was strongly dependent on the nature of the metal carbonate base used.[4] The reaction involved the co-condensation of a dithiol with an alkyl dihalide in DMF in the presence of caesium carbonate (Table 3.1). Inferior yields were observed if other metal carbonates were used. The caesium ion is not involved in any templating of the macrocyclisation step. In DMF

Table 1 Yield of macrocyclic sulfides in metal carbonate mediated cyclisation reactions

M$_2$CO$_3$	yield (%)
Li$_2$CO$_3$	0
Na$_2$CO$_3$	33
K$_2$CO$_3$	48
Rb$_2$CO$_3$	84
Cs$_2$CO$_3$	90

solution, the caesium thiolates (derived by proton abstraction of the thiol by the carbonate base) are present as solvent-separated ion pairs. The role of the caesium ion is likely to be related to improving the reaction's solubility characteristics. This may involve solubilising the Cs$_2$CO$_3$ in the DMF solvent or relate to enhancing the solubility of the caesium thiolate intermediate.

An example of the caesium carbonate-mediated cyclisation reaction is the synthesis of hexathia-18-crown-6, **2** (Scheme 3.1). Co-condensation of 3-thiapentane-1,5-dithiol with 3,6,9-trithia-1,11-dichloroundecane in DMF in the presence of caesium carbonate yielded [18]-S$_6$, **2**, in 60% overall yield. It is important to emphasise the dangerous nature of the β-halo thioethers in these reactions. Compounds that contain the –SCH$_2$CH$_2$X (X = halide or good leaving group) are powerful vesicants and as they are 'mustard gas' analogues, they are potent carcinogens. They must be handled with considerable care and appropriate precautions must be taken. As a matter of safety, therefore, a full-body laboratory coat, goggles and effective plastic gloves should be worn at all times. The reaction and all subsequent manipulations must be carried out in an efficient fume-hood. Fortunately, the products of such reactions appear to be completely innocuous.

Protocol 1.
Synthesis of 1,4,7,10,13,16-hexathiacyclooctadecane (Structure 2, Scheme 3.1)

Caution! The intermediate 2-chloroethyl sulfides are very toxic (sulfur mustards) and are very powerful vesicants and potent carcinogens. This protocol must be undertaken in an efficient hood and disposable vinyl and/or latex gloves and chemical-resistant safety goggles must be worn at all times.

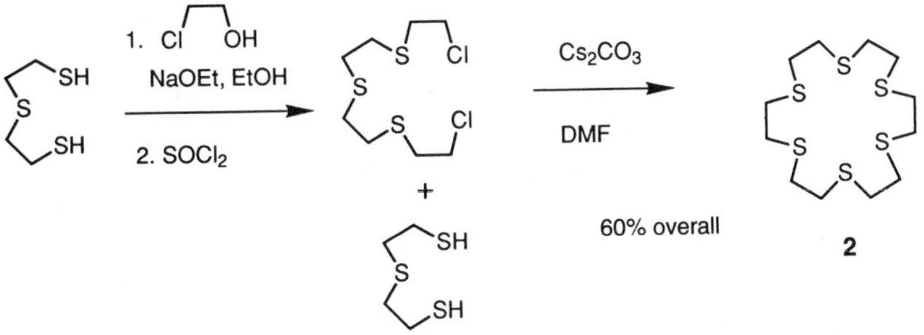

Scheme 3.1

This reaction is representative of the synthesis of cyclic thioethers from acyclic dithiols mediated by caesium carbonate in DMF.

Equipment

- Double-necked, round-bottomed flask (500 mL)
- Thermostatted hot plate stirrer
- Oil bath
- Pressure-equalising addition funnels (50 mL and 250 mL)
- Drying tube (calcium chloride)
- Double-necked round-bottomed flask (1 L)
- Separating funnel (500 mL)
- Filter papers (Whatman No. 1)
- Filter funnel
- Sintered glass filter funnel (porosity 2)
- Erlenmeyer flask (250 mL)
- Source of dry nitrogen
- Teflon-coated magnetic stirrer bar

Materials

- 3-Thiapentane-1,5-dithiol, 5.09 g, 33 mmol — stench, irritant
- 3,6,9-Trithiaundecane-1,11-diol,[a] 8 g, 33 mmol — irritant
- Thionyl chloride, 8 mL, 110 mmol — toxic, irritant
- Dry dichloromethane,[b] 600 mL — harmful by inhalation
- HPLC-grade dry dimethylformamide, 500 mL — irritant, harmful by inhalation
- Caesium carbonate,[c] 13 g, 40 mmol — harmful by inhalation
- Aqueous potassium hydroxide solution, 1 M, 180 mL — corrosive
- Anhydrous magnesium sulfate
- Hexane for recrystallisation, 80 mL — flammable, irritant
- Acetone for recrystallisation, 20 mL — flammable, irritant
- Dry methanol, 5 mL — flammable, harmful by inhalation

1. To a double-necked round-bottomed flask (500 mL) equipped with a magnetic stirrer bar and a calcium chloride drying tube, add dry dichloromethane

Protocol 1. *Continued*

(200 mL) and 3,6,9-trithiaundecane-1,11-diol (8 g, 33 mmol). Prepare a solution of thionyl chloride (8 mL, 110 mmol) in dry dichloromethane (25 mL) and transfer this to the pressure-equalising dropping funnel (50 mL) which is then stoppered.

2. Slowly add the thionyl chloride solution to the stirred suspension. A vigorous evolution of gas occurs and the diol dissolves to give a colourless solution. Stir for a further 6 h at room temperature, and then slowly add dry methanol (5 mL) to quench the excess thionyl chloride. Remove the solvent under reduced pressure using a rotary evaporator and remove traces of solvent by pumping at room temperature, under high vacuum (<0.2 mmHg). Use the toxic product dithiol (R_f = 0.77, SiO_2, CH_2Cl_2) directly in the next step without further characterisation or purification.

3. Take up the β-chloro thioether, prepared in step 2, in dry DMF (150 mL) and carefully transfer the solution to a pressure-equalising addition funnel (250 mL). Add 3-thiapentane-1,5-dithiol (5.09 g, 33 mmol) to this solution.

4. Prepare an oven-dried (electric oven, 105°C, 1 h) double-necked round-bottomed flask (1 L) equipped with a magnetic stirrer bar, fitted with a nitrogen inlet and the pressure-equalised dropping funnel prepared in step 3. Charge the flask with caesium carbonate (13 g, 40 mmol) and dry DMF (350 mL) and heat the stirred suspension to 55°C.

5. Slowly add the β-halo thioether solution from the addition funnel over a period of 18 h. Once the addition is complete, continue stirring for a further 2 h. Remove the DMF by distillation under reduced pressure (bath temperature 50°C, 0.2 mmHg) and swirl the residue with dichloromethane (2 × 150 mL) for 15 min.

6. Filter the resultant suspension through a short pad of Celite (1 cm on top of a sintered glass filter funnel) and transfer the filtrate to a separating funnel (500 mL). Wash successively with aqueous potassium hydroxide solution (1 M, 3 × 60 mL) and water (2 × 60 mL) and dry the organic layer with anhydrous magnesium sulfate (c. 3 g, 30 min).

7. Filter the 'dried' solution and remove the solvent on a rotary evaporator. Treat the residue with warm acetone–hexane (1:4, v/v, c. 100 mL), filter when warm if required, and allow the solution to cool slowly to room temperature. Separate the colourless, crystalline solid[d] and allow to dry in air, m.p. 93–4°C, 8.0 g, 56% overall, R_f = 0.34 (SiO_2, CH_2Cl_2).

[a] The diol (R_f = 0.42, silica, EtOAc) is conveniently prepared by reaction of 3-thiapentane-1,5-dithiol with 2-chloroethanol in ethanol in the presence of sodium ethoxide.
[b] Dichloromethane should be distilled from CaH_2 or P_4O_{10} prior to use.
[c] If fine-mesh (325-mesh) potassium carbonate is used, the yield is reduced to c. 40%.
[d] If necessary the product may be purified by column chromatography on silica gel (Flash: Merck Art. No. 9385, CH_2Cl_2 elutant).

3: Thia, oxa-thia and aza-thia crowns

The complexation chemistry of thia crowns has been the subject of intensive study since 1980. This has become more evident since the advent of the commercial availability of [9]-S_3, [12]-S_4, [14]-S_4, [15]-S_5, [18]-S_6 and [24]-S_8. These studies have been the subject of extensive reviews,[5,7] which have focused primarily on structural aspects of the complexes of metals such as Au, Ag and Cu and Mn, Fe, Co and Ni.

The macrocyclic effect is much less pronounced than with the corresponding oxa and aza crowns. This may be related to the very different conformation adopted by macrocycles incorporating –SCH_2CH_2S– units. As a result of the longer carbon–sulfur bond and the repulsive interaction between sulfur lone pairs adopting a *gauche* placement, the –SCH_2CH_2S– unit strongly prefers to adopt an *anti* conformation in the free ligand (Fig. 3.1). As a result, in poly-thia macrocycles that contain this unit, the sulfur lone pairs tend to point out of the ring. It follows that for metal binding to occur, considerable conformational reordering is required, which is energetically unfavourable.[5] With a trimethylene unit in the ring, the Thorpe–Ingold effect (gem dimethyl effect) may be used to advantage to predispose the sulfur lone pairs more

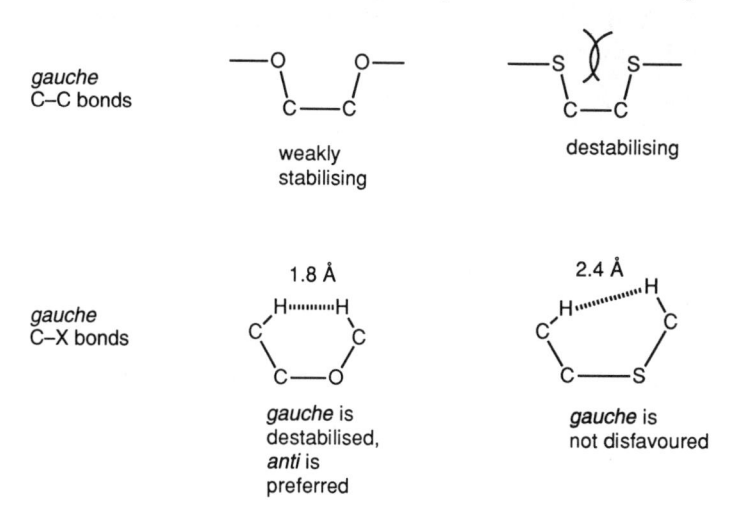

gauche C–C bonds: weakly stabilising (–O O–); destabilising (–S S–)

gauche C–X bonds: 1.8 Å, *gauche* is destabilised, *anti* is preferred (C–O); 2.4 Å, *gauche* is not disfavoured (C–S)

Fig. 3.1 1,4-Interactions at C–C and C–X bonds.

S lone pairs directed out

S lone pairs directed in

Fig. 3.2 Predisposing the [14]-S_4 macrocycle by introduction of gem-dimethyl substituents.

effectively. Thus introduction of a 2,2-dimethyl moiety into the [14]-S_4 ring changes the preferred conformation, and leads to a marked improvement in its coordination properties (Fig. 3.2).[6]

2. Cyclisation reactions forming carbon–sulfur bonds

The thiol group is relatively acidic, pK_a c. 10, so that it does not require a strong base for deprotonation. In alcohols such as methanol, ethanol or butanol, sodium carbonate may be used to generate the desired thiolate. By working under high dilution, acceptable yields of several simple thia-oxa crowns may be obtained (Scheme 3.2).[1] This reaction may be used in the synthesis of the related [12]-S_2O_2 and [15]-S_2O_3 cycles. Ligands that incorporate aromatic and/or heterocyclic subunits may also be made by this procedure.[8]

Protocol 2.
Synthesis of 1,10-dithia-4,7,13,16-tetraoxacyclooctadecane (Structure 3, Scheme 3.2)

Caution! This protocol must be undertaken in an efficient hood and disposable vinyl or latex gloves and chemical-resistant safety goggles must be worn at all times.

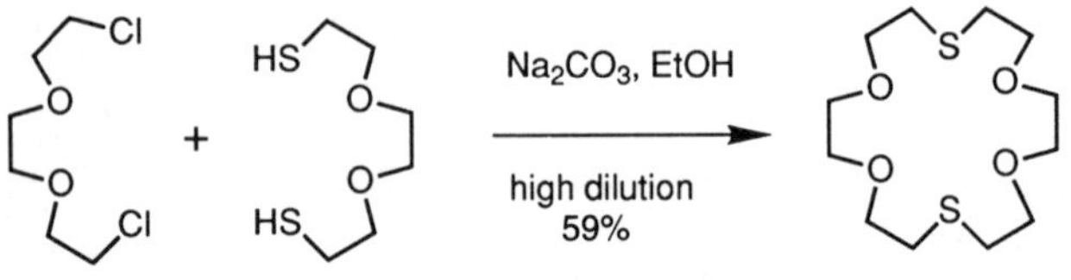

Scheme 3.2

This is a simple example of the alkylation of a dithiol at high dilution.

Equipment
- Hot plate stirrer
- Oil bath
- Pressure-equalising addition funnel (100 mL)
- Source of dry nitrogen (preferably from a nitrogen line)
- Sintered glass filter funnel (porosity 3)
- Teflon-coated magnetic stirrer bar
- Double-necked, round-bottomed flask (5 L)
- Reflux condenser
- Erlenmeyer flask (250 mL)

Materials
- 96% ethanol, 3 L — flammable
- Ethyl acetate, 180 mL — flammable
- 1,8-Dichloro-3,6-dioxaoctane,[a] 23.5 g, 0.12 mol — toxic, cancer-suspect agent
- 3,6-Dioxaoctan-1,8-dithiol,[a] 23.0 g, 0.12 mol — stench, irritant
- Sodium carbonate, 13.25 g, 0.12 mol

3: Thia, oxa-thia and aza-thia crowns

1. Pour ethanol (3 L) into a double-necked round-bottomed flask (5 L), equipped with a large magnetic stirrer bar and a reflux condenser surmounted by a nitrogen inlet. Add sodium carbonate (13.25 g, 0.125 mol) and heat the solution to 60 °C.
2. Prepare a mixture of 1,8-dichloro-3,6-dioxaoctane and 3,6-dioxaoctan-1,8-dithiol (23.0 g, 0.125 mol) in a pressure-equalising addition funnel (100 mL) and add the mixture slowly over a period of 2.5 h to the vigorously stirred boiling ethanol solution. Once the addition is complete, heat and stir the mixture for 48 h under nitrogen.
3. Remove the solvent on a rotary evaporator and treat the residue successively with three volumes of hot ethyl acetate (3 × 60 mL). Pour the extracts into a warm flask (250 mL) and allow the solution to cool to room temperature. Allow the stoppered flask to stand for 15 h at 4 °C.
4. Filter the mixture through a sintered glass filter funnel and dry the colourless crystalline solid under vacuum (0.1 mmHg, 3 h) to give the desired product, m.p. 91–2 °C, 21.6 g (59%).

[a] Commercially available.

Co-condensation of 2,6-bis(bromomethyl)pyridine, **4**, with a variety of dithiols, **5**, yields not only the [1+1] cyclisation product, e.g. **6**, but also—by working at higher concentration—the [2+2] cyclised ligand.[9,10] An example of this is provided by the formation of the tetradentate ligand **7** and the mixed donor macrocycle **8**.

The versatility of the caesium carbonate-mediated cyclisation reactions in DMF is limited only by the availability and social unacceptability of the

David Parker

precursor dithiols. Most dithiols can be synthesised readily by reaction of the related dihalide (dimesylate or ditoluenesulfonate) with thiourea followed by mild basic hydrolysis of the isothiouronium salt (see Protocol 4 for an example). In the co-condensation reaction to form the poly-thia macrocycle, the electrophilic component can be a dibromide, ditoluenesulfonate or dimesylate without significant effect on the reaction yield. The synthesis of [14]-S$_4$, **9**, for example, involves co-condensation of 1,3-dibromopropane with 3,7-dithia-1,9-thianonane and results in a 66% yield of ligand **9** (Scheme 3.3).

Protocol 3.
Synthesis of 1,4,8,11-tetrathiacyclotetradecane (Structure 9, Scheme 3.3)

Caution! Carry out all procedures in a well-ventilated hood, and wear disposable vinyl or latex gloves and chemical-resistant safety goggles.

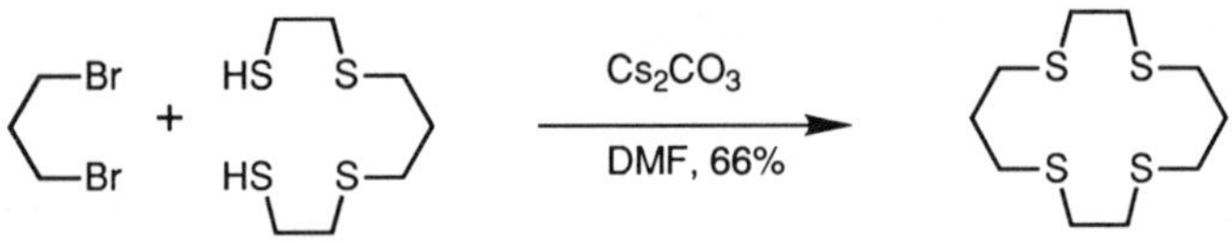

Scheme 3.3

Equipment

- Three-necked, round-bottomed flask (2 L)
- Pressure-equalising addition funnel (250 mL)
- Reflux condenser
- Mechanical stirrer equipped with a Teflon-coated stirrer paddle (7 cm)
- Source of dry nitrogen
- Thermometer (0–100 °C)

- Separating funnel (500 mL)
- Filter funnel
- Filter papers (Whatman No.1)
- Sintered filter funnel (porosity 2)
- Erlenmeyer flask (500 mL)
- Thermostatted hot plate stirrer
- Oil bath

Materials

- HPLC grade DMF,[a] 1.25 L — irritant, harmful by inhalation
- 3,7-Dithianonane-1,9-dithiol,[b] 4.56 g, 20 mmol — stench!
- Caesium carbonate, 13.04 g, 40 mmol — harmful by inhalation
- 1,3-Dibromopropane, 4.04 g, 20 mmol — irritant, harmful by inhalation, cancer-suspect agent

- Dichloromethane, 250 mL — harmful by inhalation
- Saturated aqueous sodium chloride solution, 200 mL
- Anhydrous magnesium sulfate, 10 g
- Absolute ethanol, 150 mL — flammable

1. Equip an oven-dried (electric oven, 105 °C, 1 h) three-necked round-bottomed flask (2 L) with a reflux condenser, a pressure-equalising addition funnel (250 mL) with a nitrogen inlet (see Fig. 3.3), and a mechanical stirrer fitted with a Teflon-coated stirrer paddle. Maintain the apparatus under a positive

3: Thia, oxa-thia and aza-thia crowns

pressure of dry nitrogen gas and add dry dimethylformamide (1.1 L) followed by caesium carbonate (13.04 g, 40 mmol). Heat the mixture to 60°C (bath temperature) and stir it mechanically.

2. Prepare a solution of 1,3-dibromopropane (4.04 g, 20 mmol) and 3,7-dithianonane-1,9-dithiol (4.56 g, 20 mmol) in dry DMF (150 mL) and transfer it to the pressure-equalising addition funnel. Add the solution slowly over a period of 7–9 h, and monitor the rate of addition periodically to ensure that approximately 3 mL of this solution is added every 10 min.

3. After the addition is complete, continue stirring for a further 3 h and then allow the mixture to cool to room temperature. Transfer the mixture to a single-necked round-bottomed flask (2 L), and remove the DMF by distillation under reduced pressure (bath temperature 55°C, 0.2 mmHg). Take up the residue in dichloromethane (2 × 125 mL) and transfer the solution to a separating funnel (500 mL).

4. Wash the dichloromethane solution with saturated aqueous sodium chloride solution (2 × 100 mL), dry the organic layer over anhydrous magnesium sulfate, filter and remove the organic solvent using a rotary evaporator to leave a light yellow solid residue.

5. Take up the residue in boiling absolute ethanol (100 mL) and decant the liquid into an Erlenmeyer flask (250 mL). Treat the remaining sediment with an additional volume of ethanol (50 mL) and boil the mixture for 15 min. Decant the liquid into the half-full Erlenmeyer and store the combined fractions in a fridge overnight (5°C). Separate the colourless crystalline solid by filtration (porosity 2 sintered filter funnel), and dry the product under vacuum (40°C, 0.1 mmHg, 3 h) to give the desired cycle, m.p. 119°C, 3.5 g, 66%.

[a] The HPLC grade DMF is suitable for use directly. Other grades of DMF may be distilled under reduced pressure and stored over activated molecular sieves (4 Å) prior to use.
[b] The precursor dithiol (b.p. 159–62°C, 1.2 mmHg) is conveniently prepared by reaction of propane-1,3-dithol with chloroethanol in ethanol in the presence of sodium ethoxide, followed by treatment of the resultant diol with conc. HCl, thiourea and then aqueous acid.

The electrophilic component in these co-cyclisation reactions can bear other functionality. For example, reaction of isobutenyl dichloride **10** with the appropriate linear dithiols gives good yields of polythiacycles such as **11**, **12** and **13**.[11] A related reaction allows the direct introduction of a keto group into such poly-thia macrocycles. The electrophilic component in this case is 1,3-dichloropropanone, **14**. The reaction conditions (Cs_2CO_3, DMF, 60°C), are insufficiently basic to induce a Favorskii reaction. Furthermore, under the aprotic, basic conditions, thioketal formation cannot take place, allowing successful co-cyclisation to be undertaken, such as in the formation of **15** (Scheme 3.4). Similar reactions allow the preparation of **16**, **17** and **18**.[11]

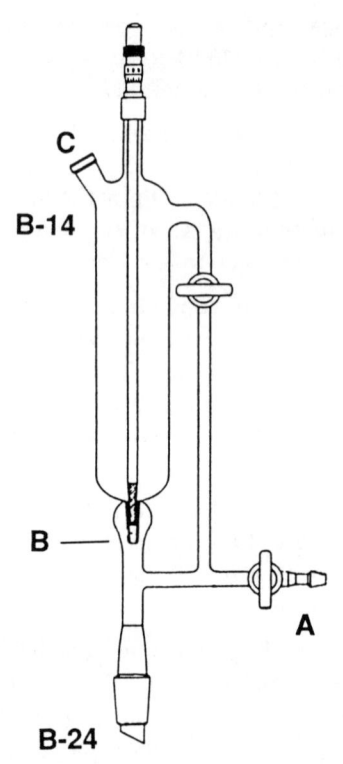

Fig. 3.3 Pressure-equalising addition funnel. Nitrogen gas is introduced via **C** and outlet **A** is connected to a gas bubbler. **B** is a ground-glass receiver for the tapered end of a 7 mm glass rod. The rod may be turned in the receiver to give the required addition rate.

10

11 n = 1,2

12 n = 1,2

13

14 + HS-S / HS-S →(Cs$_2$CO$_3$, DMF)→ 15

Scheme 3.4

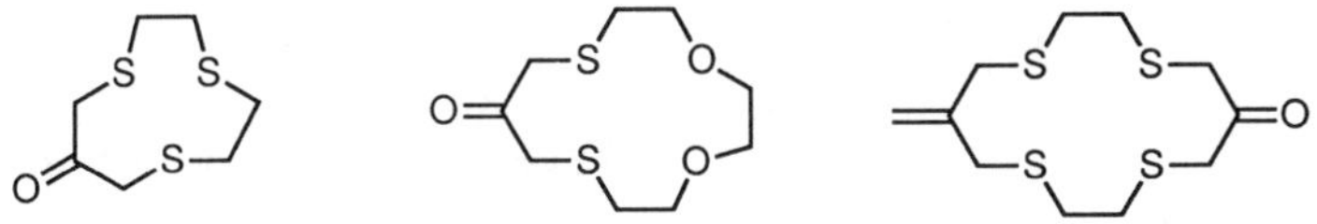

3: Thia, oxa-thia and aza-thia crowns

16 **17** **18**

In these caesium carbonate-mediated and related co-cyclisation reactions, significant yields of the larger [2 + 2] macrocycles may be obtained by working at slightly higher concentration.[8,9,12] Reaction of 1,2-dibromoethane with N-toluenesulfonyl-3-aza-pentane-1,5-dithiol, **19**, gives not only the [9]-NS$_2$ ring, **20**, but also affords a modest yield of the [2 + 2] adduct **21** (Scheme 3.5). Whilst this route probably allows most direct synthesis of the [9]-NS$_2$ macrocycle,[12] the [18]-N$_2$S$_4$ ligand (which is now commercially available) may be obtained by a variety of other methods,[13,14] usually involving formation of two C–N rather than two C–S bonds in the macrocyclisation step.

Protocol 4.
N-(p-Toluenesulfonyl)-1,4-dithia-7-azacyclononane, 20, and N,N'-bis(p-toluenesulfonyl)-1,4,10,13-tetrathia-7,16-diazacyclooctadecane (Structure 21, Scheme 3.5)

Caution! Carry out all procedures in a well-ventilated hood, and wear disposable vinyl or latex gloves and chemical-resistant safety goggles.

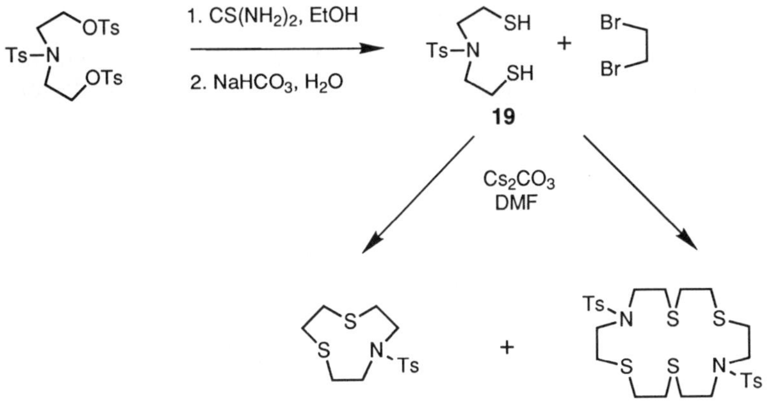

Scheme 3.5

This is an example of a macrocyclisation reaction involving formation of C–S bonds that gives a [1 + 1] major product and a [2 + 2] minor product.

Protocol 4. Continued

Equipment
- Hot plate stirrer
- Oil bath
- Column for flash chromatography (c. 60 cm × 10 cm)
- Single-necked, round-bottomed flask (1 L)
- Separating funnels (1 L and 250 mL)
- Teflon-coated magnetic stirrer bar
- Reflux condenser
- Fluted filter papers (Whatman No. 1)
- Filter funnel
- pH papers (range 0–14, BDH)
- 2 pressure equalising addition funnels (250 mL)
- Erlenmeyer flask (100 mL)
- Source of dry nitrogen

Materials
- N-(Toluenesulfonyl)-3-aza-1,5-bis (p-toluenesufonato)pentane, 75 g, 0.136 mol — cancer-suspect agent, toxic
- Silica gel for flash chromatography — harmful by inhalation
- Thiourea, 22.8 g, 0.299 mol — irritant
- Dry ethanol,[a] 500 mL — flammable
- Saturated aqueous sodium bicarbonate solution, 250 mL
- Hydrochloric acid, 6 M, 25 mL — corrosive, toxic
- Anhydrous magnesium sulfate, 10 g
- Dichloromethane for chromatography,[b] 5 L — harmful by inhalation
- Methanol for chromatography,[a] 25 mL — flammable, toxic
- Caesium carbonate, 3.6 g, 11 mmol — harmful by inhalation
- Dry dimethylformamide (HPLC grade), 900 mL — irritant, harmful by inhalation
- 1,2-Dibromoethane, 1.88 g, 10 mmol — toxic, cancer-suspect agent
- Toluene for recrystallisation, 50 mL — flammable
- Hexane for recrystallisation, 50 mL — flammable, irritant

1. Prepare a solution of N-(toluenesulfonyl)-3-aza-1,5-bis(toluenesulfonato)pentane[c] (75 g, 0.136 mol) in dry ethanol (500 mL) in a 1 L single-necked round-bottomed flask equipped with a magnetic stirrer bar and a reflux condenser attached to a nitrogen line.

2. Add thiourea (22.8 g, 0.3 mol) in portions to this solution and heat the stirred mixture under reflux, maintaining a nitrogen atmosphere for 30 h. After cooling, remove the solvent using a rotary evaporator (care! ethanol tends to bump readily so equip the rotary evaporator with a splash head). Add to the residue saturated aqueous sodium bicarbonate solution (250 mL) and boil the solution for 3 h.

3. Adjust the pH of the cooled solution to 7 using 6 M hydrochloric acid and transfer the solution to a separating funnel (1 L). Extract the aqueous layer with dichloromethane (3 × 200 mL), dry the organic extracts with magnesium sulfate (c. 10 g, 30 min), filter and remove the solvent on a rotary evaporator.

4. Prepare a silica gel chromatography column using dichloromethane, and load the column by dissolving the residue in the minimum volume of dichloromethane. Run the column with dichloromethane/methanol (199:1, v/v) and the desired dithiol elutes with an R_f of 0.5 (99:1, CH_2Cl_2–MeOH, v/v). Evaporate the eluates containing this material to yield a colourless oil

which crystallises on standing (over 24 h), 32.7 g (83%). Store the product under N_2 in a refrigerator, as oxidation causes formation of the intramolecular disulfide.

5. Equip an oven-dried, three-necked round-bottomed flask with a magnetic stirrer bar, two pressure-equalising addition funnels (250 mL) and an air condenser attached to a nitrogen line. Add caesium carbonate (3.6 g, 11 mmol) and dry DMF (500 mL) to the flask heated to 55 °C and prepare in each addition funnel solutions of 1,2-dibromoethane (1.83 g, 10 mmol) in DMF (200 mL) and N-(toluenesulfonyl)-3-azapentane-1,5-dithiol (2.91 g, 10 mmol) in DMF (200 mL), respectively.

6. Add the two solutions at equal rates over a period of 12 h with vigorous stirring at a temperature of 55°C. After addition is complete, stir for a further 3 h at 60°C, and remove the solvent by evaporation under reduced pressure (50°C, 10^{-2} mmHg). Take up the residue in dichloromethane (100 mL) and transfer the solution to a separating funnel (250 mL). Wash the organic layer with water (3 × 50 mL), dry the separated organic layer with magnesium sulfate, filter and remove the solvent on a rotary evaporator.

7. Take up the residue in warm toluene (c. 20 mL) and add hexane (c. 5 mL). Allow the solution to cool to room temperature and leave for 6 h. Filter off the crystalline solid which is the [2 + 2] 18-ring addition product, R_f = 0.4 (SiO_2, CH_2Cl_2/MeOH, 99:1), m.p. 206–7 °C, 0.31 g, 10%. Add further hexane (toluene/hexane now 2:1 v/v) and crystallise the [1 + 1], 9-ring product,[d] R_f = 0.5 (SiO_2 CH_2Cl_2/MeOH, 99:1), m.p. 122–3 °C, 1.38 g, 43%.

[a] Dry ethanol and methanol are distilled from the corresponding magnesium alkoxides.
[b] Use HPLC grade dichloromethane, or distil it from CaH_2 or P_4O_{10}.
[c] This compound (m.p. 96–7 °C, recrystallised from ethanol/toluene (5:1, v/v)) is prepared by reaction of tosyl chloride with 3-azapentane-1,2-diol in pyridine.
[d] The 9-ring product may be deprotected using HBr–AcOH–PhOH (see Protocol 2, this chapter) and the 18-ring product deprotected with Li/l. NH_3/THF–EtOH (see Protocol 5, this chapter) to give the secondary amines 1,4-dithia-7-azacyclononane, m.p. 71–2 °C, and 1,4,10,13-tetrathia-7,13-diazacyclooctadecane, m.p. 123–4 °C.

3. Cyclisation reactions forming carbon–nitrogen bonds

For the synthesis of aza-thia macrocycles, many of the methods that are available for the synthesis of aza cycles (Chapter 2) or aza-oxa crowns (Chapter 3) are also applicable. Thus the Richman–Atkins co-cyclisation of the ditoluenesulfonamide with an alkylditoluenesulfonate may be used for the synthesis of the [18]-N_4S_2 ring (Scheme 3.6). In this case, desulfonylation to afford the tetraamine 23 is best carried out with lithium in liquid ammonia.[12]

Protocol 5.
1,10-Dithia-4,7,13,16-tetraazacyclooctadecane (Structure 23, Scheme 3.6)

Caution! Carry out all procedures in a well-ventilated hood, and wear disposable vinyl or latex gloves and chemical-resistant safety goggles.

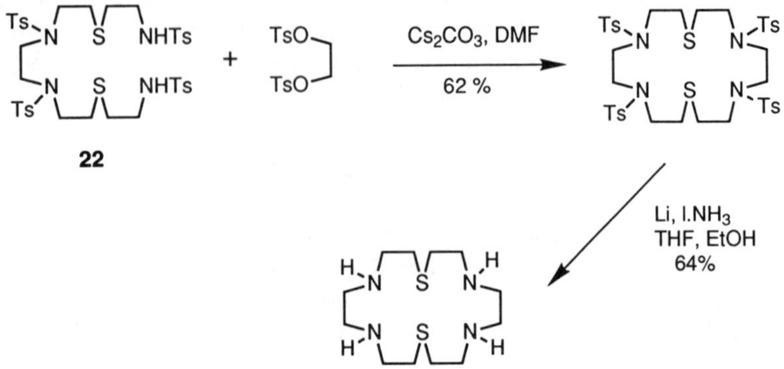

Scheme 3.6

Equipment
- Thermostatted hot-plate stirrer
- Oil bath
- Pressure-equalising addition funnel (250 mL)
- Single-necked round-bottomed flask (1 L)
- Fluted filter papers (Whatman No. 1)
- Filter funnel
- Source of dry nitrogen
- Buchner flask (250 mL)
- Buchner filter funnel
- Nitrogen inlet adaptor
- Dry ice condenser
- Double-necked, round-bottomed flask (500 mL)
- Teflon-coated magnetic stirrer bar
- Separating funnel (250 mL)

Materials
- Dry dimethylformamide,[a] 420 mL — irritant, harmful by inhalation
- Caesium carbonate, 3.13 g, 9.62 mmol — harmful by inhalation
- N,N',N'',N'''-Tetrakis (toluenesulfonyl)-3,12-dithia-6,9-diazatetradecane-1,14-diamine,[b] 4.04 g, 4.58 mmol — irritant
- 1,2-bis(toluenesulfonato)ethane, 1.69 g, 4.58 mmol — toxic, cancer-suspect agent
- Dichloromethane, 300 mL — harmful by inhalation
- Toluene for recrystallisation, 100 mL — flammable
- Diethyl ether, 80 mL — irritant, flammable
- Aqueous potassium hydroxide solution, 1 M, 30 mL — corrosive
- Ammonia gas (cylinder), 200 mL — irritant, corrosive
- Lithium, 0.75 g, 108 mmol — moisture sensitive!
- Dry ethanol, 5 mL — flammable,
- Hydrochloric acid, 6 M, 40 mL — corrosive, irritant
- Dry tetrahydrofuran,[c] 60 mL — flammable, irritant

1. Dry all glassware and stirrer bars in an electric oven (105°C) for 1 h prior to use.
2. Transfer into a single-necked round-bottomed flask (1 L), caesium carbonate

3: Thia, oxa-thia and aza-thia crowns

(3.13 g, 9.62 mmol), N,N',N'',N'''-tetrakis(toluenesulfonyl)-3-12-dithia-6,9-diazatetradecane-1,14-diamine (4.04 g, 4.58 mmol) and a magnetic stirrer bar. Pour into the flask dry DMF (300 mL) under a stream of nitrogen and equip the flask with a pressure-equalising addition funnel (250 mL) containing a solution of 1,2-bis(toluenesulfonato)ethane (1.69 g, 4.58 mmol) in dry DMF (120 mL).

3. Heat and stir the mixture at 40 °C and slowly add the contents of the addition funnel over a period of 6 h. When the addition is complete, raise the temperature to 60 °C and continue stirring for a further 4 h. Remove the DMF under reduced pressure (50 °C, 0.2 mmHg) and treat the residue with dichloromethane (50 mL). Filter the mixture and transfer the filtrate to a separating funnel (250 mL).

4. Wash the organic layer with water (3 × 25 mL) and remove the solvent using a rotary evaporator. Take up the residue in hot toluene (50 mL), filter the resultant suspension, on a Buchner funnel (washing with two further volumes of toluene, 2 × 10 mL), and remove the solvent on a rotary evaporator to yield a pale yellow solid, m.p. 255–6 °C, $R_f = 0.7$ (SiO$_2$, CH$_2$Cl$_2$/MeOH 97:3 v/v), 2.59 g, 62%. This tetratoluenesulfonamide is sufficiently pure to be used directly in the next step.

5. Take an oven-dried (electric oven, 105 °C, 1 h) double-necked round-bottomed flask (500 mL) and add a stirrer bar, the tetratosylamide (2.59 g, 2.85 mmol), dry ethanol (5 mL) and dry tetrahydrofuran (60 mL). Purge the flask with nitrogen gas for 10 min. Equip the flask with a dry ice condenser fitted with a nitrogen inlet, cool the flask to –78 °C in a dry ice/isopropanol slush bath and add dry ice and isopropanol to the dry ice condenser maintaining a strong nitrogen flow. Allow ammonia gas, from a cylinder, to condense into the cooled flask, until approximately 200 mL has been added.

6. Take a small block of lithium metal out of the oil and with a sharp knife cut approximately 1 g from it. Wash the block with dry hexane to remove adhering oil, and then cut the block into 4 smaller pieces. Dry the shiny smaller blocks with filter papers and weigh out approximately 0.75 g. Add the lithium quickly against a strong nitrogen flow to the cooled reaction flask, to generate a deep blue solution.

7. Vigorously stir the cooled mixture for 20 min at –78 °C, during which time the deep blue colour is slowly discharged and the solution becomes straw yellow in colour. Allow the reaction mixture to warm up slowly to room temperature, by removing the slush bath. During this period, the evaporating ammonia gas may be vented through a solution containing 6 M hydrochloric acid solution (at least 500 mL).

8. After all the ammonia gas has evaporated, slowly add distilled water (20 mL) to the reaction mixture and remove volatile solvents on a rotary evaporator. Treat the colourless residue with hydrochloric acid (6 M, 40 mL) and transfer the solution to a separating funnel (250 mL). Wash the aqueous

Protocol 5. *Continued*

layer with diethyl ether (2 × 40 mL), and evaporate the water using a rotary evaporator. Take up the residue in aqueous potassium hydroxide solution (1 M, 30 mL) and extract the aqueous layer exhaustively with dichloromethane (5 × 40 mL). Dry the combined organic extracts with magnesium sulfate, filter and evaporate the solvent using a rotary evaporator.

9. The resultant yellow oil solidifies on standing. Recrystallise the solid from warm toluene (*c*. 20 mL) to yield the tetraamine as a colourless solid, 370 mg, 64%, m.p. 88–9 °C. The carbon-13 NMR spectrum is diagnostic: δ_C (CDCl$_3$) 48.3 (CH$_2$N), 48.0 (CH$_2$N), 32.9 (CH$_2$S).

[a] HPLC grade DMF from Aldrich is suitable for direct use.
[b] The tetratoluenesulfonamide may be prepared in five simple steps from ethylenediamine by reaction with thiodiglycollic anhydride in CH$_2$Cl$_2$, ethanolysis, ammonolysis, borane reduction and tosylation[12].
[c] Tetrahydrofuran is distilled from lithium aluminium hydride prior to use.

The precursor tetratosylamide **22** may be prepared in five steps from ethylenediamine. The reaction sequence (Scheme 3.7) involves diacylation of ethylenediamine with 3-thiadiglycollic anhydride, followed by esterification, ammonolysis, borane reduction and toluenesulfonylation.[12] This stepwise synthesis of the desired tetratoluenesulfonamide involving a simple chain elongation strategy is reminiscent of the syntheses of large ring aza-oxa crowns (e.g. [24]-N$_4$O$_4$) discussed in Chapter 2.

The high dilution reaction involving co-condensation of a diamine with a diacid chloride has also been used to prepare several aza-thia crowns.[14–17] Reaction of 3-thiapentane-1,5-diamine with the dichloride derived from 3-thiadiglycollic acid affords a good yield of the bis-lactam which may be reduced to the diamine **24** using borane in THF (Scheme 3.8). In a similar sequence, the analogous [15]-N$_2$S$_3$ and [18]-N$_2$S$_4$ macrocycles may be prepared.[14,17]

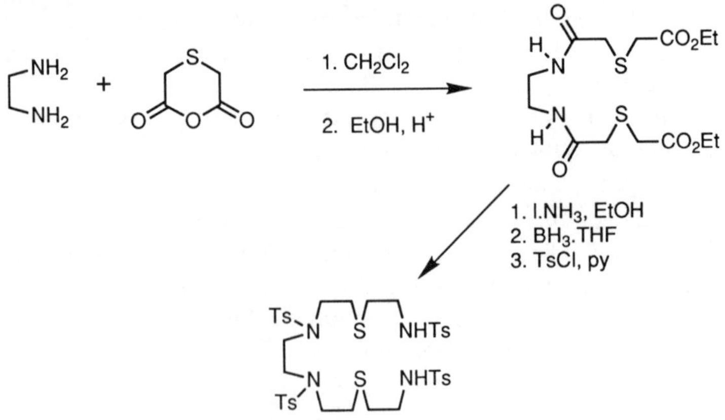

Scheme 3.7

Protocol 6.
1,7-Dithia-4,10-diazacyclododecane (Structure 24, Scheme 3.8)

Caution! Carry out all procedures in a well-ventilated hood, and wear disposable latex gloves and chemical-resistant safety goggles.

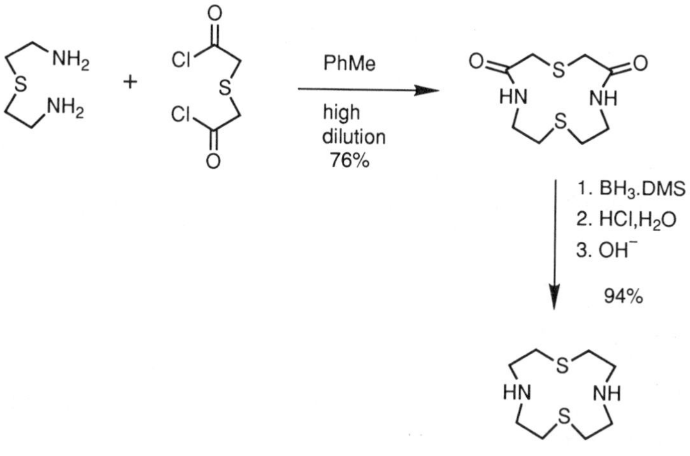

Scheme 3.8

This is an example of a high dilution condensation reaction followed by borane reduction of the intermediate lactam.

Equipment

- Three-necked, round-bottomed flask (5 L)
- Pressure-equalising addition funnels (2 × 1 L)
- Mechanical stirrer equipped with a Teflon-coated stirrer paddle (c. 10 cm)
- Buchner flask (2 L)
- Buchner filter funnel
- All glass syringe with a needle-lock Luer (50 mL)
- Calcium chloride guard tubes (2)
- Source of dry argon
- Hot plate stirrer
- Oil bath
- Sintered glass filter funnel (porosity 2)
- Column for chromatography (50 cm × 5 cm)
- Septum, B24
- Double-necked, round-bottomed flask (500 mL)

Materials

3-Thia-1,5-diaminopentane,[a] 20 g, 167 mmol	irritant
Thiadiglycollyl dichloride,[b] 16 g, 83 mmol	corrosive, toxic
Borane–dimethyl sulfide in THF, 10 M, 40 mL	flammable, toxic
Distilled toluene,[c] 3 L	flammable
Chloroform, 2 L	harmful by inhalation
Dry tetrahydrofuran,[d] 200 mL	flammable, irritant
Methanol, 50 mL	flammable
Aqueous potassium hydroxide solution, 2 M, 100 mL	corrosive
Anhydrous sodium sulfate, 20 g	
Neutral alumina for chromatography (Merck, activity II–III)	irritant
Hexane for recrystallisation, 50 mL	flammable, irritant
Hydrochloric acid, 2 M, 55 mL	corrosive, toxic

Protocol 6. *Continued*

1. Add freshly distilled toluene (2 L) to a three-necked round-bottomed flask equipped with a mechanical stirrer and two pressure-equalising addition funnels (each 500 mL; see Protocol 2, Chapter 5 for a related example).

2. Prepare a solution of 3-thia-1,5-diaminopentane (20 g, 167 mmol) in dry toluene (500 mL) in one of the addition funnels and in the other make up a solution of thiadiglycollyl dichloride (16 g, 83 mmol) in dry toluene (500 mL). Place calcium chloride guard tubes above both of these addition funnels.

3. Whilst maintaining vigorous stirring, regulate the rate of addition from each funnel to be 5 mL every 10 min. Ensure that the rate of addition is the same from both funnels. Addition is complete after about 18 h. Filter the suspension on a Buchner funnel, and wash the residue thoroughly with warm chloroform (8 × 200 mL). Combine the filtrates and remove the solvent on a rotary evaporator to yield a colourless solid.

4. Wash on a sintered glass filter funnel the product diamide with further portions of toluene (2 × 10 mL) and then chloroform (5 × 5 mL). Dry the solid residue under high vacuum (0.1 mmHg, 2 h) to give a colourless solid, m.p. 196–7 °C, 14.7 g, 76%. Use this diamide directly in the following step without further purification.

5. Allow an oven-dried two-necked round-bottomed flask (500 mL) equipped with a magnetic stirrer bar to cool down under dry nitrogen. Fit a septum to one joint and a reflux condenser attached to a nitrogen line to the other. Under a stream of nitrogen, add to the flask dry tetrahydrofuran (200 mL) and the diamide (5 g, 21.4 mmol) prepared in step 4.

6. Add slowly by syringe, a solution of borane–dimethyl sulfide in THF (10 M, 40 mL) to the stirred solution. Replace the septum with a glass stopper and heat the solution to reflux under nitrogen for 16 h. Allow the solution to cool to room temperature, replace the stopper by a septum, and add dry methanol (20 mL) slowly by syringe whilst maintaining stirring. There follows a vigorous evolution of hydrogen.

7. Transfer the mixture to a single-necked round-bottomed flask (500 mL) and remove the solvent using a rotary evaporator. Treat the residue with methanol (50 mL) and hydrochloric acid (2 M, 55 mL) and boil the solution under reflux for 4 h. Remove the solvent under reduced pressure on a rotary evaporator and take up the residue in chloroform (100 mL) and aqueous potassium hydroxide solution (2 M, 100 mL).

8. Transfer the mixture to a separating funnel, separate the organic layer and wash the aqueous layer with chloroform (4 × 50 mL). Combine the organic extracts and dry them over anhydrous sodium sulfate (*c.* 10 g, 2 h). Filter the mixture using a sintered filter funnel and evaporate the filtrate under reduced pressure.

9. Prepare a column for chromatography using neutral alumina (activity II–III,

3: Thia, oxa-thia and aza-thia crowns

Merck) and toluene as the elutant. Dissolve the residue in the minimum volume of warm toluene (add a little dichloromethane if necessary) and elute the column with toluene. Evaporate the toluene eluates on a rotary evaporator to yield a residue which may be recrystallised from chloroform/hexane (1:2, v/v; c. 80 mL) to give the product diamine as colourless needles, m.p. 78–9 °C, 4 g, 94%.

[a] 3-Thia-1,5-diaminopentane (b.p. 113–16 °C, 12 mmHg) is most directly prepared by reaction of 2-aminoethanethiol with aziridine in absolute ethanol at room temperature.
[b] Prepare this acid chloride by reaction of thiadiglycollic acid with thionyl chloride.[17] Do not distil it as it may decompose to a toxic mustard gas analogue.
[c] Toluene is distilled from calcium hydride under nitrogen.
[d] Tetrahydrofuran is distilled from lithium aluminium hydride under nitrogen.

Metal ion-templated cyclisation reactions have been pursued since the early work of Busch.[18–20] One of the first examples of this type of reaction involved the alkylation of the nickel complex of the N_2S_2 ligand **25** (Scheme 3.9). Reaction with 1,2-bis(bromomethyl)benzene in a non-polar solvent gave the *trans*-dibromo–nickel complex quantitatively. No reaction occurred, however, with dibromoethane or dibromopropane. The juxtaposition of three five-membered ring chelates in **25** imposes some degree of strain on the complex which is relieved by the adoption of a trapezoidal (rather than a strained square planar) structure. In the reaction shown in Scheme 3.9, a relatively unstrained seven-ring chelate, with a large S–M–S bond angle, is formed in the cyclisation step. Smaller ring chelates prefer more acute bite angles (e.g. for a five-ring chelate, the ideal bite angle is c. 70°) and four such five-ring chelates cannot be accommodated around a square planar nickel centre.

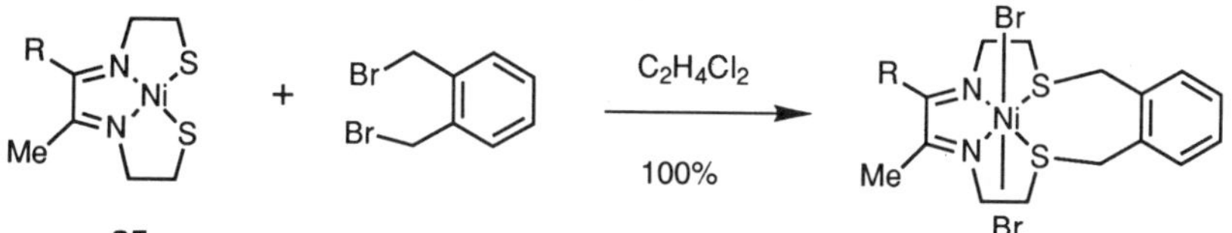

Scheme 3.9

More versatile reactions are being discovered in metal-templated reactions involving C–N bond formation.[21–23] Co-condensation of formaldehyde, nitroethane and the aminothioether **26** in the presence of copper(II) yields an intermediate di-imino nitro macrocycle. Subsequent dissolving metal reduction with zinc in HCl yields the saturated aza-thia macrocycle **27** in which the hydroxy group and amino groups are *trans*-related (Scheme 3.10).[22,23]

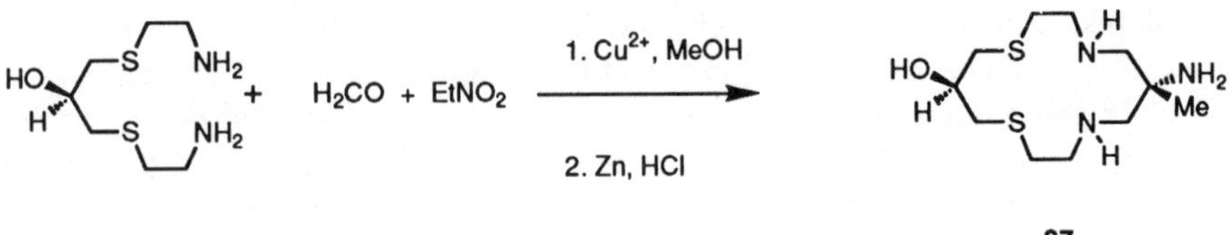

Scheme 3.10

26 + H₂CO + EtNO₂ →(1. Cu²⁺, MeOH; 2. Zn, HCl)→ **27**

This reaction (a related example is given in Protocol 7) is representative of a family of Cu(II)-directed syntheses employing formaldehyde and a range of carbon acids which yield saturated macrocycles with additional functional groups. The reaction also works well for a range of commercially available linear tetraamines, yielding in that case, C-substituted N_4 macrocycles.

Protocol 7.
Synthesis of 6-methyl-1,11-dithia-4,8-diaza-cyclotetradecane-6-amine[a] (Scheme 3.11)

Caution! Carry out all procedures in a well-ventilated hood, and wear disposable vinyl or latex gloves and chemical-resistant safety goggles.

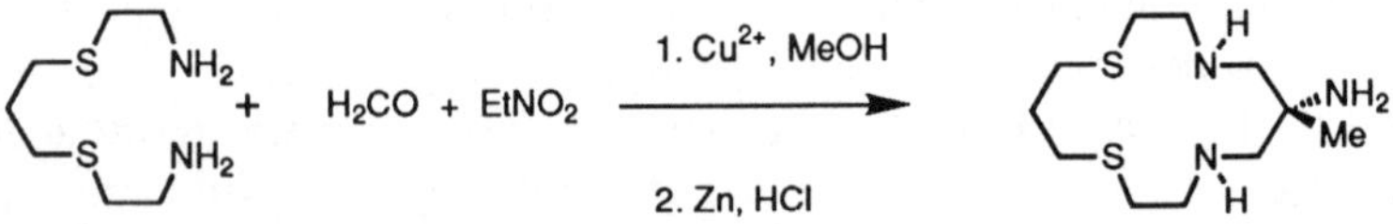

Scheme 3.11

This reaction is representative of the synthesis of copper(II)-templated syntheses using formaldehyde, a carbon acid and a linear substituted diamine.

Equipment

- Hot plate stirrer
- Erlenmeyer flask (1 L)
- Water-cooled condenser
- Glass chromatography columns (c. 30 cm × 4 cm) fitted with a sintered glass disc (porosity 1) at the base of the column
- Separating funnel (2 × 500 mL)
- Beaker (2 L)
- Separating funnel (1 L)
- Large conical flask (c.2 L)
- Rotary evaporator
- Magnetic stirrer bar

Materials

- 3,7-Dithia-1,9-diaminononane,[b] 5.0 g, 26 mmol — irritant
- Copper(II) nitrate hydrate, 6.0 g, 26 mmol — harmful by ingestion
- Methanol, 50 mL — flammable, harmful by inhalation
- Nitroethane, 3.0 g, 41 mmol (50% excess) — flammable, toxic, irritant
- Formaldehyde (aq. solution, 38%), 25 mL, 300 mmol — irritant
- Aqueous sodium hydroxide solution (2.5 M) — corrosive
- Zinc powder, 25 g — pyrophoric
- Hydrochloric acid (1 M and 5 M) — corrosive

3: Thia, oxa-thia and aza-thia crowns

- SP Sephadex C-25 cation exchange resin (Na⁺ form), 100 g
- Dowex 50 W × 2 cation exchange resin (H⁺ form), 100 g
- Aqueous sodium chloride solution (0.25 M)

1. Dissolve the copper(II) nitrate solid (6.0 g) in water (300 mL) in a 1 L conical flask containing a stirrer bead and add a solution of the 3,7-dithianonane-1,9-diamine (5.0 g) in methanol (50 mL) slowly. Warm (to 50°C) and stir for 1 h. Cool and filter through Kieselguhr (or an alternative filter aid) under suction to yield a dark purple filtrate.

2. Adjust the pH of the filtrate to approximately 10 with NaOH solution, and add nitroethane (3 g) and 38% aqueous formaldehyde (25 mL). Readjust the pH to 10.

3. Reflux the mixture overnight, then filter under suction through Kieselguhr to remove any brown solids.

4. Prepare two columns, one containing Sephadex and the other Dowex resins, by forming a slurry of the resin in water which is then poured into the empty column. Allow the resin to settle, and wash well with water prior to use.[c]

5. Dilute the filtrate to c.6 L with water, and adsorb on to the column of SP-Sephadex C-25 cation exchange resin, employing the 1 L separating funnel as the column reservoir. Wash the column with water (c.500 mL), then elute with 0.25 M NaCl solution.

6. Collect only the major (or sole) purple band.

7. Place the purple solution into one of the separating funnels, and place an equal volume of 2 M HCl in a second separating funnel. Arrange both funnels over a large beaker containing excess zinc powder (25 g) on a hot plate stirrer. Add a magnetic stirrer bar and commence stirring the zinc powder, while adding the solution of copper complex and the acid solution slowly at approximately the same rates. Stir and warm the final suspension until the aqueous phase is colourless (add extra zinc powder if necessary), then filter to remove solids.

8. Dilute the colourless filtrate at least five-fold (preferably more), and adsorb on to the column of Dowex 50W×2 (H⁺ form) cation exchange resin.

9. Wash the column with water (c. 500 mL), then wash with 1.0 M HCl (c. 1 L) to remove zinc ion, and elute with 5 M HCl to remove the macrocycle. [Test for presence of the colourless macrocycle by taking small aliquots from time to time, adjusting the pH to c. 2, adding some copper(II) ion, and observing the formation of a purple acid-stable colour.]

10. Take the fractions containing the ligand and remove the water on a rotary evaporator to yield a white powder of the hydrochloride salt. To assist in drying, add a small amount of methanol and rotary evaporate to dryness again. Dry the solid finally in a vacuum desiccator (3 h, 0.05 mmHg) to yield the hydrochloride salt which melts at high temperature with decomposition.

Protocol 7. *Continued*

The ^{13}C NMR spectrum is diagnostic (D_2O): 19.5 (Me), 24.0 (CCH_2C), 28.0 and 36.5 (CH_2S), 47.0 and 50.5 (CH_2N and CHN), 53.5 (quaternary C).

[a] The reaction can be performed on any reasonable scale and the yield is in the range 30 to 50%. The main limitations on the scale of the reaction are the capacities of the chromatography columns which are governed by the size and resin volume and/or the number of repetitions of the chromatography steps employed.
[b] The precursor diamine 3,7-dithia-1,9-diaminononane can be conveniently prepared by reaction of the sodium salt of 2-aminoethanethiol with 1,3-dibromopropane in absolute ethanol.[24]
[c] After use, these columns can be reused very extensively following washing with sodium ion (concentrated NaCl) or hydrogen ion (5 M HCl) respectively, followed by water.

References

1. Ray, P. C. *J. Chem. Soc.* **1920**, 1090.
2. Meadow, J. R.; Reid, E. E. *J. Am. Chem. Soc.* **1934**, *56*, 2177.
3. Dann, J. R.; Chiesa, P. P.; Gates, J. W. *J. Org. Chem.* **1961**, *26*, 1991.
4. Buter, J.; Kellogg, R. M. *J. Org. Chem.* **1981**, *46*, 4481.
5. Cooper, S. R. *Acc. Chem. Res.* **1988**, *21*, 141; Cooper, S. R.; Rawle, S. C. *Struct. Bonding* **1990**, *72*, 1.
6. Desper, J. M.; Gellman, S. H.; Wolf, R. E.; Cooper, S. R. *J. Am. Chem. Soc.* **1991**, *113*, 8663.
7. Blake, A. J.; Schroder, M. *Adv. Inorg. Chem.* **1990**, *35*, 1.
8. Vogtle, F. *Tetrahedron* **1969**, *25*, 3231; Vogtle, F.; Weber, E. *Angew. Chem., Int. Ed. Engl.* **1974**, *13*, 149.
9. Parker, D.; Lehn, J. M.; Rimmer, J. *J. Chem. Soc., Dalton Trans.* **1985**, 1517.
10. Helps, I. M.; Matthes, K. E.; Parker, D.; Ferguson, G. *J. Chem. Soc., Dalton Trans.* **1989**, 915.
11. Buter, J.; Kellogg, R. M.; van Bolhuis, F. *J. Chem. Soc., Chem. Commun.* **1990**, 282.
12. Craig, A. S.; Kataky, R.; Matthews, R. C.; Parker, D.; Ferguson, G.; Lough, A.; Adams, H.; Bailey, N. R.; Schneider, H. *J. Chem. Soc., Perkin Trans. 2* **1990**, 1523.
13. Black, D. St. C.; McLean, J. A. *J. Chem. Soc., Chem. Commun.* **1968**, 1004.
14. Alberts, A. H.; Annunziata, R.; Lehn, J. M. *J. Am. Chem. Soc.* **1977**, *99*, 8502.
15. Agnus, Y.; Louis, R. *Nouv. J. Chim.* **1981**, *5*, 305.
16. Black, D. St. C.; McLean, J. A. *Aust. J. Chem.* **1971**, *24*, 1401.
17. Alberts, A. H.; Lehn, J. M.; Parker, D. *J. Chem. Soc., Dalton Trans.* **1985**, 2311.
18. Busch, D. H.; Lindoy, L. F. *J. Am. Chem. Soc.* **1969**, *91*, 4690.
19. Kotovic, V.; Taylor, L. T.; Busch, D. H. *J. Am. Chem. Soc.* **1969**, *91*, 2122.
20. Busch, D. H.; Thompson, M. C. *J. Am. Chem. Soc.* **1964**, *86*, 3651.
21. Lawrance, G. A.; Maeder, M.; Hambley, T. W.; Wilkes, E. N. *J. Chem. Soc., Dalton Trans.* **1992**, 1283.
22. Comba, P.; Lawrance, G. A.; Rossignoli, M.; Skelton, B. W.; White, A. H. *Aust. J. Chem.* **1988**, *41*, 773.
23. Wei, G.; Allen, C. C.; Hambley, T. W.; Lawrance, G. A.; Maeder, G. *Aust. J. Chem.* **1995**, *48*, in press.
24. Hay, R. W.; Gidney, P. M.; Lawrance, G. A. *J. Chem. Soc., Dalton Trans.* **1975**, 779.

4

Crown ethers

DAVID B. AMABILINO, JON A. PREECE and
J. FRASER STODDART

1. Introduction

Crown ethers[1] are perhaps the most widely used family of host compounds in supramolecular chemistry[2]—the chemistry of the non-covalent bond. The fortuitous discovery[3] of the macrocyclic polyethers by Pedersen in 1967 laid the foundation for an exhaustive study of their preparation and complexing abilities, primarily with metal cations[4] but also with neutral and even anionic species.[5] Moreover, the metal ions which are complexed by crown ethers can also be utilised as templates[6] for their formation.

The syntheses of crown compounds invariably rely upon the Williamson ether synthesis, a dated but reliable reaction which is extemely useful in the synthesis of these medium-ring and large-ring compounds.[7] The syntheses are usually not discussed in much detail in the literature, since the emphasis in the general area of supramolecular chemistry is on the properties of the target compounds, not on their preparation. It is frequently the case, however, that these apparently conventional syntheses are far from straightforward.

2. Synthesis of 1,1′,4,4′-tetra-*O*-benzyl-2,2′:3,3′-bis-*O*-oxydiethylene-di-L-threitol, a chiral crown ether

The interest in chiral crown ethers arises not only from their ability to differentiate between the enantiomers of racemic substrates[8] containing substituted primary ammonium centres by virtue of N–H· · ·O hydrogen bonding interactions, but also because they can behave as chiral reagents or catalysts when enantioselective reactions are performed on appropriate substrates.[8,9]

The synthesis of chiral crown ethers is generally approached by exploiting the chiral pool, that is, naturally occurring materials which are available in enantiomerically pure form.

The chiral 18-crown-6 derivative **1** (Scheme 4.1), and similar structures derived from it, are capable of exhibiting enantiomeric differentiation in the complexation of racemic primary ammonium salts. Compound LL-**1** has been

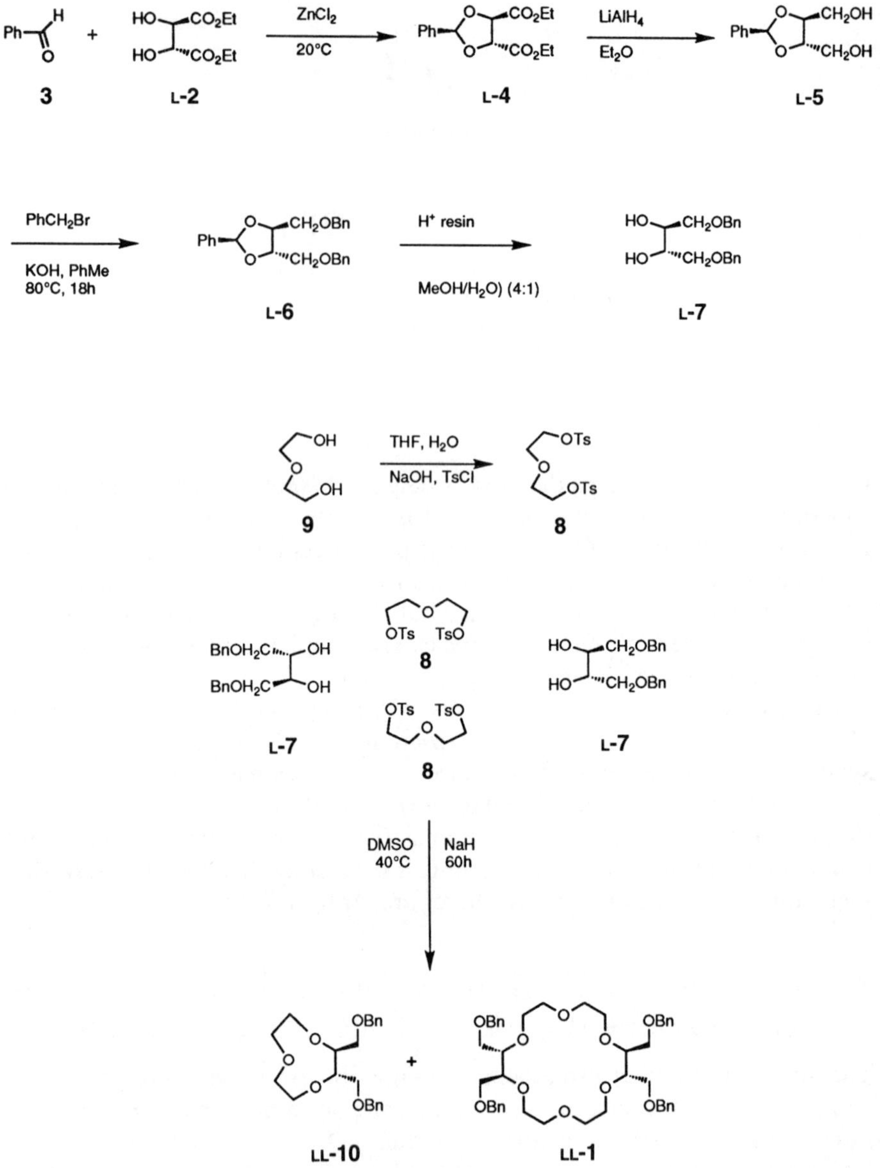

David B. Amabilino et al.

Scheme 4.1

synthesised[10] according to the route illustrated in Scheme 4.1. The chiral starting material employed[2-5] in the synthesis of this crown ether is diethyl tartrate **2**, which is commercially available in both its D or L forms. The route described relates to the synthesis of the LL crown ether. Diethyl L-tartrate, L-**2**, is reacted with benzaldehyde, **3**, in the presence of zinc chloride to yield

4: Crown ethers

the *O*-benzylidene derivative L-**4**, which is readily reduced with lithium aluminium hydride to give the diol L-**5**. The primary hydroxyl groups are converted into the corresponding benzyl ethers by treatment with an excess of potassium hydroxide and benzyl bromide. This reaction affords the dibenzyl ether L-**6**. A problem is encountered in the synthesis when the 1,3-dioxolan ring of L-**6** is cleaved during an acid-catalysed reaction. Although some hydrolysis of the benzyl ethers also occurs, the pure diol L-**7** can be isolated in 38% yield. This diol is converted into the disodium salt by treatment with sodium hydride in dimethyl sulfoxide. Addition of the ditoluenesulfonate **8** of diethylene glycol **9**, followed by separation of the reaction products, yields the 9-crown-3 analogue L-**10** in 3% yield and the desired 18-crown-6 derivative LL-**1** in 11% yield.

Protocol 1.
Synthesis of 2,3-*O*-benzylidene-L-tartrate (Structure L-4, Scheme 4.2)

Caution! Carry out all procedures in a well-ventilated hood, and wear disposable vinyl or latex gloves and chemical-resistant safety goggles.

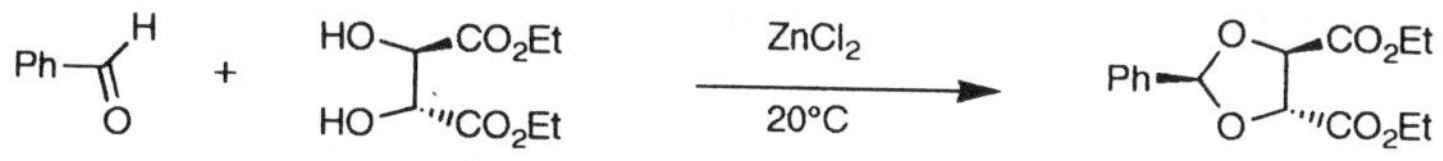

Ph–CH(=O) + HO–CH(CO₂Et)–CH(OH)–CO₂Et →(ZnCl₂, 20°C)→ Ph–CH(O–)(O–)–CH(CO₂Et)–CH–CO₂Et

3   **L-2**   **L-4**

Scheme 4.2

This procedure has been adapted from the method of Ehrlenmeyer.[11]

Equipment
- Double-necked, round-bottomed flask (1 L)
- Mechanical stirrer
- Separating funnel (2.5 L)
- Erlenmeyer flask (1 L)
- Filter funnel

Materials
- Diethyl L-tartrate, **2**, 250 g
- Benzaldehyde, **3**, 250 g   **harmful**
- Fused zinc chloride, 250 g   **causes burns**
- Chloroform for extraction, 500 mL   **toxic**
- Sodium bicarbonate solution
- Ethanol for crystallisation   **highly flammable**
- Magnesium sulfate, anhydrous, 10 g

1. Combine diethyl L-tartrate, L-**2** (250 g), benzaldehyde, **3** (250 g) and fused zinc chloride (250 g) in a dry round-bottomed flask, fitted with a mechanical stirrer.
2. Stir the mixture for 20 h at room temperature.

Protocol 1. *Continued*

3. Extract the viscous mixture with chloroform (500 mL). Transfer the mixture to a separating funnel (2.5 L) and extract the organic layer with water (3 × 500 mL) and then with aqueous sodium bicarbonate solution (2 × 500 mL).
4. Separate the chloroform layer and dry it over anhydrous magnesium sulfate (10 g).
5. Filter the organic layer, and remove the solvent on a rotary evaporator at 75–80 °C. A yellow oil remains that solidifies on standing.
6. Recrystallise the material from an ethanol–water mixture to afford **4** (122 g, 35%), m.p. 45 °C, $[\alpha]_D^{20}$ −33.8° (c = 1.5, chloroform).

Protocol 2.
Synthesis of 2,3-benzylidene-L-threitol (Structure L-5, Scheme 4.3)

Caution! Carry out all procedures in a well-ventilated hood, and wear disposable vinyl or latex gloves and chemical-resistant safety goggles. Lithium aluminium hydride is a powder which reacts VIOLENTLY with water and alcohols. Care should always be taken to avoid moisture when handling this material.

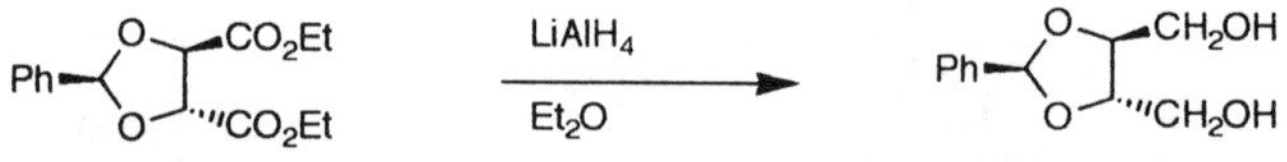

Scheme 4.3

Equipment

- Double-necked, round-bottomed flask (1 L)
- Water-cooled condenser
- Nitrogen bubbler
- Filter funnel
- Pressure-equalising addition funnel (100 mL)
- Magnetic heater and stirrer
- Magnetic stirrer bar
- Rotary evaporator
- Soxhlet extractor and thimble

Materials

- 2,3-O-Benzylidene-L-tartrate, L-4, 64 g, 225 mmol
- Lithium aluminium hydride, 16 g, 421 mmol   **flammable and reacts violently with water**
- Dry diethyl ether,[a] 350 mL   **highly flammable**
- Ether for extraction, 200 mL   **highly flammable**
- Magnesium sulfate, anhydrous, for drying
- Source of dry nitrogen gas
- Whatman No.1 filter paper

1. Dry all glassware for at least 2 h in an oven at 150 °C, assemble the round-bottomed flask, the condenser and the pressure-equalising addition funnel while still warm, immediately attach a calcium chloride guard tube

4: *Crown ethers*

to the top of the condenser and stopper the pressure-equalising addition funnel.

2. Carefully add the lithium aluminium hydride to the reaction vessel, via one of the necks, and immediately replace the pressure-equalising addition funnel or the condenser.

3. Pour the dry diethyl ether (350 mL) into the pressure-equalising addition funnel, replace the stopper at the top of the funnel, and add the solvent cautiously to the lithium aluminium hydride.

4. Dissolve 2,3-*O*-benzylidene-L-tartrate, L-**4** (64 g, 225 mmol) in diethyl ether (200 mL) in a dry stoppered flask, and transfer the resulting solution to the pressure-equalising addition funnel.

5. Add the solution of 2,3-*O*-benzylidene-L-tartrate, L-**4**, dropwise over a period of 1 h to the lithium aluminium hydride suspension with stirring (do not heat the mixture at this stage).

6. When addition is complete, heat the mixture to reflux for 2 h with stirring, and then allow it to cool.

7. Carefully add dropwise an aqueous sodium hydroxide solution (15% w/w, 60 mL) to the reaction mixture in order to destroy the excess of lithium aluminium hydride.

8. Filter the inorganic precipitate through a fluted filter paper under gravity, and wash the solid well with diethyl ether (4 × 50 mL)—do not discard the solid residues.

9. Dry the organic solution over magnesium sulfate (10 g), filter the suspension and remove the solvent by evaporation under reduced pressure (rotary evaporator). The resulting oil solidifies on standing, affording L-**5**.

10. Transfer the inorganic solid residues from step 8 to a Soxhlet thimble, and place this thimble into a Soxhlet extractor. Extract for 3 d with diethyl ether (50 mL).

11. Evaporate the diethyl ether from the extraction to leave more of the crystalline product, which can be combined with the material obtained from step 9, to afford L-**5**. This product can be recrystallised from diethyl ether to yield L-**5** (42.1 g, 92%), m.p. 69–70°C, $[\alpha]_D^{20}$ −11.4°C (c = 2.1, MeOH).

[a] Diethyl ether should be predried over sodium wire, then distilled from lithium aluminium hydride in an inert atmosphere (nitrogen or argon), and should be used immediately.

Protocol 3.
Synthesis of 1,4-di-O-benzyl-2,3-O-benzylidene-L-threitol (Structure L-6, Scheme 4.4)

Caution! Carry out all procedures in a well-ventilated hood, and wear disposable vinyl or latex gloves and chemical-resistant safety goggles. Benzyl bromide is a potent lachrymator.

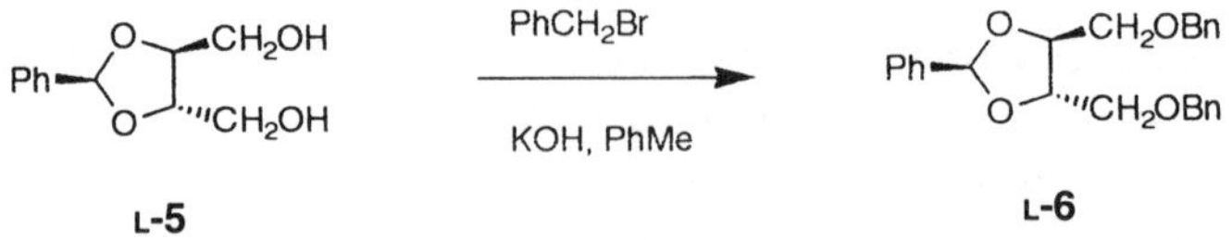

L-5   PhCH₂Br / KOH, PhMe   L-6

Scheme 4.4

Equipment

- Double-necked, round-bottomed flask (500 mL)
- Water-cooled condenser
- Guard tube
- Powder funnel
- Magnetic heater and stirrer
- Magnetic stirrer bar
- Filter funnel (500 mL)
- Separating funnel
- Vacuum distillation equipment
- Vacuum pump

Materials

- 2,3-O-Benzylidene-L-threitol, L-5, 15 g, 71 mmol
- Benzyl bromide, 71.5 g, 420 mmol   **irritant**
- Powdered anhydrous potassium hydroxide,[a] 30 g, 536 mmol   **causes burns**
- Dry toluene,[b] 150 mL   **highly flammable**
- Calcium chloride for guard tube   **irritant**
- Magnesium sulfate, anhydrous, for drying
- Whatman No. 1 filter papers
- Solid carbon dioxide for cooling   **danger of burns**
- Acetone for cooling   **flammable**

1. Dry all glassware for at least 2 h in an oven at 150 °C, assemble the round-bottomed flask and condenser while still warm, immediately attach a calcium chloride guard tube to the top of the condenser and stopper the second neck.

2. Dissolve the 2,3-O-benzylidene-L-threitol, L-5 (51 g, 71 mmol) in dry toluene (150 mL) in the reaction vessel by heating and stirring the mixture at 60 °C.

3. Add the powdered potassium hydroxide through a powder funnel to the toluene solution, via the second neck. Immediately afterwards add benzyl bromide (71.5 g, 420 mmol).

4. Raise the temperature of the reaction mixture to 80 °C with vigorous stirring, and maintain these conditions for 18 h.

5. Allow the reaction mixture to cool, then filter it under gravity through a fluted filter paper.

4: Crown ethers

6. Transfer the toluene solution to a separating funnel (500 mL), and wash the organic layer with distilled water (3 × 200 mL).
7. Dry the toluene solution over magnesium sulfate, and filter it through a fluted filter paper.
8. Set up a vacuum distillation apparatus so that the container vessel is placed in a water bath and the receiver vessel is cooled in an acetone/CO$_2$ bath.
9. Distil the toluene and residual benzyl bromide under reduced pressure (0.05 mmHg) at a water bath temperature of 80°C.
10. The remaining yellow oil L-6 (26.8 g, 96%) is used in the next step of the synthesis without further purification.

[a] Powdered potassium hydroxide should be prepared by grinding the commercially available material in a mortar and pestle.
[b] Toluene should be distilled from sodium–benzophenone prior to use.

Protocol 4.
Synthesis of 1,4-di-*O*-benzyl-L-threitol (Structure 7, Scheme 4.5)

Caution! Carry out all procedures in a well-ventilated hood, and wear disposable vinyl or latex gloves and chemical-resistant safety goggles.

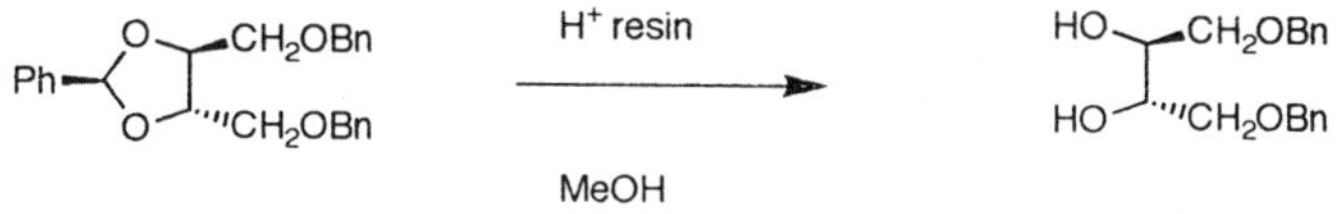

Scheme 4.5

Equipment
- Single-necked, round-bottomed flask (500 mL)
- Water-cooled condenser
- Magnetic heater and stirrer
- Oil bath
- Magnetic stirrer bar
- Column for chromatography

Materials
- 1,4-di-*O*-benzyl-2,3-*O*-benzylidene-L-threitol, L-**6**, 57 g, 146 mmol
- Zeo-Karb 325 ion-exchange resin, H⁺ form, 30 g
- Methanol, 300 mL   **toxic and flammable**
- Distilled water, 50 mL
- Silica gel for column chromatography, 1.5 kg   **toxic by inhalation**
- Diethyl ether for column chromatography   **highly flammable**
- Chloroform for recrystallisation   **toxic by inhalation**
- Light petroleum ether for recrystallisation   **highly flammable**

1. Combine 1,4-di-*O*-benzyl-2,3-*O*-benzylidene-L-threitol, L-**6**, Zeo-Karb 325 ion-exchange resin, H⁺ form and a 4:1 mixture of methanol:water (250 mL).

Protocol 4. Continued

2. Raise the temperature of the mixture to reflux with stirring, and maintain these conditions for 1 h.
3. Filter the reaction mixture through a fluted filter paper, and wash the solid resin with methanol (30 mL).
4. Concentrate the filtered solution on a rotary evaporator. A brown oil remains.
5. Chromatograph the oil on silica gel using diethyl ether as the eluant.
6. Recrystallise the product obtained by chromatography from chloroform–light petroleum (b.p. 60–80 °C) affording a white solid L-7 (17.4 g, 38%), m.p. 66 °C, $[\alpha]_D^{20}$ –5.5° (c = 5.1, chloroform).

Protocol 5.
Synthesis of 1,1′,4,4′-tetra-*O*-benzyl-2,2′:3,3′-bis-*O*-oxydiethylenedi-L-threitol (Structure LL-1, Scheme 4.6)

Caution! Carry out all procedures in a well-ventilated hood, and wear disposable vinyl or latex gloves and chemical-resistant safety goggles.

Scheme 4.6

Equipment
- Triple-necked, round-bottomed flask
- Water-cooled condenser
- Pressure-equalising addition funnel (50 mL)
- Separating funnel (250 mL)
- Magnetic heater and stirrer
- Stirrer bar
- Oil bath
- Column for chromatography

Materials
- Sodium hydride (60% suspension in mineral oil), 0.64 g, 16 mmol
- 1,4-Di-*O*-benzyl-L-threitol, L-7, 2.0 g, 6.6 mmol
- Diethylene glycol ditoluenesulfate, 8, 3.2 g, 7.7 mmol — irritant
- Dry dimethyl sulfoxide (DMSO),[a] 50 mL — toxic
- Silica gel for column chromatography, 100 g — toxic by dust inhalation
- Ethyl acetate for column and thin-layer chromatography, 500 mL — highly flammable
- Preparative thin-layer chromatography plate, 20 cm × 20 cm × 2 mm
- Magnesium sulfate, anhydrous, for drying
- Whatman No. 1 filter paper
- Chloroform, 100 mL — toxic by inhalation
- Distilled water, 100 mL
- Nitrogen gas

4: Crown ethers

1. Dry all glassware for at least 2 h in an oven at 150 °C, and assemble flask, pressure-equalising addition funnel, condenser and nitrogen bubbler on top of condenser while still warm, passing nitrogen through the bubbler whilst apparatus cools, and stopper the third neck.
2. Dissolve 1,4-di-*O*-benzyl-L-threitol, L-**7** (2.0 g, 6.6 mmol) in dry dimethyl sulfoxide (25 mL) in the reaction vessel, under an atmosphere of nitrogen, adding via the third neck, and then restopper.
3. Add sodium hydride (60% suspension in oil) (0.64 g, 16 mmol) to the solution, adding via the third neck, and then restopper.
4. Stir the resulting suspension for 1 h at room temperature and then raise the temperature of the solution to 40 °C.
5. Dissolve diethylene glycol ditoluenesulfonate, **8** (3.2 g, 7.7 mmol) in dry dimethyl sulfoxide (25 mL), and transfer the solution to the pressure-equalising addition funnel.
6. Add the solution of ditoluenesulfonate, **8**, to the reaction mixture dropwise over a period of 30 min.
7. Stir the reaction mixture under nitrogen at 40 °C for 6 h, then allow the mixture to cool.
8. Add water (50 mL) dropwise to the reaction mixture to quench the excess of sodium hydride.
9. Transfer the resulting solution to a separating funnel, and extract with chloroform (3 × 30 mL). Combine the organic layers, transfer them back into the separating funnel and extract them with water (3 × 30 mL).
10. Dry the separated organic layer over magnesium sulfate, and filter the mixture under gravity through a fluted filter paper.
11. Concentrate the organic solution on a rotary evaporator. The resulting oil may retain dimethyl sulfoxide, therefore drying on a vacuum line equipped with a nitrogen trap is advised.
12. Chromatograph the oil on silica gel in a column using ethyl acetate as the eluant. This column yields a mixture of two compounds.
13. Apply the mixture of compounds as a dichloromethane solution to a preparative thin-layer chromatography plate using a syringe. Allow the plate to air-dry, and then elute it with ethyl acetate.
14. The faster running material on the plate (R_f = 0.5) corresponds to the nine-membered ring 1:1 cyclic 'adduct' L-**10**. Remove the silica from the plate (**caution:** wear a mask as absorbed materials on silica dust are extremely hazardous) using a scalpel, and dissolve the product in chloroform (20 mL).
15. Filter the chloroform–silica gel suspension under gravity through a fluted filter paper, and wash the silica with chloroform (5 mL). Removal of the solvent from the filtrate on a rotary evaporator leaves an oil which

Protocol 5. Continued

solidifies on cooling. This material is the 9-crown-3 derivative L-**10** (81 mg, 3%), $[\alpha]_D^{20}$ +19.3° (c 3.5, chloroform).

16. The slower running material from the preparative thin layer chromatography plate (R_f = 0.4) corresponds to the 2:2 cyclic 'adduct', that is, the desired 18-crown-6 derivative, 1,1',4,4'-tetra-O-benzyl-2,2':3,3'-oxydiethylenedi-L-threitol LL-**1**. Removal and isolation of this material from the silica (remember, caution, mask necessary when using silica) as in steps 14 and 15 for the smaller crown, affords a colourless oil LL-**1** (274 mg, 11%), $[\alpha]_D$ +5.8° (c = 3.5, chloroform).

[a] Dry dimethyl sulfoxide can be obtained by distilling the solvent under reduced pressure from calcium hydride and subsequently storing over 4 Å molecular sieves.

Higher yields of crown ethers are generally obtained if the cyclisation reaction can be templated by a metal ion of appropriate size. Thus, potassium ions are well known to enhance the yields of 18-crown-6 derivatives,[1,3] while addition of the smaller lithium ion enhances the yields of reactions giving 13- and 14-membered tetraoxa rings.[4b,4c] Such a reaction has been used to good effect in the synthesis of the chiral 14-crown-4 dibenzyl ether, L-**11**, which can be prepared in a yield of 65% (Scheme 4.7) (cf. 11% for the untemplated 18-crown-6 analogue) by co-condensation of L-**7** with the appropriate linear ditoluenesulfonate.

Protocol 6.
Synthesis of (2*S*,3*S*)-2,3-bis(benzyloxymethyl)-1,4,8,11-tetraoxacyclotetradecane (Structure L-11, Scheme 4.7)

Caution! Carry out all procedures in a well-ventilated hood, and wear disposable vinyl or latex gloves and chemical-resistant safety goggles.

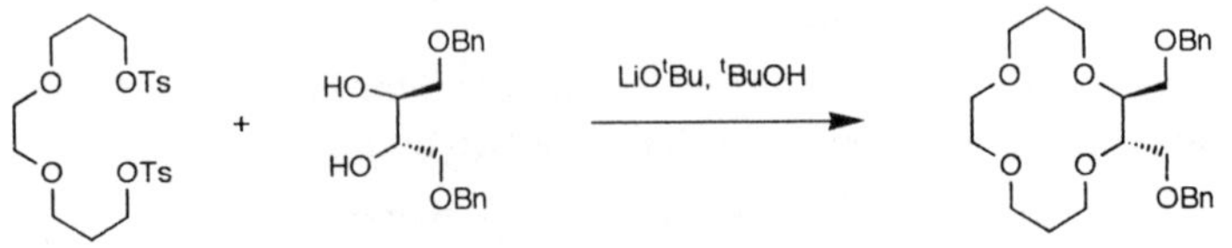

Scheme 4.7

Equipment
- Double-necked, round-bottomed flask (50 mL)
- Single-necked, round-bottomed flask (50 mL)
- Single-necked flask (250 mL)
- Water-cooled condenser
- Separating funnel (100 mL)
- Hot plate stirrer
- Oil bath
- Stirrer bar
- Column for chromatography
- Filter funnel
- pH indicator strips (BDH pH 0–6)
- Gas line adaptor
- Filter papers (Whatman No. 1)
- Conical flask (50 mL)
- Source of dry nitrogen

4: Crown ethers

Materials

• Dry t-butanol	flammable and irritant
• Li metal, 40 mg, 6.0 mmol	flammable and moisture sensitive
• (2S,3S)-(−)-1,4-Bis(benzyloxymethyl)butane-2,3-diol, 0.62 g, 2.1 mmol	
• 1,10-Bis(toluenesulfonato)-4,7-dioxadecane, 1.00 g, 2.1 mmol	irritant and toxic
• Lithium bromide, anhydrous, 0.18 g, 2.1 mmol	irritant and moisture sensitive
• 6 M HCl	corrosive and irritant
• Dichloromethane, 50 mL	Irritant, harmful by inhalation
• Chloroform, 50 mL	irritant, harmful by inhalation
• Hexane, for chromatography	flammable
• Ethyl acetate, for chromatography	flammable
• Potassium carbonate, anhydrous, for drying	irritant
• Alumina (neutral, activity II-III)[b]	irritant
• Nitrogen gas	

1. Dry all glassware in an oven. Assemble the two-necked flask, pressure-equalising addition funnel, condenser, gas line adaptor, stirrer bar and stopper on the hot plate stirrer while warm and allow to cool under nitrogen.

2. Pour 35 mL of dry t-butanol[a] into the flask.

3. Take a lithium metal block out of the oil and cut approximately 40 mg from it. Ensure all the surfaces of the cut piece are shiny and metallic. Wash the lithium thoroughly in dry hexane (25 mL) in a 50 mL conical flask. Dry the lithium on a paper and add to the t-butanol as quickly as possible to avoid oxidation of the metal.

4. Stir the t-butanol and lithium under nitrogen until all the lithium has dissolved, resulting in a fine pale white suspension.

5. Add the diol L-7 (0.62 g), the ditoluenesulfate (0.5 g) and lithium bromide (0.18 g) in sequence and stir at 60°C under nitrogen for 36 h. The reaction mixture is a fairly viscous white suspension.

6. After 36 h add a further 0.5 g of the ditoluenesulfonate and stir the mixture under nitrogen at 60°C for 72 h.

7. Allow the mixture to cool, remove the stirrer bar and pour the contents into a 50 mL round-bottomed flask. Remove the solvent on a rotary evaporator.

8. Treat the residue with 6 M hydrochloric acid until pH paper indicates the pH of the solution is 2.

9. Pour the solution into a 100 mL separating funnel and extract the aqueous layer with dichloromethane (2 × 25 mL) and chloroform (2 × 25 mL).

10. Dry the organic solution with anhydrous potassium carbonate (1 g) and filter it through fluted filter paper.

11. Transfer the organic solution to a 250 mL round-bottomed flask and remove the solvent on a rotary evaporator.

Protocol 6. Continued

12. Apply the residue to an alumina column (2.5 cm × 20 cm) and elute with hexane:ethyl acetate (5:1) to yield a colourless oil L-**11** (R_f = 0.5), (0.61 g, (65%),[c] $[\alpha]_D^{20}$ = –10.6° (c = 1.0, dichloromethane).

[a] t-Butanol was used as received and was stored over 4 Å molecular sieves.
[b] Alumina was prepared by soaking for 24 h in ethyl acetate and then eluting with 5:1 hexane:ethyl acetate until the excess ethyl acetate was washed off.
[c] The reaction gives similar yields at 10 times the stated scale.

3. Bis-*p*-phenylene-34-crown-10 synthesis—a receptor for π-electron-deficient aromatics

Although relatively small benzo crown ethers have been used widely in studies of metal ion complexation, larger crown ethers incorporating two π-electron-rich aromatic residues are capable of forming complexes with a range of π-electron-deficient aromatic units within their cavities.[12] An interesting example[13] is bis-*p*-phenylene-34-crown-10, **12**, (BPP34C10 is an acronym used to describe it), which forms a strong complex with the bishexafluorophosphate salt of the π-electron-deficient herbicide, paraquat. The complex formed between these two complementary species, in which the paraquat dication is threaded through the centre of the macrocycle, has been used as the starting point for the self-assembly of a wide range of catenanes and rotaxanes.[14]

The receptor BPP34C10, **12**, is probably best prepared using the route described in Scheme 4.8. The ditoluenesulfonate **13** of commercially available tetraethylene glycol **14** can be readily prepared (see Protocol 7, this chapter) on the hundred-gram scale using a modified procedure described by Ouchi *et al.*[15] in which the diol is reacted with *p*-toluenesulfonyl chloride in a mixture of water and THF at 0°C. Reaction of the resulting ditoluenesulfonate with commercial 4-benzyloxyphenol yields the dibenzyl ether derivative **16**, which can be subjected to catalytic hydrogenolysis to afford the bisphenol **17**, a compound which can be used, without purification, in the final step of the synthesis. The bisphenol **17** is reacted with the ditoluenesulfonate **13** of tetraethylene glycol **14** in dry DMF under pseudo high dilution conditions, by employing caesium carbonate as the base, with caesium toluenesulfonate present as a 'template' and tetrabutylammonium iodide included as a phase-transfer catalyst. The use of caesium carbonate as the base is known to enhance the yields of the macrocycles formed in similar reactions. The product BPP34C10, **12**, can be isolated in approximately 40% yield. In addition, the 'dimeric' macrocyclic polyether, tetrakis-*p*-phenylene-68-crown-20 (TPP68C20), is obtained in 5% yield.

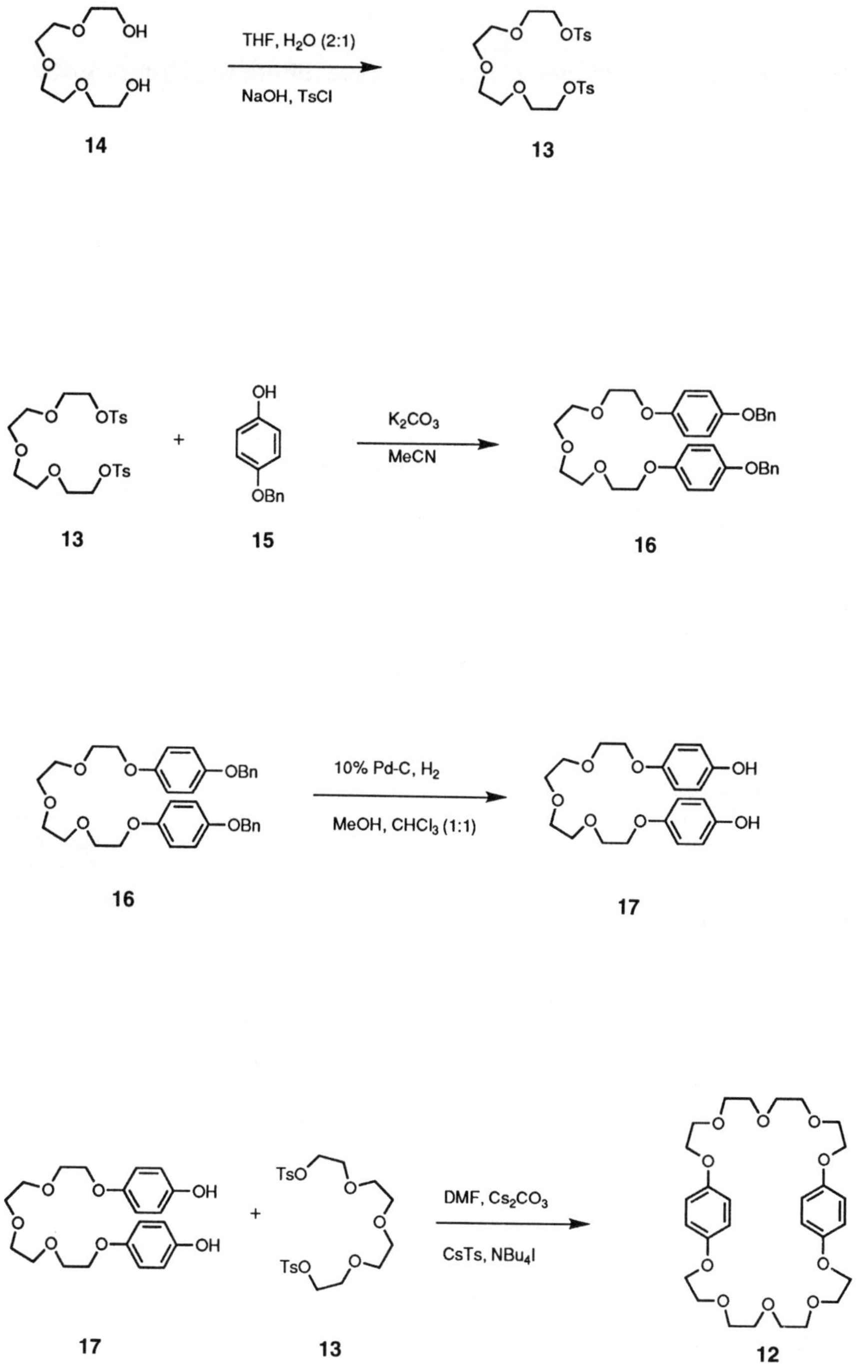
OH
OH
THF, H_2O (2:1)
NaOH, TsCl
OTs
OTs
14
13
OTs
OTs
+
OH
OBn
K_2CO_3
MeCN
OBn
OBn
13
15
16
OBn
OBn
10% Pd-C, H_2
MeOH, $CHCl_3$ (1:1)
OH
OH
16
17
OH
OH
+
TsO
TsO
DMF, Cs_2CO_3
CsTs, NBu_4I
17
13
12

Scheme 4.8

Protocol 7.
Synthesis of tetraethylene glycol ditoluenesulfonate (Structure 13, Scheme 4.9)

Caution! Carry out all procedures in a well-ventilated hood, and wear disposable vinyl or latex gloves and chemical-resistant safety goggles.

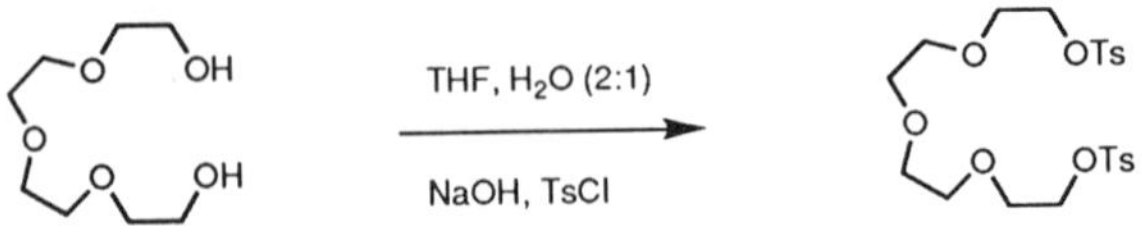

Scheme 4.9

This procedure has been adapted from the method of Ouchi et al.,[15] and can be applied widely to the synthesis of analogous ditoluenesulfonates (see **8**, Scheme 4.6).

Equipment
- Double-necked, round-bottomed flask (2 L)
- Pressure-equalising addition funnel (250 mL)
- Separating funnel (2 L)
- Thermometer
- Stirrer bar
- Filter funnel
- Magnetic stirrer
- Ice bath

Materials
- Tetrahydrofuran (THF), 400 mL — highly flammable
- Distilled water, 200 mL
- Sodium hydroxide, 40 g, 1 mol — causes burns
- Tetraethylene glycol, **14**, 68.3 g, 0.35 mol — irritant
- p-Toluenesulfonyl chloride, 145 g, 0.76 mol — causes burns
- Toluene, 300 mL — highly flammable
- Calcium chloride, anhydrous, for drying, 50 g — irritant

1. Prepare an ice-salt cooling bath. Assemble the round-bottomed flask, and pressure-equalising addition funnel with a stirrer bar and a thermometer. Place the round-bottomed flask into the ice-salt bath.

2. Dissolve the sodium hydroxide (40 g) in water (150 mL) and pour this mixture through a funnel into the round-bottomed flask. Allow it to cool to 0 °C. While this solution is cooling, prepare a solution of tetraethylene glycol, **14** (68.3 g, 0.35 mmol) in THF (200 mL) and water (50 mL) and pour this solution into the round-bottomed flask.

3. Dissolve the p-toluenesulfonyl chloride (145 g) in THF (200 mL) and transfer it to the pressure-equalising addition funnel. Add this solution dropwise to the mixture contained in the round-bottomed flask over a period of 4 h. *Important.* The temperature of the reaction mixture must not be allowed to rise above 5 °C. It should be maintained at around 0 °C. Great patience is necessary!

4: Crown ethers

4. When the addition of the *p*-toluenesulfonyl chloride is complete, stir the reaction mixture for a further 3 h, keeping the temperature between 0 and 5 °C. It is important that the reaction mixture is not allowed to stand for prolonged periods.

5. Pour the reaction mixture on to an ice-water mixture (200 g in 100 mL). Add toluene (300 mL) and allow the ice to thaw. Separate the two layers in a separating funnel, and wash the organic phase with distilled water (2 × 100 mL).

6. Dry the organic phase over calcium chloride (50 g) for 1 h. Other drying agents (*e.g.* magnesium sulfate) can also be included, but calcium chloride is important to remove any monotoluenesulfate. Filter the mixture and remove the solvent from the filtrate on a rotary evaporator, being careful to keep the temperature below 60 °C.

7. The product, which is usually obtained as an oil, should be washed with several portions of light petroleum (5 × 100 mL) by shaking the petroleum vigorously with the oil. This procedure removes unreacted *p*-toluenesulfonyl chloride.

8. The product **13** is obtained as a clear oil (154.2 g, 87%), R_f = 0.31 with hexane: acetone (3:2, v/v) as eluant.

Protocol 8.
Synthesis of 1,11-bis[4-(benzyloxy)phenoxy]-3,6,9-trioxaundecane (Structure 16, Scheme 4.10)

Caution! Carry out all procedures in a well-ventilated hood, and wear disposable vinyl or latex gloves and chemical-resistant safety goggles.

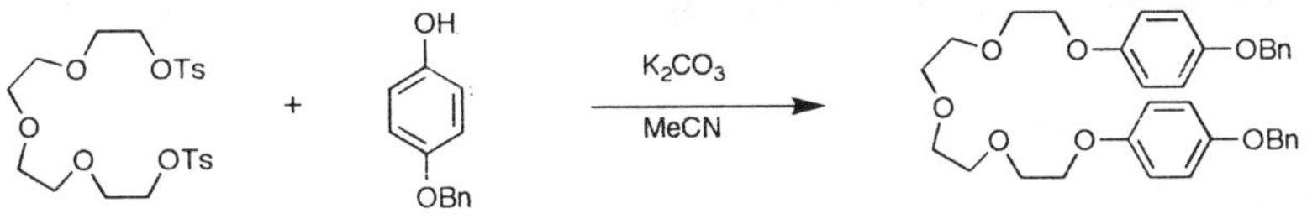

Scheme 4.10

This procedure is a general one for the reaction of phenol derivatives with aliphatic toluenesulfates.

Equipment

- Double-necked, round-bottomed flask (1 L)
- Water-cooled condenser
- Pressure-equalising addition funnel (500 mL)
- Nitrogen bubbler
- Ultrasonic bath
- Bullet magnetic stirrer bar
- Efficient magnetic stirrer
- Source of nitrogen
- Buchner funnel
- Buchner flask (1 L)
- Rotary evaporator

Protocol 8. Continued

Materials
- Dry acetonitrile, 300 mL   toxic and flammable
- Potassium carbonate, anhydrous, 19.4 g, 140 mmol
- Tetraethylene glycol ditoluenesulfonate, **12**, 20.1 g, 40 mmol   irritant
- 4-Benzyloxyphenol, **15**, 16.0 g, 80 mmol   irritant
- Magnesium sulfate, anhydrous, for drying, 10 g
- Nitrogen gas

1. Dry all glassware for 2 h in an oven at 105°C, assemble the round-bottomed flask, pressure-equalising addition funnel, condenser, magnetic stirrer bar and nitrogen bubbler while still warm, and immediately connect the nitrogen supply so that a flow of the gas passes through the equipment.

2. Place the potassium carbonate (19.4 g, 140 mmol) in the reaction vessel and then add the acetonitrile (200 mL).

3. Sonicate the suspension in an ultrasonic bath for approximately 30 min, agitating the mixture every few minutes. Although this procedure is not essential, it increases greatly the efficiency of the reaction. Dry the exterior of the vessel thoroughly before transferring it to the oil bath.

4. Place the stirrer bar into the suspension. Degas the suspension further by passing a stream of nitrogen through the solvent for 15 min.

5. Add the 4-benzyloxyphenol **15** (16.0 g, 80 mmol) as a solid to the suspension and warm it to near refluxing temperatures with vigorous stirring. Maintain the reaction mixture under these conditions for 30 min.

6. Dissolve the tetraethylene glycol bistosylate **13** in dry acetonitrile (100 mL). Then degas the resulting solution by bubbling nitrogen through the solution. Transfer the solution to the pressure-equalising addition funnel.

7. Add the acetonitrile solution of the ditoluenesulfonate to the warm reaction mixture over a period of 30 min.

8. Boil the reaction mixture under reflux with vigorous stirring for 2 d.

9. Filter the reaction mixture through a sintered Buchner funnel at a water pump whilst it is still warm. Wash the solid residue with aliquots of ethyl acetate (100 mL) and dichloromethane (100 mL).

10. Remove all the solvents on a rotary evaporator, and partition the residue between dichloromethane (150 mL) and 1 M aqueous sodium hydroxide solution (150 mL). Pour the mixture into a separating funnel (500 mL), remove the top aqueous phase, and extract the bottom organic phase with water (100 mL), and then dry it over magnesium sulfate (10 g). Filter under gravity and remove the solvent on a rotary evaporator.

11. Recrystallise the white solid from boiling methanol to afford **16** (13.38 g, 60%), m.p. 78–80°C.

4: Crown ethers

Protocol 9.
Synthesis of 1,11-bis(4-hydroxyphenoxy)-3,6,9-trioxaundecane (Structure 17, Scheme 4.11)

Caution! Carry out all procedures in a well-ventilated hood, and wear disposable vinyl or latex gloves and chemical-resistant safety goggles. Hydrogen is a highly flammable gas. No naked flames should be in the area of this experiment while hydrogen is present.

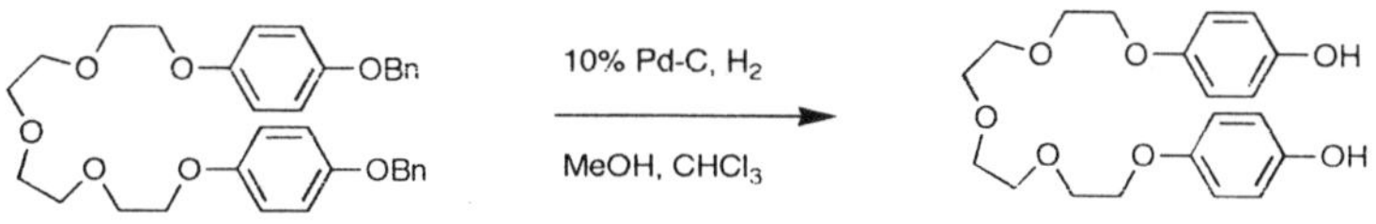

Scheme 4.11

This protocol can also be applied for the deprotection of analogous benzyl ether derivatives.

Equipment
- Double-necked, round-bottomed flask (500 mL)
- Magnetic stirrer bar
- Magnetic stirrer
- Sintered Hirsch funnel

Materials
- 1,11-Bis[4-(benzyloxy)phenoxy]-3,6,9-trioxaundecane, **16**, 6.52 g, 11.7 mmol
- Chloroform, 150 mL   harmful by inhalation
- Methanol, 150 mL   toxic, flammable
- 10% palladium on charcoal, 0.65 g   flammable solid
- Celite filter agent, 20 g   harmful by inhalation
- Hydrogen gas   explosive gas

1. Place the 10% palladium on charcoal in the reaction vessel and wash it with chloroform–methanol (1:1, 50 mL) by swirling the suspension. Allow the mixture to settle and decant off the solvent. Repeat the same procedure one more time.

2. Dissolve the dibenzyl ether **16** (6.52 g, 11.7 mmol) in chloroform–methanol (1:1, 100 mL) and pour it into the reaction vessel. Put the stirrer bar into the suspension, attach the bubbler and line connected to the source of hydrogen to the sintered inlet, and begin to stir.

3. Open the source of hydrogen so that a very gentle flow of the gas passes through the reaction mixture. A rapid flow of hydrogen is *not necessary*. The reaction only requires the presence of a hydrogen atmosphere above it in order to proceed efficiently.

4. After stirring for 2 h, stop the flow of hydrogen and discontinue stirring the reaction mixture. Check the contents of the solution by TLC on silica gel using acetone:hexane (1:1, v/v) as eluent.

Protocol 9. Continued

5. When the reaction is complete, suspend the Celite filter agent (approximately 20 g) in chloroform–methanol (1:1, 50 mL), and apply it to a sintered Hirsch funnel, without applying a vacuum. Keep a layer of solvent over the surface of the filter agent before performing the next step.

6. Pour the reaction mixture through the bed of the filter agent, after applying water pump suction to the vessel below the Hirsch funnel. Wash the solid residue in the funnel with more chloroform–methanol (1:1, 50 mL).

7. The solvent is removed on a rotary evaporator, affording an oil **17** (4.31 g, 98%). This oil can be used in the next stage of synthesis without further purification. *Important.* This material should be used relatively soon (days) after its preparation. It should be stored under a nitrogen atmosphere. If it is stored for prolonged periods, it should be kept in a refrigerator or freezer.

Protocol 10.
Synthesis of bis-*p*-phenylene-34-crown-10 (Structure 12, Scheme 4.12)

Caution! Carry out all procedures in a well-ventilated hood, and wear disposable vinyl or latex gloves and chemical-resistant safety goggles. Crown ethers are known to be highly toxic by absorption through the skin.

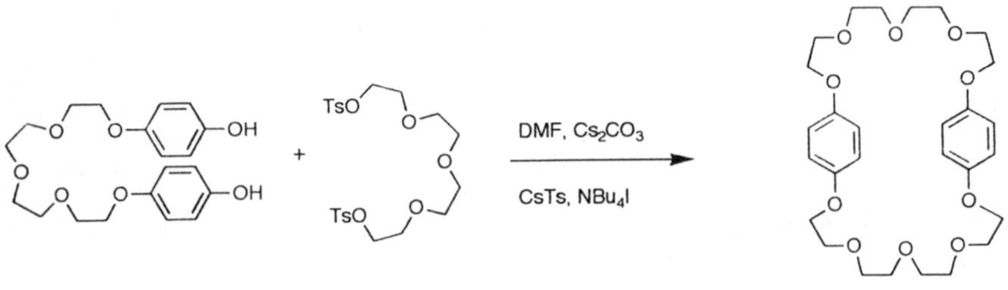

Scheme 4.12

The procedure for the synthesis of the crown ether BPP34C10, **12**, that is outlined below is representative of caesium-templated cyclisations of phenol derivatives with ditoluenesulfonates.

Equipment

- Two-necked, round-bottomed flask (2 L)
- Pressure-equalising addition funnel (250 mL)
- Bullet-shaped magnetic stirrer bar
- Powder funnel
- Column for chromatography
- Efficient hot plate magnetic stirrer
- Oil bath
- Buchner flask and funnel (sinter)
- Separating funnel (1 L)
- Thermometer

4: Crown ethers

Materials

- Dry dimethylformamide (DMF), 1.1 L   harmful by inhalation and skin contact
- Caesium carbonate, anhydrous, 31.3 g, 96 mmol   harmful
- Caesium toluenesulfonate, 7.29 g, 24.0 mmol
- Tetrabutylammonium iodide, 1.13 g, 3.1 mmol   toxic
- Tetraethylene glycol ditoluenesulfonate, **13**, 6.05 g, 12.0 mmol   irritant
- Bisphenol **17**, 4.55 g, 12.0 mmol   irritant
- Magnesium sulfate, anhydrous, for drying, 5 g
- Dichloromethane for extraction and chromatography, 2.5 L   toxic
- Ethyl acetate for washing and chromatography, 2 L   flammable
- Silica gel for column chromatography   toxic by dust inhalation
- Nitrogen gas

1. Dry all glassware for 2 h in an electric oven at 105°C and assemble the flask and the pressure-equalising addition funnel while still warm. Immediately connect the nitrogen supply so that a flow of the gas passes through the equipment.

2. Place the caesium carbonate (31.3 g), caesium toluenesulfonate (3.65 g) and tetrabutylammonium iodide (1.13 g) in the reaction vessel through a powder funnel and then pour dry DMF (800 mL) through the same funnel. Degas the resulting suspension with a flow of nitrogen for 15 min. Transfer the nitrogen supply to the bubbler, so that the gas is flowed over the reaction mixture.

3. Place the reaction vessel in the oil bath and warm the suspension to 80°C with vigorous stirring.

4. Dissolve the bisphenol **17** (4.55 g) in dry degassed DMF (100 mL) and transfer it to the pressure-equalising addition funnel attached to the reaction vessel. Add the solution dropwise over a period of 30 min to the reaction mixture, maintaining the temperature at 80°C with vigorous stirring for a period of 30 min.

5. Dissolve tetraethylene glycol ditoluenesulfonate, **13** (6.05 g) and caesium toluenesulfonate (3.64 g) in dry degassed DMF (100 mL) and transfer it to the pressure-equalising dropping funnel attached to the reaction vessel. Add the solution over a period of 30 min to the reaction mixture, maintaining the reaction conditions as described in step 4.

6. Raise the temperature of the reaction mixture to 100°C and stir vigorously under the nitrogen atmosphere for 2 d.

7. Filter the reaction mixture at the pump while it is still warm. Wash the solid residue with DMF (100 mL) and ethyl acetate (100 mL).

8. Remove all the solvents on a rotary evaporator at 80°C.

9. Partition the residue between 1 M aqueous sodium hydroxide (400 mL) and dichloromethane (400 mL). Transfer the mixture to a separating funnel.

Protocol 10. *Continued*

10. Separate the organic layer and extract with water (2 × 200 mL). Dry the organic phase over magnesium sulfate (5 g), filter the mixture through a fluted filter paper and remove the solvent on a rotary evaporator.
11. Dissolve the residue in the minimum amount of dichloromethane and apply the solution to a silica gel column (3 cm × 40 cm). Use a gradient elution, starting with dichloromethane and increasing the polarity of the eluant to dichloromethane:ethyl acetate (9:1, v/v).
12. When the BPP34C10 has been eluted from the column, increase the polarity of the eluant to dichloromethane:ethyl acetate (5:1, v/v) to obtain the 'dimeric' TPP68C20.
13. Combine the eluants and remove the solvents *in vacuo*. BBP34C10, **12**, (2.48 g, 40%) m.p. 87–8°C and the 2:2 adduct TPP68C20 (0.32 g, 5%) m.p. 96–8°C are isolated as solids.

References

1. (a) Weber, E.; Vögtle, F. *Top. Curr. Chem.* **1981**, *98*, 1–41. (b) Gokel, G. W. *Monograph in Supramolecular Chemistry: Crown Ethers and Cryptands*; The Royal Society of Chemistry, London, **1991**.
2. Lehn, J.-M. *Nature* **1993**, *260*, 1762–1763.
3. Pedersen, C. J. *J. Am. Chem. Soc.* **1967**, *89*, 7017–7036.
4. (a) Pedersen, C. J.; Frensdorff, H. K. *Angew. Chem., Int. Ed. Engl.* **1972**, *11*, 16–25. (b) Kataky, R.; Nicholson, P. E.; Parker, D. *J. Chem. Soc., Perkin Trans 2* **1990**, 321. (c) Collie, L.; Dennes, J. E.; Parker, D.; O'Carroll, F. *J. Chem. Soc., Perkin Trans 2* **1993**, 1747.
5. Vögtle, F.; Sieger, H.; Müller, W. M. *Top. Curr. Chem.* **1981**, *98*, 107–161.
6. (a) Anderson, S.; Anderson, H. L.; Sanders, J. K. M. *Acc. Chem. Res.* **1993**, *26*, 469–478. (b) Busch, D. H. *J. Inc. Phenom.* **1992**, *12*, 389–395.
7. (a) Blasius, E.; Janzen, K.-P. *Top. Curr. Chem.* **1981**, *98*, 163–189. (b) Montanari, F.; Landini, D.; Rolla, F. *Top. Curr. Chem.* **1982**, *101*, 147–200. (c) Vögtle, F.; Knops, P. *Angew. Chem., Int. Ed. Engl.* **1991**, *30*, 958–960.
8. Stoddart, J. F. *Top. Stereochem.* **1987**, *17*, 207–287.
9. Kellogg, R. M. *Top. Curr. Chem.* **1982**, *101*, 111–148.
10. Curtis, W. D.; Laidler, D. A.; Stoddart, J. F.; Jones, G. H. *J. Chem. Soc., Perkin Trans. I* **1977**, 1756–1769.
11. Ehrlenmeyer, E. *Biochem. Z.* **1915**, *68*, 351.
12. Ashton, P. R.; Slawin, A. M. Z.; Spencer, N.; Stoddart, J. F.; Williams, D. J. *J. Chem. Soc., Chem. Commun.* **1988**, 1066–1068.
13. Allwood, B. L.; Spencer, N.; Shariari-Zavareh, H.; Stoddart, J. F.; Williams, D. F. *J. Chem. Soc., Chem. Commun.* **1988**, 1064–1066.

14. (a) Anelli, P. L.; Ashton, P. R.; Ballardini, R.; Balzani, V.; Delgado, M.; Gandolfi, M. T.; Goodnow, T. T.; Kaifer, A. E.; Philp, D.; Pietraszkiewicz, M.; Prodi, L.; Reddington, M. V.; Slawin, A. M. Z.; Spencer, N.; Stoddart, J. F.; Williams, D. J. *J. Am. Chem. Soc.* **1992**, *114*, 193–218. (b) Ashton, P. R.; Preece, J. A.; Stoddart, J. F.; Tolley, M. S. *Synlett* **1994**, 789–792. (c) Ashton, P. R.; Preece, J. A.; Stoddart, J. F.; Tolley, M. S.; Williams, D. J.; White, A. J. P. *Synthesis* **1994**, 1344–1352.
15. Ouchi, M.; Inoue, Y.; Liu, Y.; Nagamune, S.; Nakamura, S.; Wada, K.; Hakushi, K. *Bull. Chem. Soc. Jpn.* **1990**, *63*, 1260–1262.

5

Cryptands

BERNARD DIETRICH

1. Introduction

The cryptands were first prepared in 1969 and form a series of well-defined complexes (cryptates) with alkali and alkaline-earth cations. In this chapter the synthesis of the first cryptand, **8**, a macrobicyclic ligand, will be described[1,2]. The schematic representation (Fig. 5.1) shows that one deals with a multi-step synthesis. The major drawback of this approach is the rather large number of synthetic steps, but the route offers the advantage of being able to construct unsymmetrical compounds (A ≠ B ≠ C).

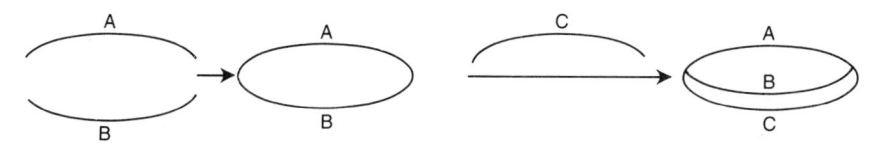

Fig. 5.1

In order to gain more direct access to these bicyclic ligands, methods have been developed which allow the formation of the macrobicycle in one step. The tripod–tripod coupling is an example of this method (Fig. 5.2) which may be performed without the need for high-dilution conditions. The synthesis of the bis-tren macrobicycle, **17**, illustrates this approach.[3]

Fig. 5.2

2. Synthesis of the cryptand [2.2.2]

The construction of the macrobicyclic system is shown in Scheme 5.1. The acyclic starting materials are the diacid chloride **2** and the diamine **3**.[4] An alternative synthesis of the intermediate monocycle is given in Chapter 2 (Protocol 5).

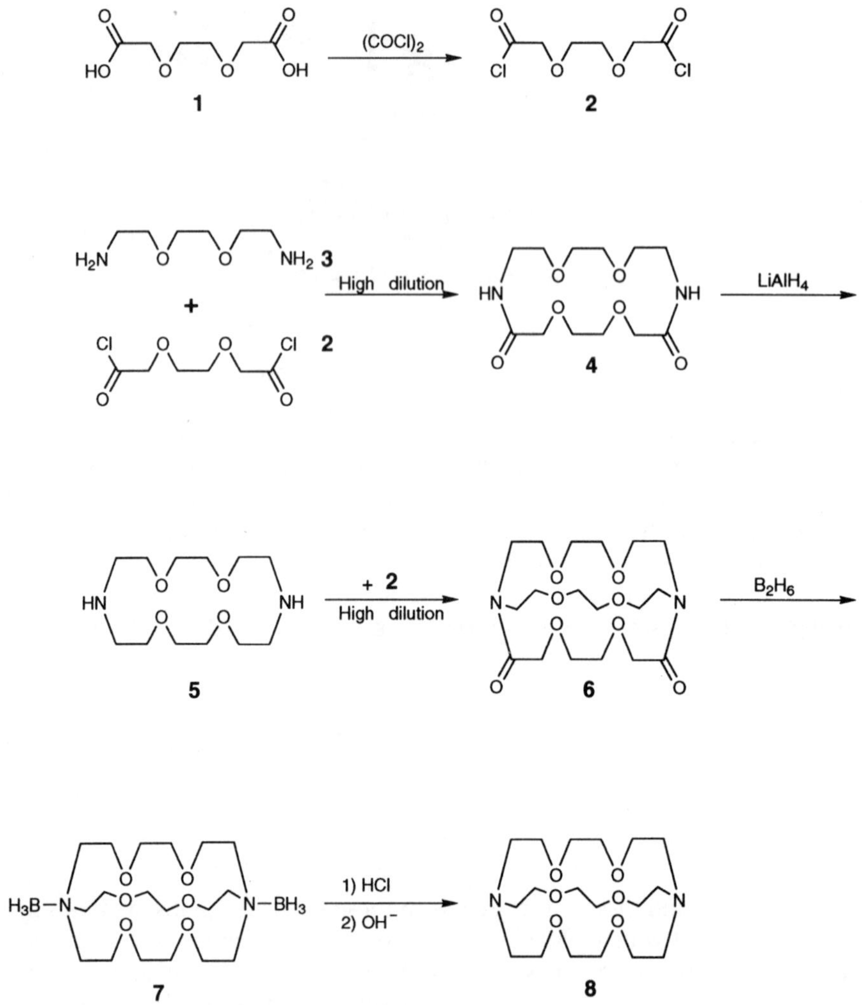

Scheme 5.1

The macrocyclization step involving the two bifunctional compounds **2** and **3** is performed under high-dilution conditions. For this reaction there is a need to use a specially adapted apparatus illustrated in Fig. 5.3. The crucial features of this reaction are the precision addition funnels which have to deliver the solutions at a low and constant rate and the vigorous stirring ensured by a high-speed motor and by the creased flask.

One can note that for this step the ratio of reagents **3:2** is 2:1, the excess diamine, **3**, playing the role of base which takes up the hydrogen chloride liberated by the coupling reaction. The separation of the monocyclic diamide **4** from the reaction mixture containing amide oligomers is easy as the latter

5: Cryptands

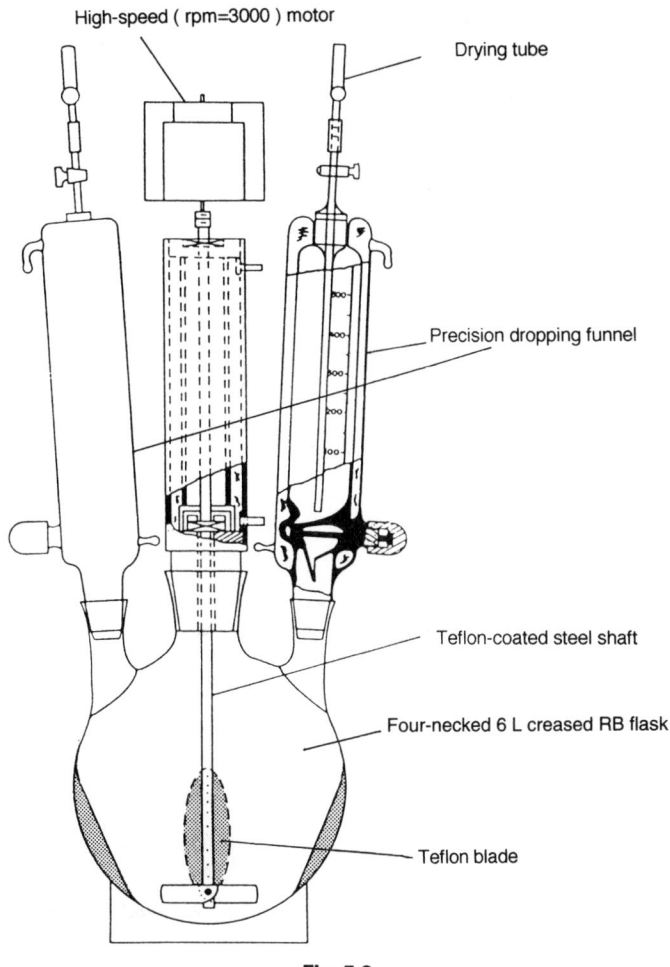

Fig. 5.3

(Labels: High-speed (rpm=3000) motor; Drying tube; Precision dropping funnel; Teflon-coated steel shaft; Four-necked 6 L creased RB flask; Teflon blade)

compounds do not elute on chromotographic separation using alumina. For the reduction step, **4 → 5** using LiAlH$_4$, a mechanical stirrer is strongly recommended. This is required because following the destruction of the excess of LiAlH$_4$ by water addition, the reaction mixture becomes very sticky. For a good recovery of the amine **5** the solid obtained after filtration has to be washed thoroughly with hot toluene.

The bicyclization step is accomplished under identical conditions to the first macrocyclization step, the base being in this case added triethylamine. The reduction of the bicyclic amide **6** has to be realized with diborane (LiAlH$_4$ cleaves the macrobicyclic system). Note that there is no need to isolate the intermediate aminoborane **7**: the crude mixture obtained after destruction of the excess of diborane (step 14 in Protocol 5) can be treated directly with 6 M HCl (step 2 in Protocol 6).

Protocol 1.
Synthesis of the diacid chloride 2 of 3,6-dioxaoctanedioic acid (Scheme 5.2)

Caution! Carry out all procedures in a well-ventilated hood, and wear disposable vinyl or latex gloves and chemical-resistant safety goggles. Oxalyl chloride is a very dangerous material.

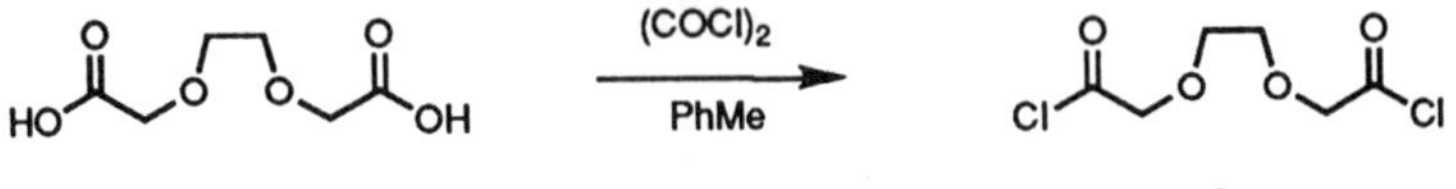

Scheme 5.2

Equipment
- Single-necked, round-bottomed flask (500 mL)
- Graduated pipette (50 mL)
- Pipette filler
- Graduated cylinder (100 mL)
- Drying tube
- Magnetic stirrer
- Teflon-coated magnetic stirrer bar
- Rotary evaporator
- Sintered glass conical filter funnel
- Filtering flask (500 mL)
- Vacuum pump
- Micro-filter candle

Materials
- 3,6-Dioxaoctanedioic acid,[a] 15 g, 84.2 mmol — corrosive, irritant
- Oxalyl chloride, 30 g, 236 mmol — poisonous, severely irritating to skin, eyes, respiratory tract
- Dry toluene,[b] ~300 mL — flammable, highly toxic, carcinogen
- Dry pyridine, 3 drops — flammable, irritant, toxic
- Dry petroleum ether 30–60°, ~400 mL — flammable
- Dry ether, ~200 mL — flammable, explosive, harmful by inhalation
- Drierite, ~50 g — irritant
- Dry ice — harmful to skin
- Liquid nitrogen — harmful to skin

1. Dry all glassware in an electric oven at 110°C for 2 h.
2. Introduce with care the stirring bar in the 500 mL round-bottomed flask. Fill the drying tube with drierite.
3. Pour 100 mL of dry toluene in the flask, and add 3 drops of dry pyridine, the 3,6-dioxaoctanedioic acid (15 g) and the oxalyl chloride (30 g). Protect the reaction from moisture with the drying tube.
4. Stir the suspension at room temperature for 20 h. The solid dissolves slowly.
5. Rapidly filter the reaction mixture with suction through a sintered glass conical filter funnel, rinse the flask and the filter with three portions of dry toluene (3 × 30 mL).

5: Cryptands

6. Transfer the solution to a 500 mL round-bottomed flask and remove the solvent on a rotary evaporator with a vacuum pump. To the dry residue add 100 mL of dry toluene and evaporate the solvent, this operation should be repeated twice. *Caution!* The vacuum pump should be protected by an efficient trap containing liquid nitrogen. *Caution!* The collected solvent still contains oxalyl chloride which should be destroyed before discharge.
7. Crystallize **2** from dry ether/dry petroleum ether at −70 °C. Remove the solvent with the micro-filter candle. Recrystallise the solid under the same conditions.
8. Dry the crystalline solid under vacuum to obtain pure **2** (16.3 g, 90%) m.p. 19–20 °C. This compound should be used forthwith or stored under N_2 at −30 °C.

[a] The diacid is commercially available but the purity has to be checked. Purity can be improved by the preparation of a dianilide derivative (see an example in Ref. 2). This diacid can also be synthesised from triethylene glycol.[2]
[b] The solvent used in the original preparation was benzene which can certainly be replaced by the less harmful toluene.

Protocol 2.
Synthesis of 2,9-dioxo-1,10-diaza-4,7,13,16-tetraoxacyclooctadecane (Structure 4, Scheme 5.3)

Caution! Carry out all procedures in a well-ventilated hood, and wear disposable vinyl or latex gloves and chemical-resistant safety goggles.

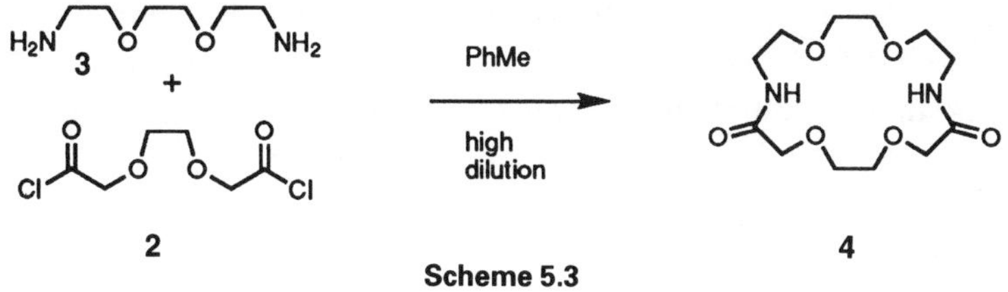

Scheme 5.3

The following procedure for the synthesis of macrocycle **4** is representative of a high dilution reaction.

Equipment

- Four-necked, creased round-bottomed flask (6 L)
- Two precision dropping funnels (500 mL) [for example: NORMAG 'Dosiertrichter' DBP 1 259 728 (BASF)]
- Three drying tubes
- Stirrer guide with cooling jacket
- Teflon-coated steel shaft with Teflon blade
- Mechanical stirrer 3000 r.p.m.

- Strong lab jack
- Rubber round-bottomed flask support
- Three rubber protected clamps
- Two conical filters
- Two graduated cylinders (500 mL)
- Rotary evaporator
- Column for chromatography
- Sintered glass filter funnel

Protocol 2. Continued

Materials
- 3,6-Dioxa-1,8-diaminooctane, 3, 14.8 g, 100 mmol — irritant to skin
- Diacid chloride 2, 10.8 g, 50.2 mmol — irritant to skin
- Dry benzene, ~7 L — flammable, highly toxic, carcinogen
- Chloroform, 600 mL — irritant, harmful by inhalation
- Alumina (neutral, activity II-III), 200 g — irritant
- Heptane for crystallization, ~300 mL — flammable
- Drierite — irritant

1. Dry all glassware in an electric oven at 105°C for 2 h (except the addition funnels).
2. Assemble the apparatus as shown in Fig. 5.3. Do not forget the water cooling of the stirrer guide and of the addition funnels. Flush the apparatus with dry nitrogen for 10 min.
3. Prepare separate solutions of the diamine 3 (14.8 g) in 500 mL dry benzene and the diacid chloride 2 (10.8 g) in 500 mL of dry benzene.
4. Pour 1200 mL of dry benzene into the 6000 mL round-bottomed flask.
5. Filter[a] the diamine solution in funnel A and the acid chloride solution in funnel B. Place the drying tubes on the two addition funnels and on the fourth neck of the reaction flask.
6. Start the vigorous stirring and slowly open the addition funnels: a constant rate of addition is required; ~1 drop per second.
7. Ensure frequently that the volumes in the two addition funnels are identical. The total time of addition is about 8 h.[b]
8. Filter the reaction mixture through a sintered glass filter funnel. Rinse the flask with benzene (3 × 50 mL).
9. Concentrate the filtrate to ~100 mL by evaporation under reduced pressure (rotary evaporator).
10. Pass this solution through a column of 200 g alumina and elute with benzene,[c] ~5 L.
11. Evaporate the solvent and crystallise the diamide from benzene–heptane to yield 4 (10.9 g, 75%), m.p. 110–11 °C.

[a] This avoids the presence of small particles which could obstruct the addition funnels.
[b] During the addition a precipitate appears which is a mixture of low molecular weight polymers and the diammonium chloride salt.
[c] Toluene may be used instead of benzene in this protocol.

5: Cryptands

Protocol 3.
Synthesis of 1,10-diaza-4,7,13,16-tetraoxacyclooctadecane (Structure 5, Scheme 5.4)

Caution! Lithium aluminium hydride is a dangerous metal. Carry out all procedures in a well-ventilated hood, and wear disposable vinyl or latex gloves and chemical-resistant safety goggles.

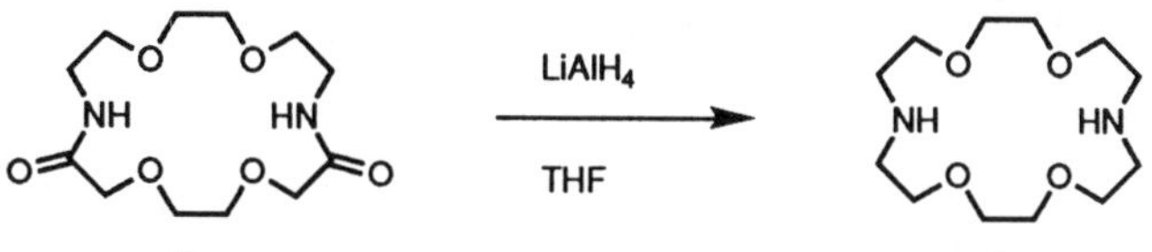

Scheme 5.4

Equipment

- Three-necked, round-bottomed flask (1 L)
- Pressure-equalising additional funnel (500 mL)
- Water-cooled condenser
- Two drying tubes
- Explosion-proof mechanical stirrer
- Stirrer guide with cooling jacket
- Teflon-coated steel shaft with Teflon blade
- Hot plate
- Oil bath
- Lab jack
- Graduated cylinder (500 mL)
- Sintered filter
- Filter funnel
- Rotary evaporator
- Source of dry nitrogen

Materials

- Diamide **4**, 13.8 g, 47.5 mmol — irritant
- Lithium aluminium hydride, 12 g, 316 mmol — flammable on contact with water, highly corrosive for skin, eyes, lungs
- Dry tetrahydrofuran, 500 mL — highly flammable, can contain peroxide (explosion!), irritant
- Toluene, 300 mL — flammable, toxic
- Petroleum ether 30–60°, ~300 mL — flammable
- Drierite, ~80 g — irritant
- Sodium hydroxide 15%, aqueous solution 20 mL — corrosive
- Benzene,[a] ~100 mL — flammable, highly toxic, carcinogenic

1. Dry all glassware in an electric oven at 110°C for 2 h.
2. Assemble the three-necked flask, mechanical stirrer, condenser, addition funnel and drying tubes, and purge the assembly with dry nitrogen.
3. Pour 100 mL dry tetrahydrofuran in the flask and add under gentle stirring LiAlH$_4$ (12 g).
4. Introduce a solution of the diamide **4** (13.8 g) in THF (~300 mL) in the addition funnel.
5. Stir the LiAlH$_4$ suspension and add slowly the diamide solution (a moderate heating is observed). The time of addition is c. 1 h.
6. Reflux the mixture for 24 h, under nitrogen.
7. Remove the oil bath and allow the mixture to cool.

Protocol 3. *Continued*

8. Stir the suspension well and add slowly and successively: (i) a mixture of 20 mL H$_2$O in 50 mL THF; (ii) 20 mL of a 15% solution of NaOH; and finally (iii) a mixture of 50 mL H$_2$O in 50 mL THF.
9. Filter the mixture, wash the solid residue thoroughly with hot toluene (3 × 100 mL).
10. Evaporate the combined solvents and crystallise the diamine from benzene[a]–petroleum ether to yield **5** (9.4 g, 75%), m.p. 114–16 °C.

[a] Toluene may be used in place of benzene.

Protocol 4.
Synthesis of 2,9-dioxo-1,10-diaza-4,7,13,16,21,24-hexaoxabicyclo[8.8.8]hexacosane (Structure 6, Scheme 5.5)

Caution! Carry out all procedures in a well-ventilated hood, and wear disposable vinyl or latex gloves and chemical-resistant safety goggles.

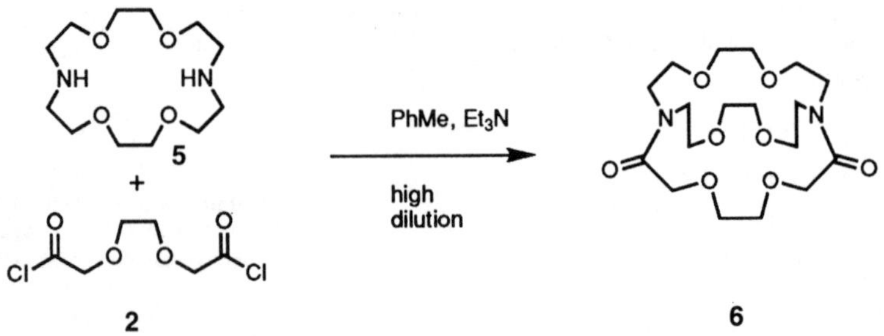

Scheme 5.5

The following procedure for the synthesis of macrobicycle **6** is representative of a high dilution reaction.

Equipment
- Four-necked, creased round-bottomed flask (6 L)
- Two precision dropping funnels (500 mL) [for example: NORMAG 'Dosiertrichter' DBP 1 259 728 (BASF)]
- Three drying tubes
- Stirrer guide with cooling jacket
- Teflon-coated steel shaft with Teflon blade
- Mechanical stirrer (3000 r.p.m.)
- Strong lab jack
- Rubber round-bottomed flask support
- Three rubber-protected clamps
- Two conical filters
- Two graduated cylinders (500 mL)
- Rotary evaporator
- Column for chromatography
- Sintered glass filter funnel

Materials
- Diamine **5**, 5.24 g, 20 mmol
- Diacid chloride **2**, 4.4 g, 20.46 mmol   corrosive, irritant to skin
- Triethylamine, 4.4 g, 43.5 mmol   highly flammable, irritant vapour

5: Cryptands

- Dry toluene, ~5 L — flammable, highly toxic, carcinogen
- Petroleum ether 30–60°, ~200 mL — flammable
- Alumina (neutral, activity II-III), 200 g — irritant
- Drierite — irritant

1. Dry all glassware in an electric oven at 110°C for 2 h (except the addition funnels).
2. Assemble the apparatus as shown in Fig. 5.3. Do not forget the water cooling of the stirrer guide and the addition funnels. Flush the apparatus with dry nitrogen.
3. Prepare the solutions of the diamine **5** (5.24 g) and triethylamine (4.4 g) in 500 mL dry toluene and of the diacid chloride **2** (4.4 g) in 500 mL of dry toluene.
4. Pour 1200 mL of dry toluene in the 6 L flask.
5. Filter[a] the diamine triethylamine solution in funnel A and the acid chloride solution in funnel B. Place the drying tubes on the two addition funnels and on the fourth neck of the reaction flask.
6. Start the vigorous stirring and slowly open the addition funnels: constant rate of addition ~1 drop per second.
7. Ensure throughout the addition that the volumes in the two addition funnels are identical. Time of addition is 10 h.[b]
8. Filter the reaction mixture through a sintered glass filter funnel. Rinse the flask with toluene (3 × 50 mL).
9. Concentrate the filtrate to ~100 mL using a rotary evaporator.
10. Pass this solution through a column of 200 g alumina and elute with toluene (c. ~3 L).
11. Evaporate the solvent and crystallise the macrobicyclic diamide from toluene/petroleum ether (c. 100 mL, 2:1 v/v) to yield **6** (3.64 g, 45%), m.p. 114–15°C.

[a] This avoids the presence of small particles which could obstruct the addition funnels.
[b] During the addition a precipitate appears which is a mixture of polyamides and of triethylammonium chloride salt.

Protocol 5.
Synthesis of 1,10-borane-1,10-diaza-4,7,13,16,21,24-hexaoxabicyclo[8.8.8]hexacosane (Structure 7, Scheme 5.6)

Caution! Carry out all procedures in a well-ventilated hood, and wear disposable vinyl or latex gloves and chemical-resistant safety goggles. Diborane is a flammable and highly irritating compound.

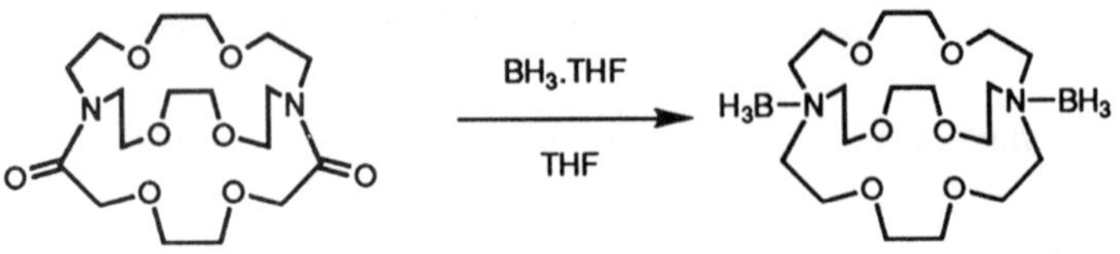

Scheme 5.6

Equipment

- Three-necked, round-bottomed flask (1 L)
- Water-cooled condenser
- Addition funnel with pressure-equalization arm (200 mL)
- Stopper
- Septum
- Three-way stopcock
- Nitrogen balloon
- Graduated cylinder (250 mL)
- Teflon-coated magnetic stirrer bar
- All-glass syringe with a needle-lock Luer (100 mL)
- Stainless steel needle (gauge no 18 or 20)
- Cooling bath
- Hot plate stirrer
- Oil bath

Materials

- Diamide **6**, 10 g, 24.7 mmol
- 1.0 M solution of borane in tetrahydrofuran, 100 mL, 100 mmol — flammable, highly irritating
- Dry tetrahydrofuran, 300 mL — highly flammable, can contain peroxide (explosion!), irritant
- Chloroform, 500 mL — irritant, harmful by inhalation
- Hexane, 300 mL — flammable
- Source of dry nitrogen

1. Dry all glassware in an electric oven at 110 °C for 2 h.
2. Prepare a solution of diamide **6** (10 g) in 150 mol of dry THF.
3. Assemble the apparatus as shown in Fig. 5.4.
4. Introduce the diamide solution to the addition funnel.
5. Purge the apparatus with N_2 using the three-way stopcock: (i) vacuum; (ii) nitrogen. Repeat this operation three times.
6. Cool the reaction flask with an ice cooling bath and start the stirring.
7. Fill the syringe with 100 mL diborane in THF from the bottle containing this solution (see Fig. 5.5). *Caution!* (a) support the bottle containing the diborane solution by a clamp; (b) this operation should be made with great

5: Cryptands

care, the plunger has to be held firmly in order to avoid the spillage of diborane (*fire hazard!*).

8. Transfer the solution in the syringe to the reaction flask by puncturing the flask septum.
9. Add the diamide solution contained in the addition funnel dropwise to the stirred diborane solution. The time of addition is 30 min. During the addition: (i) a violent reaction occurs after addition of each drop; (ii) gas is liberated (keep an eye on the nitrogen balloon); (iii) a white and abundant precipitate appears in the flask.
10. Remove the ice bath and allow the temperature to rise to room temperature.
11. Boil the mixture under reflux for 2 h. Allow the mixture to cool.
12. Destroy the excess of diborane by very slow addition of water (*c.* 20 mL) through the septum.
13. Flush the apparatus with nitrogen.
14. Remove the solvent on a rotary evaporator.
15. Wash the residue with hot chloroform (3 × 100 mL) and filter the combined chloroform solution.
16. Concentrate the chloroform solution to 100 mL and add hexane (300 mL). Crystals of the aminoborane **7** (10 g, 100%) m.p. 164–6°C precipitate from the solution.

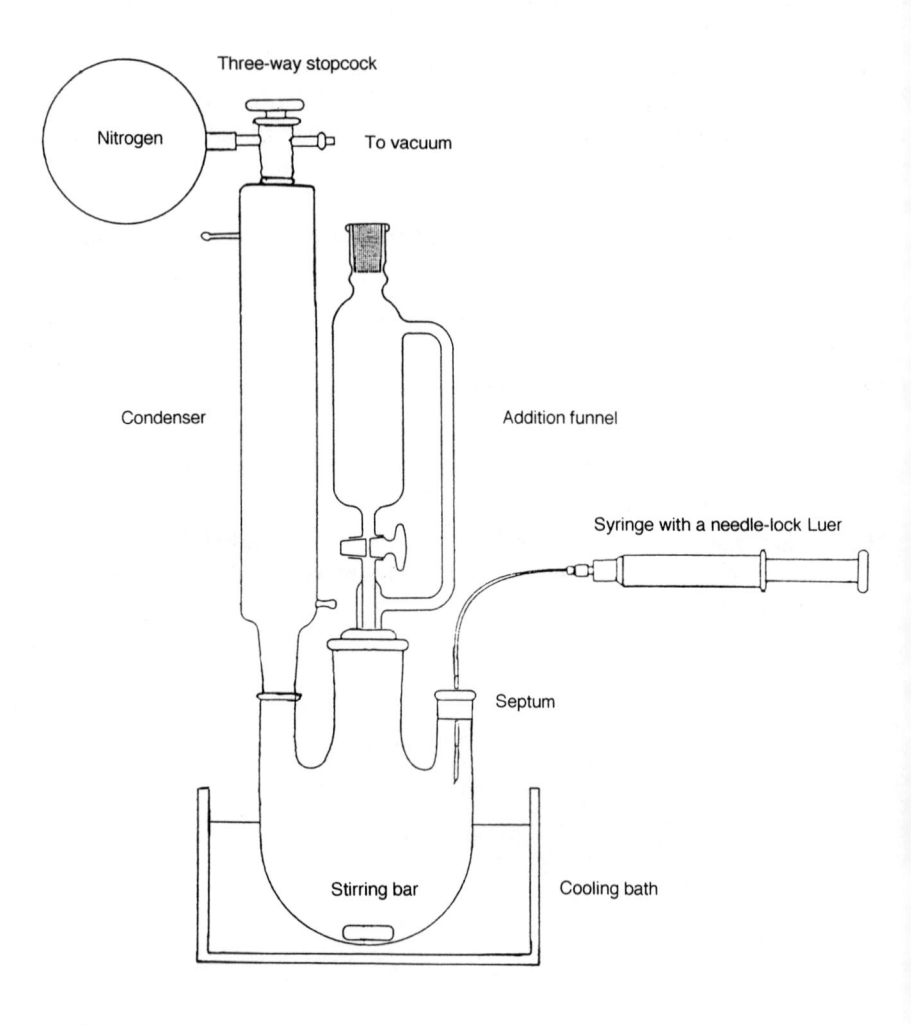

Fig. 5.4

5: *Cryptands*

Nitrogen

Syringe with a needle-lock Luer

Diborane in THF

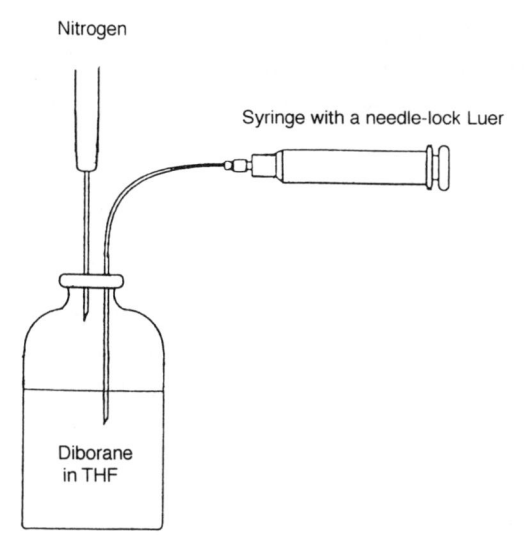

Fig. 5.5

Protocol 6.
Synthesis of 1,10-diaza-4,7,13,16,21,24-hexaoxabicyclo[8.8.8]hexacosane (Structure 8, Scheme 5.7)

Caution! Carry out all procedures in a well-ventilated hood, and wear disposable vinyl or latex gloves and chemical-resistant safety goggles.

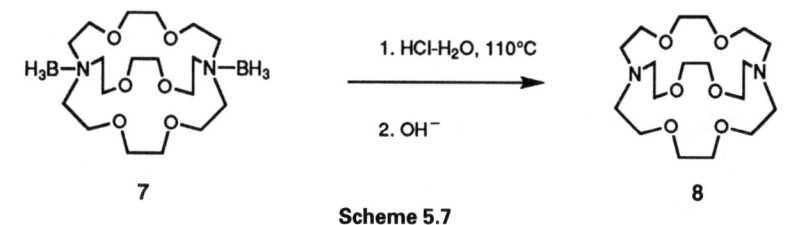

H_3B-N [structure] $N-BH_3$ 1. HCl-H$_2$O, 110°C   → N [structure] N

2. OH$^-$

      7   8

Scheme 5.7

Equipment

- Single-necked, round-bottomed flask (500 mL)
- Super water-cooled condenser
- Hot plate stirrer
- Oil bath
- Graduated cylinder (250 mL)
- Teflon-coated stirring bar
- Column for ion-exchange resin
- pH indicator strip (1–10)
- Single-necked round-bottomed flask (1 L)
- Rotary evaporator
- Glass wool

Materials

- Aminoborane derivative 7, 10 g, 24.7 mmol
- 6 M hydrogen chloride, 200 mL   **corrosive, irritant**
- Dowex 1x8–100 ion-exchange resin, 150 g
- Hexane, 300 mL   **flammable**

Protocol 6. *Continued*

1. Introduce the aminoborane derivative **7** (10 g) to the 500 mL reaction flask.
2. Add the stirrer bar and 200 mL of 6 M HCl and place the condenser on the flask.
3. Boil the stirred solution under reflux for 3 h (bath temperature 110 °C). *Caution!* A large quantity of foam develops during the heating. Ensure that there is no overflow on the top of the condenser (a piece of glass wool on the top of the condenser can prevent this).
4. Remove the acid solution under vacuum on a rotary evaporator. Dry the residue thoroughly on a vacuum pump (0.1 mmHg, 3 h).
5. Prepare the ion-exchange column: (i) introduce a small piece of glass wool above the stopcock; (ii) add the slurry of Dowex resin (150 g) and water (300 mL); (iii) exchange the Cl$^-$ of the commercial resin to OH$^-$ by treatment of the resin with 200 mL of 10% NaOH aqueous solution; and (iv) wash the resin with water until the eluent is neutral (pH indicator strip).
6. Dissolve the residue obtained in step 4 in 50 mL water.
7. Pass this solution over the Dowex 1 column. Wash the column with water. The eluate stays basic (control with pH indicator strip) as long as **8** comes out. About 1 L of water is necessary.
8. Evaporate the aqueous solution on a rotary evaporator.
9. Dry the residue under high vacuum (60 °C, 0.1 mmHg).
10. Crystallise the product **8** from hexane (*c.* 200 mL) (8.85 g, 95%) m.p. 68–9 °C.

---

3. Synthesis of bis-tren

The bis-tren ligand, **17**, incorporates two triethylenetetramine sub-units, and forms a series of well-defined 2:1 (M$_2$L) complexes with transition metal ions such as Cu(II), Ni(II) and Co(II).

The first reported synthesis of this compound involved the stepwise procedure described in Scheme 5.8.[5] One can note that there are two cyclization reactions which were accomplished in 59% and 60% yields. The synthesis required ten steps and the overall yield was 7%. In order to gain more direct access to the macrobicyclic compound a shorter route involving a tripod–tripod coupling sequence was developed (Scheme 5.9).

Treatment of the monochloro diethylene glycol **9** with dihydropyran yields the protected alcohol **10**. Reaction of 2,2′,2″-triaminotriethylamine (tren) **11** with toluenesulfonyl chloride gives the tritoluenesulfonyl derivative **12**. The pyranyl ether **10** may be condensed with the tris(sodium) salt of **12** leading to **13**. Removal of the tetrahydropyranyl group is achieved in high yield under

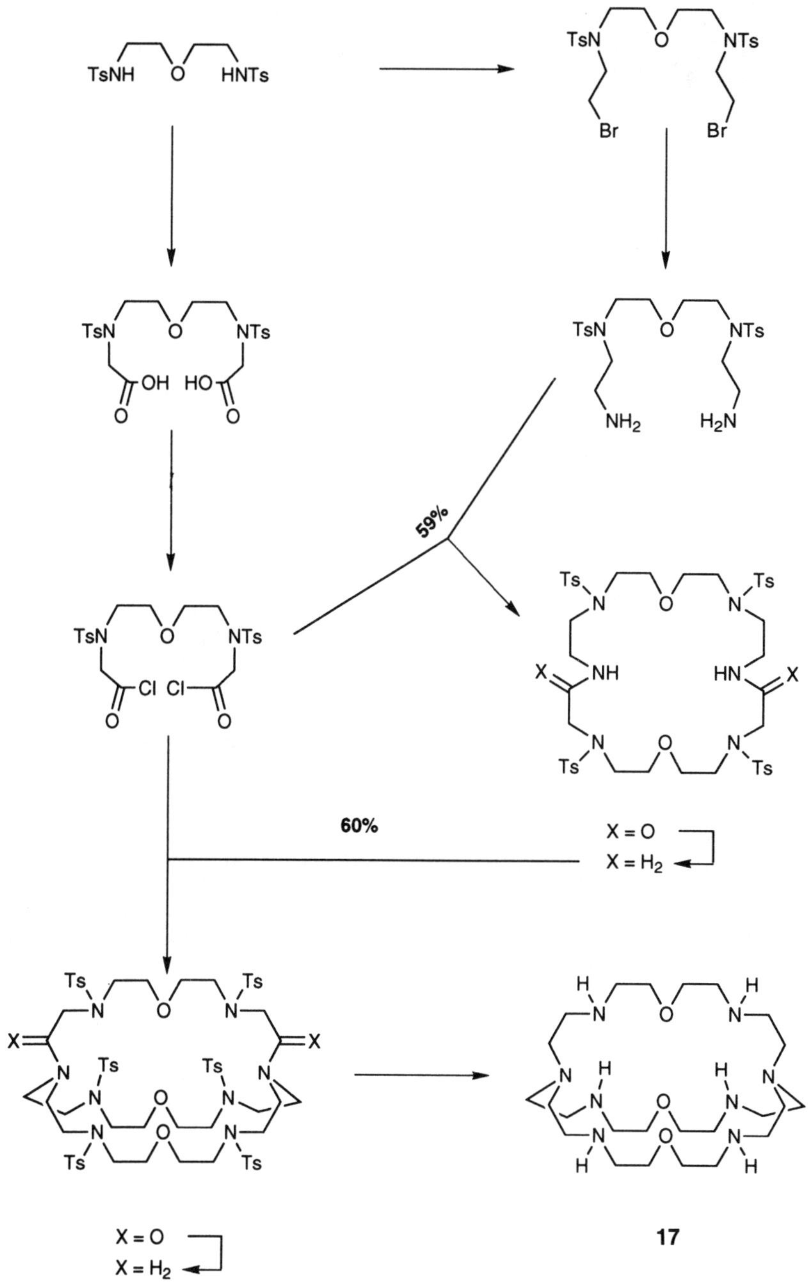
TsNH
O
HNTs
TsN
O
NTs
Br
Br
TsN
O
NTs
OH
HO
O
O
TsN
O
NTs
NH2
H2N
59%
TsN
O
NTs
Cl
Cl
O
O
Ts
N
O
N
Ts
X
NH
HN
X
N
O
N
Ts
Ts
60%
X = O
X = H2
Ts
N
O
N
Ts
X
X
N
Ts
Ts
N
N
O
N
N
O
N
Ts
Ts
X = O
X = H2
H
N
O
N
H
N
H
H
N
N
O
N
N
O
N
H
H

17

Scheme 5.8

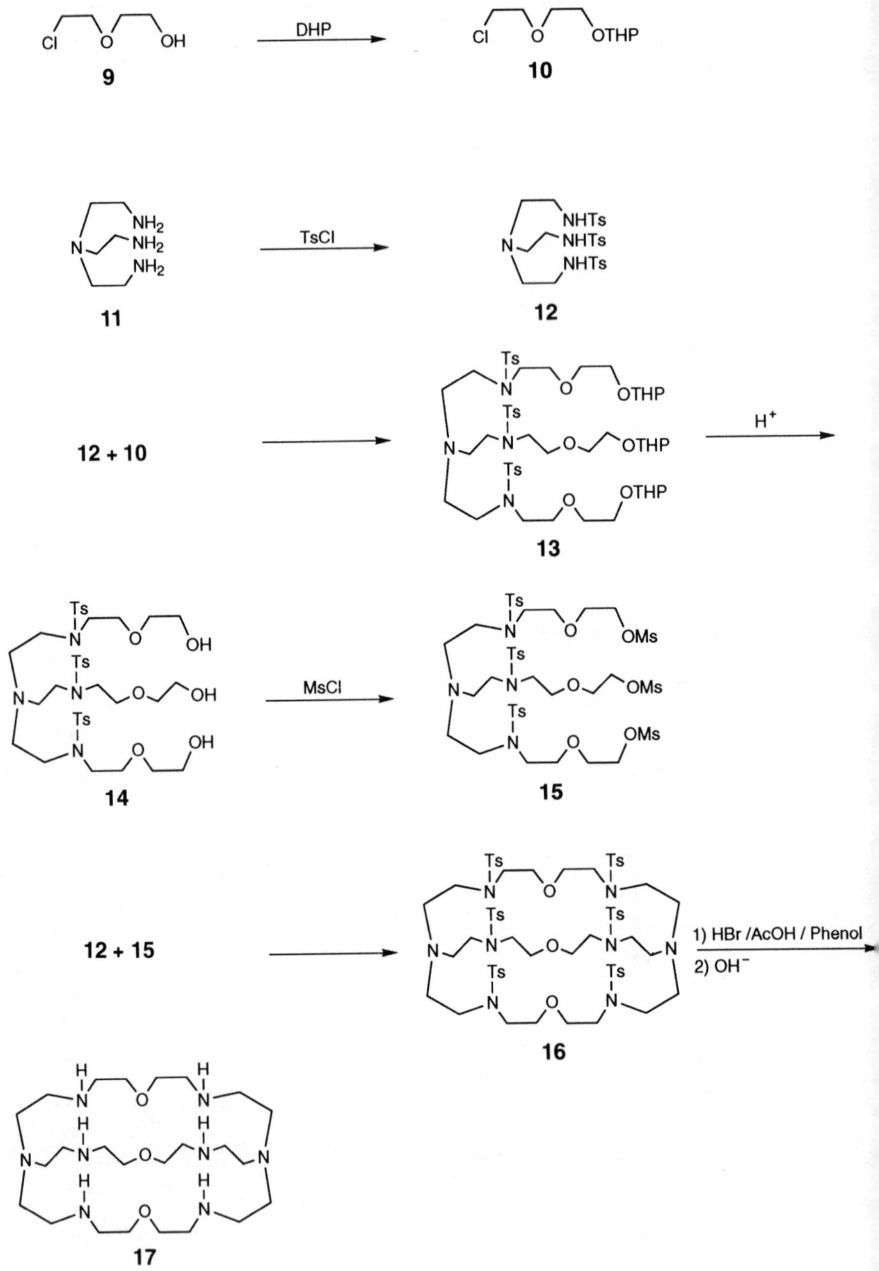

Scheme 5.9

5: Cryptands

toluenesulfonic acid catalysis. Reaction with methanesulfonyl chloride converts the triol **14** into the trimesylate **15**. Tripod–tripod coupling of **12** and **15** was performed in hot DMF in the presence of a large excess of K_2CO_3 yielding the hexatoluenesulfonyl macrobicycle **16** in 31% yield. The toluenesulfonyl groups may be removed with HBr/AcOH/phenol giving **17** as its octabromide salt. The free macrobicyclic octaamine **17** was obtained by passing **17**·8HBr over Dowex 1x8 resin in its basic (OH⁻) form.

By this procedure the final compound **17** may be obtained in only seven steps with an overall yield of 12%, which is a significant improvement over the original lengthy route.

Protocol 7.
Synthesis of 1-chloro-5-(tetrahydro-2H-pyran-2-yloxy)-3-oxapentane (Structure 10, Scheme 5.10)

Caution! Carry out all procedures in a well-ventilated hood, and wear disposable vinyl or latex gloves and chemical-resistant safety goggles.

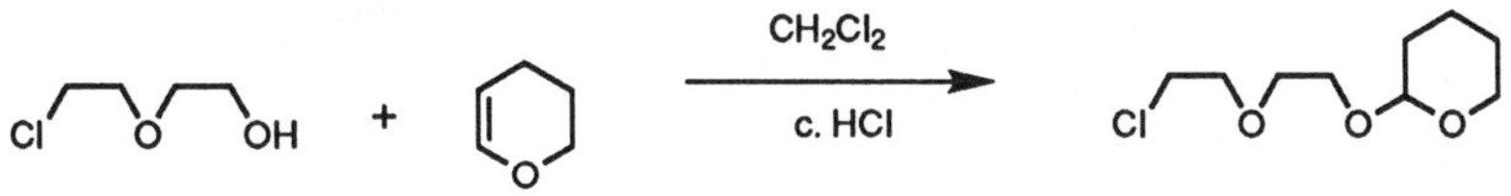

Scheme 5.10

Equipment
- Double-necked, round-bottomed flask (500 mL)
- Water-cooled condenser
- Addition funnel (200 mL)
- Hot plate stirrer
- Teflon-coated magnetic stirrer bar
- Graduated cylinder (250 mL)

Materials
- Monochloro diethylene glycol **9**, 56 g, 450 mmol — corrosive
- Freshly distilled 3,4-dihydro-2H-pyran, 44.8 g, 532 mmol — irritant
- Dichloromethane, 200 mL — irritant, harmful by inhalation
- Concentrated hydrogen chloride, 12 drops — corrosive, irritant
- Potassium carbonate, 10 g, 72 mmol — irritant

1. Equip the round-bottomed flask with a stirrer and attach the condenser and the addition funnel.
2. Introduce into the round-bottomed flask the monochloro diethylene glycol **9** (56 g) and 150 mL CH_2Cl_2. Start the stirring.
3. Through the addition funnel add within 30 min the solution of 3,4-dihydro-2H-pyran (44.8 g) in 50 mL CH_2Cl_2.
4. After the addition is complete add 12 drops of conc. HCl.
5. Heat the mixture at 40 °C for 1 h. Allow to cool.
6. Add 10 g of potassium carbonate.[a]

Protocol 7. Continued

7. Evaporate the solvent and dry the residue on a vacuum pump for 12 h to obtain **10** as an oil,[b] in quantitative yield. This material may be used directly in the next step without further purification.

[a] Potassium carbonate is added to maintain a basic medium necessary for the stability of the protecting group.
[b] The recovery of **10** for the next step is made by its solubilization in DMF and filtration of K_2CO_3.

Protocol 8.
Synthesis of N,N',N''-tritoluenesulfonyl-2,2',2''-nitrilotriethylamine (Structure 12, Scheme 5.11)

Caution! Carry out all procedures in a well-ventilated hood, and wear disposable vinyl or latex gloves and chemical-resistant safety goggles Toluenesulfonyl chloride is a corrosive compound with a persistent odour.

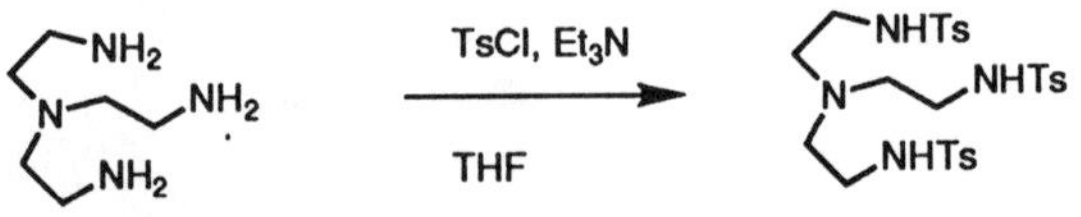

Scheme 5.11

This protocol is typical of the sulfonylation of a primary amine. The tritoluenesulfonylated derivative can also be obtained by an alternative method.[3]

Equipment

- Single-necked, round-bottomed flask (500 mL)
- Magnetic stirrer
- Teflon-coated magnetic stirrer bar
- Separating funnel (1 L)
- Spatula
- Graduated cylinder (100 mL)
- Column for chromatography
- Rotary evaporator

Materials

- Tris(2-aminoethyl)amine, **11**, 10 g, 68 mmol — corrosive, irritant
- Toluenesulfonyl chloride, 44 g, 232 mmol — highly irritating, corrosive
- Triethylamine, 40 mL, 288 mmol — irritant
- Tetrahydrofuran, 200 mL — highly flammable, can contain peroxide (explosion!), irritant
- Dichloromethane, 3000 mL — irritant, harmful by inhalation
- Ethanol, 500 mL — flammable
- Silica gel, 200 g — irritant
- Sodium sulfate, 50 g — irritant

1. Introduce into the round-bottomed flask the stirring bar, the tetraamine **11** (10 g), the triethylamine (40 mL) and 200 mL THF.

5: Cryptands

2. To this stirred solution, add toluenesulfonyl chloride (44 g) in small portions. Maintain the stirring for 12 h at room temperature.
3. Evaporate the solvent, using a rotary evaporator.
4. To the residue add 400 mL CH_2Cl_2 and 200 mL H_2O.
5. Extract the aqueous phase with CH_2Cl_2 (2 × 50 mL).
6. Dry the combined organic phases with Na_2SO_4.
7. Concentrate the CH_2Cl_2 solution, using a rotary evaporator (c. 50 mL).
8. Chromatograph this solution on a large silica gel column with CH_2Cl_2 as eluent.
9. Evaporate the solvent and crystallise the residue from hot ethanol (c. 500 mL) to obtain **12** (41.4 g, 90%), m.p. 108°C.

Protocol 9.
Synthesis of N,N',N''-tris[8-(tetrahydro-2H-pyran-2-yl) oxy-3-toluenesulfonyl-6-oxa-3-azaoctyl] amine (Structure 13, Scheme 5.12)

Caution! Carry out all procedures in a well-ventilated hood, and wear disposable vinyl or latex gloves and chemical-resistant goggles. Sodium reacts violently with water (fire hazard) and is very corrosive to skin and eyes.

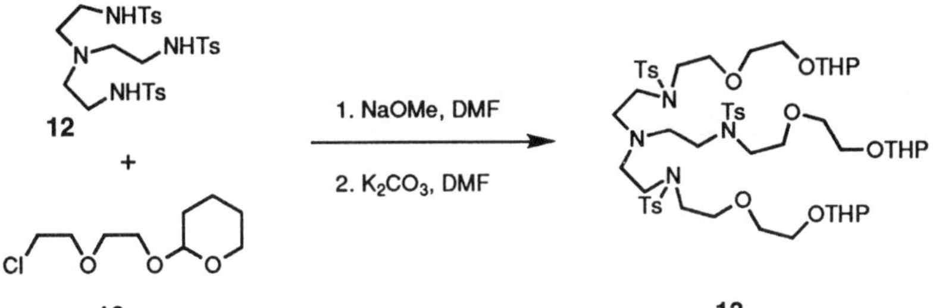

Scheme 5.12

Equipment

- Double-necked, round-bottomed flask (1 L)
- Water-cooled condenser
- Addition funnel with pressure-equalization arm (200 mL)
- Hot plate stirrer
- Teflon-coated magnetic stirring bar
- Graduated cylinder (500 mL)
- Source of nitrogen
- Separating funnel (2000 mL)
- Column for chromatography
- Tweezers
- Sharp knife
- Two crystallising dishes (100 mL)
- Oil bath
- Two drying tubes
- Rotary evaporator

Protocol 9. Continued

Materials

- Sodium, 9.6 g, 420 mmol — reacts violently with water, corrosive
- Dry methanol, 200 mL — flammable, harmful by inhalation
- Tren, tritoluenesulfonylated derivative **12**, 85 g, 139 mmol
- Dry petroleum ether 30–60°, 200 mL — flammable
- Dry dimethylformamide, 600 mL — harmful vapour, irritant
- *O*-protected monochloro diethylene glycol **10**, 118 g, 560 mmol
- Potassium carbonate, 15 g — irritant
- Dichloromethane, ~1000 mL — irritant, harmful by inhalation
- Sodium sulfate for drying, 100 g — irritant
- Dry hexane, 4000 mL — flammable, irritant
- Toluene, 2000 mL — flammable, toxic
- Alumina (neutral, activity II–III), 1000 g — irritant
- Drierite, 50 g — irritant

1. Introduce the magnetic stirrer to the 1 L round-bottomed flask, surmounted by the condenser.
2. Locate the nitrogen source adaptor on the top of the condenser and allow a slow stream of nitrogen to go through the apparatus.
3. Remove with a knife the oxidized part of the sodium lump. Wash the clean piece in petroleum ether and weigh the sodium rapidly (9.6 g). Cut the sodium in small pieces (~500 mg each). *Caution!* Before cleaning with water, *all* the materials which have been in contact with sodium should be treated with dry isopropanol.
4. Pour 200 mL of dry methanol into the 1 L flask, start the stirring and maintain the stream of nitrogen.
5. Add the sodium pieces slowly, in small portions with the tweezers, to the methanol. (*Caution!* do not add too rapidly.)
6. When *all* sodium has reacted add the tritoluenesulfonyl derivative **12** (85 g). Stop the nitrogen stream and protect the reaction with drying tubes.
7. Heat the mixture at 50°C for 2 h.
8. Evaporate the solvent. Dry the residue on a vacuum pump (50°C, 0.1 mmHg) for several hours.
9. To the dry sodium salt add 400 mL of dry DMF and assemble the addition funnel.
10. Prepare the solution of **10** (118 g) in 200 mL dry DMF (see step 7, (and footnote,) in Protocol 7). Introduce this solution into the addition funnel. Protect the reaction with drying tubes.
11. To the stirred suspension of the sodium salt of **12** slowly add the solution of **10**. Time of addition: ~1 h.
12. Add potassium carbonate (10 g) to the mixture.

5: Cryptands

13. Heat the mixture at 110 °C for 8 h.
14. Evaporate the solvent on the rotary evaporator under vacuum.
15. To the residue add 500 mL CH_2Cl_2 and 1000 mL H_2O. Transfer this mixture into the separating funnel.
16. Separate the organic layer, extract the water layer with 3 × 200 mL of CH_2Cl_2. Combine the organic layers and dried over 100 g Na_2SO_4; add 5 g of K_2CO_3.[a]
17. Evaporate the solvent and triturate the residue four times with 500 mL hexane.[b]
18. Dissolve the residue in the minimum volume of a 1:1 mixture of toluene:hexane and allow the solution to be adsorbed on to an alumina column (1 kg). Elute the column with dry toluene (c. 2 L).
19. After evaporation of the solvent and drying on a vacuum pump, the product 13 (106 g, 68%) is obtained as a viscous oil.

[a] Potassium carbonate is added to maintain a basic medium necessary for a good stability of the protecting group.
[b] This trituration is made in order to eliminate the unreacted 10.

Protocol 10.
Synthesis of 6,6′,6″-tritoluenesulfonyl-8,8′,8″-nitrilotri (3-oxa-6-azaoctanol) (Structure 14, Scheme 5.13)

Caution! Carry out all procedures in a well-ventilated hood, and wear disposable vinyl or latex gloves and chemical-resistant safety goggles.

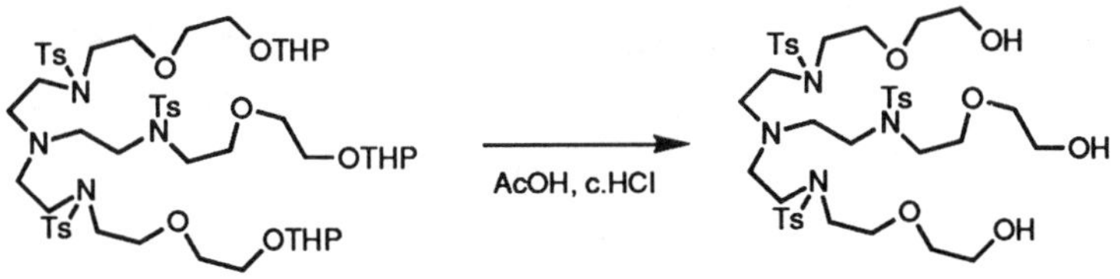

13 → 14
AcOH, c.HCl

Scheme 5.13

Equipment
- Single-necked, round-bottomed flask (2 L)
- Water-cooled condenser
- Separating funnel (1000 mL)
- Graduated cylinder (1000 mL)
- Teflon-coated magnetic stirrer bar
- Hot plate stirrer
- Oil bath
- Column for chromatography
- Conical filter funnel
- Filtering flask (1000 mL)

113

Protocol 10. Continued

Materials
- Triprotected compound **13**, 106 g, 94 mmol
- Concentrated hydrogen chloride, 50 mL   corrosive, irritant
- Acetic acid, 800 mL   corrosive, irritant
- Toluene, 600 mL   flammable, toxic
- Dichloromethane, 300 mL   irritant, harmful by inhalation
- Potassium carbonate, 10 g   irritant
- Sodium hydroxide, 1 M, 150 mL   highly corrosive, very dangerous for eyes
- Methanol, 600 mL   flammable, harmful by inhalation
- Silica gel, 100 g   irritant
- Alumina (neutral, activity II–III), 1.5 kg   irritant

1. Introduce into the round-bottomed flask compound **13** (106 g), concentrated HCl (50 mL) and acetic acid (800 mL). Place the condenser on top of the flask.
2. Heat the mixture at 50°C for 12 h (the solution turns progressively darker).
3. Evaporate the solvent under reduced pressure using a rotary evaporator.
4. Add toluene (200 mL) and evaporate to dryness.[a] Repeat this operation three times.
5. Dissolve the residue in CH_2Cl_2 (500 mL), add K_2CO_3 (10 g) and stir for 12 h.
6. Filter the reaction mixture and evaporate the solvent using a rotary evaporator. Dry on a vacuum pump (0.1 mmHg) for several hours. The triacetate is obtained as a colourless oil (87 g, 93%).
7. To the residue add 1 M NaOH (150 mL) and methanol (500 mL).
8. Heat the mixture at 50°C for 1 h.
9. Evaporate the solvent using a rotary evaporator.
10. Add water and extract the mixture with CH_2Cl_2 (3 × 200 mL).
11. Evaporate the combined extracts, and reduce the volume to c. 40 mL.
12. Filter the solution through 100 g of silica gel. Use 2% $MeOH/CH_2Cl_2$ as eluant.
13. Evaporate the solvent using a rotary evaporator.
14. Dissolve the residue in the minimum volume of CH_2Cl_2 and pass the solution over an alumina column (1.5 kg). Use 1% $MeOH/CH_2Cl_2$ as eluant.
15. Evaporate the solvent. The triol is obtained as a viscous oil, and may be used directly in the next step of the synthesis.

[a] Toluene is added and evaporated in order to eliminate all the acids.

5: Cryptands

Protocol 11.
Synthesis of 6,6',6''-tritoluenesulfonyl-8,8',8''-nitrilotri(3-oxa-6-azaoctyl)tris(methanesulfonate) (Structure 15, Scheme 5.14)

Caution! Carry out all procedures in a well-ventilated hood, and wear disposable vinyl or latex gloves and chemical-resistant safety goggles. Methanesulfonyl chloride is a corrosive compound.

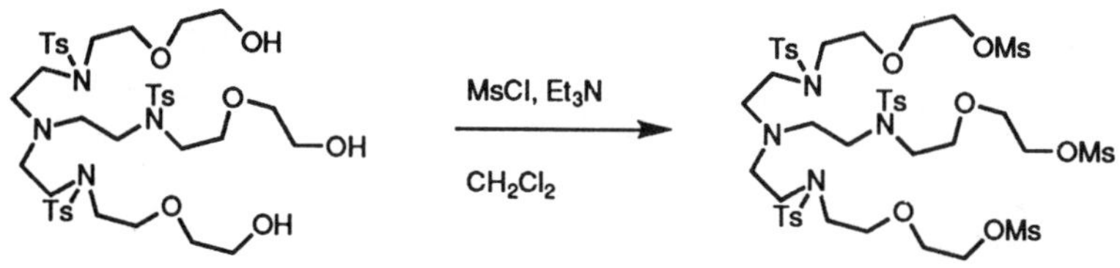

Scheme 5.14

Equipment

- Single-necked, round-bottomed flask (500 mL)
- Addition funnel (50 mL)
- Graduated cylinder (500 mL)
- Magnetic stirrer
- Teflon-coated magnetic stirrer bar
- Separating funnel (1000 mL)
- Ice bath

Materials

- Triol **14**, 7.7 g, 88 mmol
- Triethylamine, 22 mL, 160 mmol — irritant
- Methanesulfonyl chloride, 2.8 g, 36 mmol — highly irritant
- Dichloromethane, 300 mL — irritant, harmful by inhalation
- 1 M hydrochloric acid, 50 mL — corrosive, irritant
- 1 M sodium hydroxide, 50 mL — highly corrosive, very dangerous for eyes
- Sodium sulfate, 50 g — irritant
- Ice, 1 kg

1. Introduce to the round-bottomed flask triol **14** (7.7 g), triethylamine (22 mL) and CH_2Cl_2 (300 mL). Cool the solution at 0 °C with an ice bath. Start the stirring and place the addition funnel on the reaction flask.
2. Add methanesulfonyl chloride slowly (2.8 mL).
3. Stir the mixture for 1 h at 0 °C and 2 h at room temperature.
4. Transfer the solution to a separating funnel.
5. Wash the CH_2Cl_2 solution successively with cold HCl (1 M, 50 mL), cold NaOH (1 M, 50 mL) and cold water (50 mL).
6. Dry the organic solution with sodium sulfate.
7. Filter and evaporate the solvent. The trimesylated compound **15** (9.7 g, 98%) is a viscous oil which may be used directly for the cyclisation step.

Protocol 12.
Synthesis of 4,10,16,22,27,33-hexatoluenesulfonyl-7,19,30-trioxa-1,4,10,13,16,22,27,33-octaazabicyclo[11.11.11]pentatriacontane (Structure 16, Scheme 5.15)

Caution! Carry out all procedures in a well-ventilated hood, and wear disposable vinyl or latex gloves and chemical-resistant safety goggles.

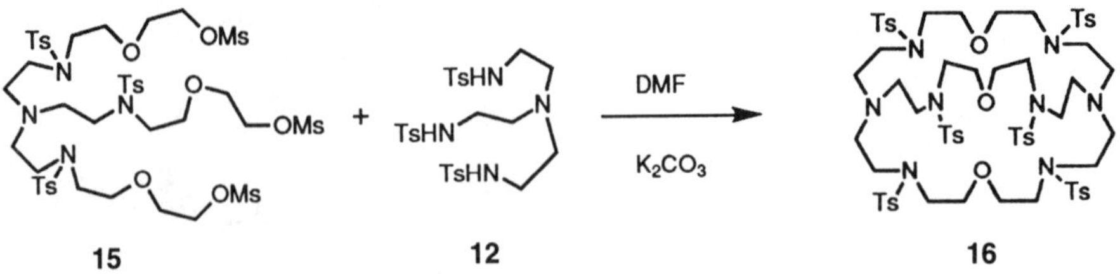

Scheme 5.15

The following procedure for the synthesis of macrobicycle **16** is representative of the tripod–tripod coupling reaction.

Equipment
- Double-necked, round-bottomed flask (1 L)
- Water-cooled condenser
- Column for chromatography
- Hot plate stirrer
- Oil bath
- Teflon-coated magnetic stirring bar
- Source of dry nitrogen
- Graduated cylinder (500 mL)
- Nitrogen inlet adaptor
- Rotary evaporator

Materials
- 6,6′,6″-Tritoluenesulfonyl-8,8′,8″-nitrilotri(3-oxa-6-azaoctyl)trismethanesulfonate, **15**, 10.95 g, 9.9 mmol
- N,N′,N″-Tritoluenesulfonyl-2,2′,2″-nitrilotriethylamine, **12**, 6.05 g, 9.9 mmol
- Dry DMF, 500 mL   **highly irritant**
- Fine mesh potassium carbonate, 50 g, 360 mmol   **irritant**
- Dichloromethane, ~4 L   **irritant, harmful by inhalation**
- Sodium hydroxide, 1 M, 100 mL   **highly corrosive, very dangerous for eyes**
- Magnesium sulfate, 50 g   **irritant**
- Toluene, ~1 L   **flammable, toxic**
- Alumina (neutral, activity II–III), 400 g   **irritant**

1. Introduce to the round-bottomed flask tritoluenesulfonylated-tren, **12** (6.05 g), 500 mL DMF, potassium carbonate,[a] (50 g) and trimesyl derivative **15** (10.95 g). Add the condenser and the nitrogen inlet adaptor.
2. Heat the well-stirred mixture at 80 °C for 30 h.
3. After cooling, filter the mixture and wash the solid residue with CH$_2$Cl$_2$ (3 × 50 mL).
4. Evaporate the solvent on a rotary evaporator.
5. Treat the residue with 300 mL CH$_2$Cl$_2$ and 100 mL of 1 M NaOH.
6. Transfer the mixture to a separating funnel.

7. Wash the organic phase with water (2 × 100 mL).
8. Dry the CH_2Cl_2 solution over $MgSO_4$, filter and evaporate the solvent under reduced pressure.
9. Dissolve the residue in the minimum volume of a 1/1 mixture of toluene/CH_2Cl_2.
10. Pass this solution through an alumina column (400 g). Use CH_2Cl_2/toluene 3/1 as eluant. Pure **16** (4.34 g, 31%) is obtained as an amorphous solid.

[a] In most of the described syntheses of this type, K_2CO_3 is replaced by Cs_2CO_3. The yields are usually somewhat better.

Protocol 13.
Synthesis of 7,19,30-trioxa-1,4,10,13,16,22,27,33-octaazabicyclo [11.11.11]pentatriacontane (Structure 17, Scheme 5.16)

Caution! Carry out all procedures in a well-ventilated hood, and wear disposable vinyl or latex gloves and chemical-resistant safety goggles. Hydrogen bromide in acetic acid gives corrosive and dangerous fumes.

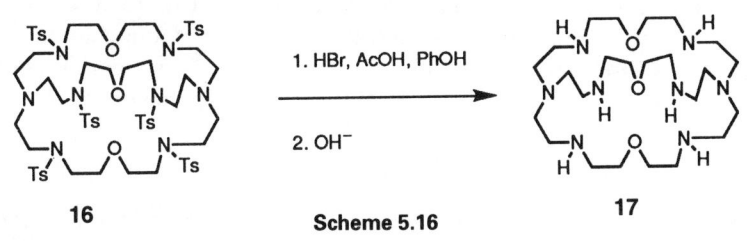

Scheme 5.16

Equipment
- Double-necked, round-bottomed flask (250 mL)
- Graduated cylinder (100 mL)
- Hot plate stirrer
- Teflon-coated magnetic stirrer bar
- Water-cooled condenser
- Right-angle tubing adaptor
- Polyethylene tubing, ~1 m
- Separating funnel (500 mL)
- Column for Dowex-1 resin
- pH indicator strip (1–10)

Materials
- Hexatoluenesulfonylated macrobicycle **16**, 2.6 g, 1.8 mmol
- Phenol, 4 g, 34 mmol **poisonous, caustic**
- Hydrogen bromide, 33% in acetic acid, 70 mL **highly irritant**
- Dichloromethane, 500 mL **irritant, harmful by inhalation**
- Dowex 1 × 8-100 ion exchange resin, 70 g
- Sodium hydroxide, 1 M, 100 mL **highly corrosive**

1. To the 250 mL flask introduce the hexatoluenesulfonylated macrobicycle **16** (2.6 g), phenol (4 g) and 70 mL of a 33% HBr/AcOH solution.
2. Place the condenser and at its outlet the tubing which should be directed to the back part of the hood.
3. Heat the stirred mixture at 80 °C for 16 h.

4. After cooling evaporate the mixture. *Caution!* avoid any contact with the HBr vapours.
5. To the residue add 50 mL of H_2O and 100 mL of CH_2Cl_2 and transfer the mixture to the separating funnel.
6. To eliminate phenol, extract the aqueous phase with 4×100 mL of CH_2Cl_2.
7. Evaporate the aqueous phase and dry the residue on a vacuum pump for 12 h.
8. Prepare the ion-exchange column: (i) introduce a small piece of glass wool above the stopcock; (ii) add the slurry of Dowex resin (70 g) and 200 mL of water; (iii) exchange the Cl^- of the commercial resin to OH^- by treatment of the resin with 100 mL or 10% NaOH; and (iv) wash the resin with water until the eluent is neutral (pH indicator strip).
9. Dissolve the residue obtained in 30 mL water.
10. Pass this solution over the Dowex-1 column, wash the column with water. The eluate stays basic (control with pH indicator strip) as long as **17** comes out; about 600 mL of water is necessary.
11. Evaporate the water solution on a rotary evaporator.
12. Dry the residue on a good vacuum pump for 24 h. The free octaamine **17** (0.78 g, 87%) is obtained as a viscous oil. This compound turns yellow rapidly and should not be stored in this form but as its polyammonium salt (chloride or bromide as desired). Further details are given in Ref. 3.

4. Concluding remarks

The two methods detailed here have been used widely in many related macrocycle syntheses, but various other approaches have also been investigated. Several reviews and books give a more thorough discussion of the synthetic strategies that may be used.[6]

References

1. Dietrich, B.; Lehn, J.-M.; Sauvage, J. P. *Tetrahedron Lett.* **1969**, 2885.
2. Dietrich, B.; Lehn, J.-M.; Sauvage, J. P. *Tetrahedron* **1973**, *29*, 1629.
3. Dietrich, B.; Hosseini, M. W.; Lehn, J.-M.; Sessions, R. B. *Helv. Chim. Acta* **1985**, 289.
4. At the time of the original [2.2.2] synthesis the compounds **1** and **3** were not commercially available. They had to be prepared from a common starting material, the triethylene glycol, which was transformed by oxidation to the diacid **1** or by bromination followed by Gabriel reaction to the diamine **3**.[2]
5. Lehn, J.-M.; Pine, S. H.; Watanabe, E. I.; Willard, A. K. *J. Am. Chem. Soc.* **1977**, *99*, 6766.
6. Gokel, G. W.; Korzeniowski, S. H. *Macrocyclic Polyether Syntheses*; Springer-Verlag: Berlin, **1982**; Rossa, L.; Vögtle, F. *Top. Curr. Chem.* **1983**, *113*, 1; Knops, P.; Sendhoff, N.; Mekelburger, H.-B.; Vögtle, F. *Top. Curr. Chem.* **1991**, *161*, 1; Krakowiak, K. E.; Bradshaw, J. S. *Isr. J. Chem.* **1992**, *32*, 3; Dietrich, B.; Viout, P.; Lehn, J.-M. *Macrocyclic Chemistry*; VCH: Weinheim, **1993**.

6

Torands

THOMAS W. BELL and JULIA L. TIDSWELL

1. Introduction

Torands are a class of macrocyclic ligands having perimeters that are completely formed by smaller, fused rings.[1-7] This chapter describes the synthesis of the simplest known example of the class, dodecahydrohexaazakekulene 1.[8] This fused-ring analogue of sexipyridine 3[9] is also the parent system for a series of substituted torands, such as tributyldodecahydrohexaazakekulene 2 (TBDK).[1,2,4-7] Larger ring, 'expanded' torands are also known, such as 4.[3] Syntheses of torand 1 and sexipyridine 3 were reported in the original communications,[8,9] but little is known about their complexation properties. On the other hand, sufficiently large quantities of tributyltorand 2 have been synthesized to permit investigation of the structures[4,5] and stabilities of alkali metal complexes as well as the spectroscopic effects of complexation.[10] Torand 2 forms extremely stable complexes with metal cations,[2,6,7] while expanded torand 4 binds the larger, triangular guanidinium cation.[3]

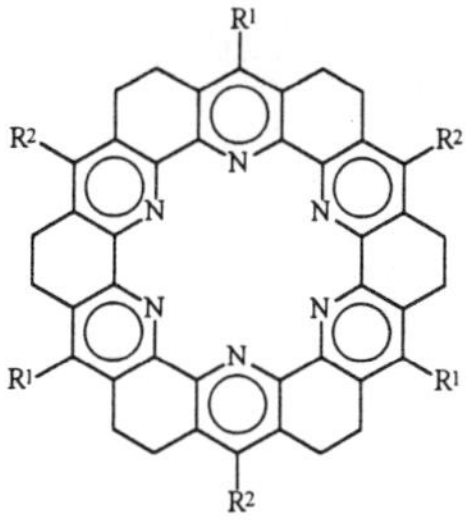

1 R1 = R2 = H
2 R1 = H; R2 = butyl

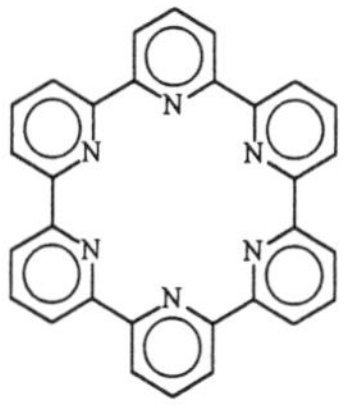

3

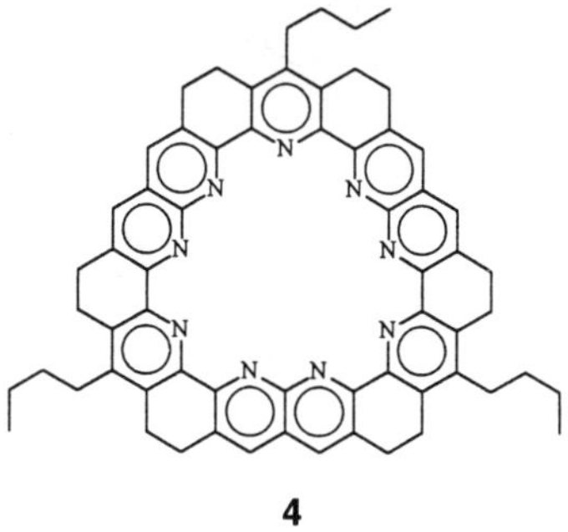

4

This chapter illustrates torand synthesis with the preparation of unsubstituted torand 1 by the route developed for synthesis of tributyltorand 2.[1,6,7] The synthesis of 1 is two steps shorter than that of 2 because commercially available 1,2,3,4,5,6,7,8-octahydroacridine, 5, replaces the 9-butyl derivative, which is prepared in two steps.[11,12] Many of the intermediates could also be used in the synthesis of other unsubstituted torands by methods analogous to those reported for preparation of the butyl derivatives.[10,13–15]

In Scheme 6.1, outlining the synthesis of torand **1**, each of the numbered arrows corresponds to one of the eleven protocols. This approach to **1** sequentially combines three octahydroacridine subunits, whereas the previously

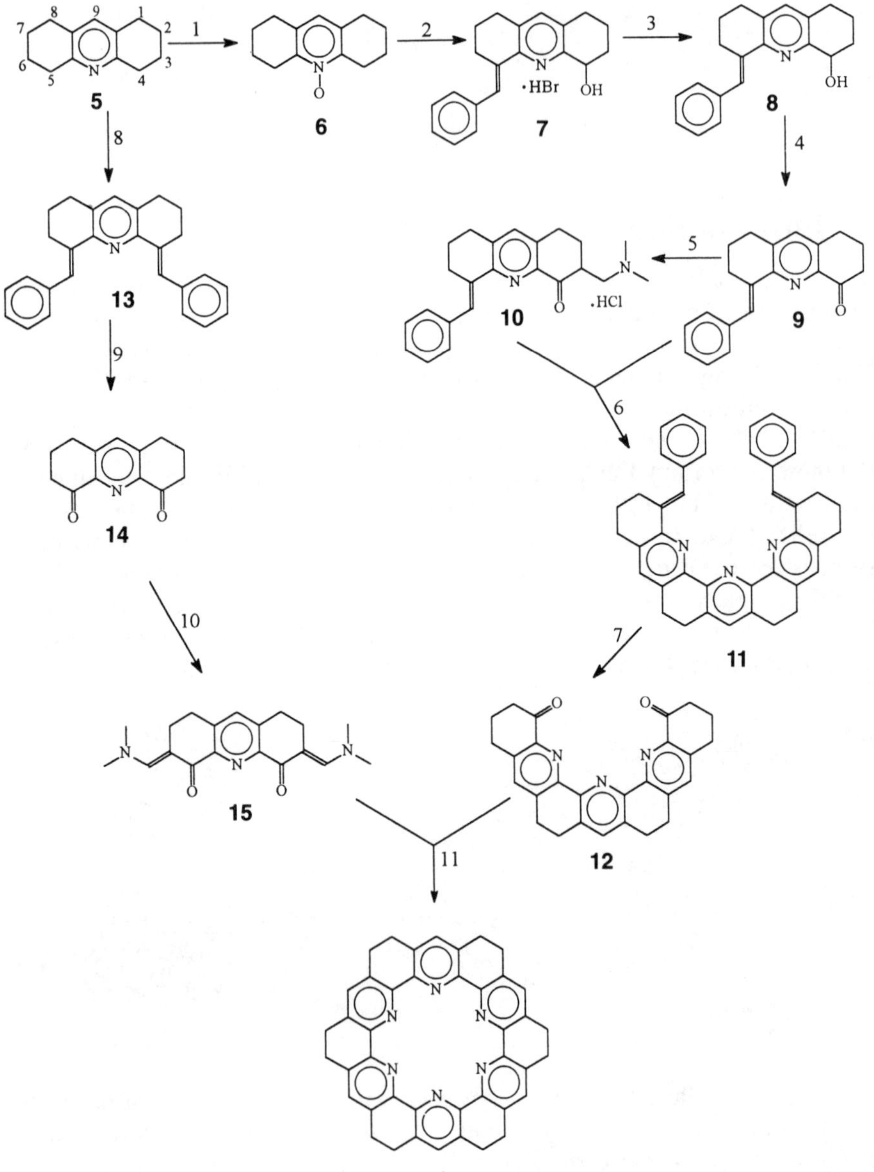

Scheme 6.1

6: Torands

described synthesis[8] involves dimerization of decahydrodibenzophenanthroline derivatives. Each of the three octahydroacridine subunits can bear a substituent in the 9-position (4-pyridyl position); hence the synthesis can be used to prepare symmetrically or unsymmetrically trisubstituted torands (e.g. **2**). Variation of steps 5, 6, 10, and 11 also allows the incorporation of up to three aryl groups.[6,16] Thus a wide variety of substituted torands can be prepared by modification of the protocols in this chapter, but the reader is cautioned that attendant changes in solubilities would require that new procedures be developed for the purification of substituted products.

2. Oxidation of octahydroacridine

The first step (Protocol 1) is analogous to the N-oxidation of 9-butyl-1,2,3,4,5,6,7,8-octahydroacridine described in *Organic Syntheses*.[12] Various peroxy acids may be used to synthesize N-oxides from pyridines, but the method described here is especially convenient for large-scale preparations. The oxidizing agent, potassium peroxymonosulfate (Caroate), is employed as the inexpensive reagent Oxone[R], which has the composition $2KHSO_5 \cdot KHSO_4 \cdot K_2SO_4$. The reactants are not completely soluble in aqueous methanol, but they are sufficiently soluble for completion of the reaction in 24 h when an excess of oxidant is used. The buffering action of sodium bicarbonate is also critical to completion of the reaction; if omitted, the acidity of the reaction mixture gradually increases, reducing the rate of reaction because the starting material becomes protonated.[17]

Protocol 1.
Synthesis of 1,2,3,4,5,6,7,8-octahydroacridine N-oxide (Structure 6, Scheme 6.2)

Caution! Carry out all procedures in a well-ventilated hood, and wear disposable vinyl or latex gloves and chemical-resistant safety goggles.

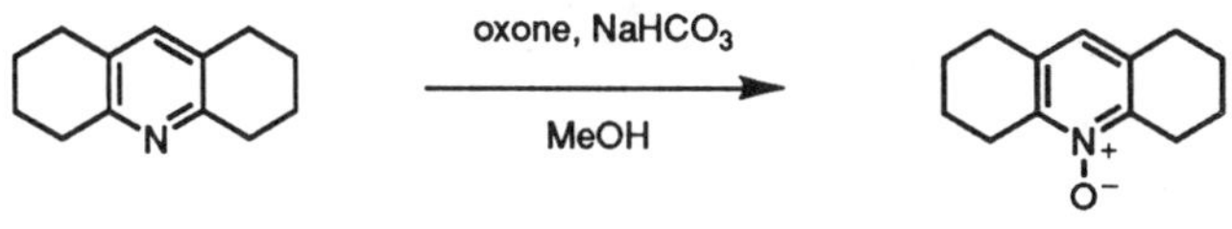

Scheme 6.2

Equipment
- Three-necked, round-bottomed flask (5 L)
- Variable-speed mechanical stirrer
- Glass stirring rod with Teflon stirring blade
- Adapter for stirring rod
- Water-cooled condenser
- Nitrogen inlet adapter
- Source of dry nitrogen
- Thermometer with adaptor
- Wide-mouth short-stem funnel
- Glass filter funnel
- Fluted filter paper
- Heating mantle

121

Protocol 1. Continued

- Variable transformer
- Sintered glass funnels (150 and 350 mL, med. porosity)
- Vacuum filter flasks (500 mL and 4 L)
- Separating funnel (1 L)
- Round-bottom flask (1 L)
- Two Erlenmeyer flasks (500 mL)
- Rotary evaporator
- Insulated cylindrical glass vessel (19 cm in diameter × 14 cm h)

Materials

- 1,2,3,4,5,6,7,8-Octahydroacridine, 50.0 g, 0.267 mol — irritant
- Sodium bicarbonate, 73.0 g, 0.868 mol — moisture sensitive
- Oxone^R (2KHSO$_5$·KHSO$_4$·K$_2$SO$_4$), 185 g, 0.30 mol — oxidizer, irritant
- Methanol, 1.8 L — highly toxic, flammable liquid
- Dichloromethane, 400 mL — toxic, irritant
- Ethyl acetate, 250 mL — flammable liquid, irritant
- Hexane, 300 mL — flammable liquid, irritant
- Sodium sulfate, anhydrous, for drying, c. 75 g — irritant, hygroscopic
- Dry ice, 500 g
- Acetone, 500 mL — flammable liquid, irritant

1. Assemble the three-necked flask and glass stirring rod with blade and adapter. The curved part of the blade should rest gently against the bottom of the round-bottomed flask. Attach the glass rod to the shaft of the stirrer by means of a short piece of rubber tubing and make certain that the mechanical stirrer, glass stirring rod, and round-bottomed flask are aligned so that there is no strain on any of the parts. Test the assembly by turning on the mechanical stirrer to determine that the stirring rod and blade rotate freely.

2. Place the condenser on a second neck and place the nitrogen inlet adaptor on top of the condenser. Turn on the source gas to flush the apparatus with N$_2$.

3. Use the wide-mouth funnel to add the 1,2,3,4,5,6,7,8-octahydroacridine, sodium bicarbonate, and 1.5 L of methanol. Start the mechanical stirrer and stir the mixture for 5–10 min.

4. Add the Oxone^R and 600 mL of distilled water. Replace the wide-mouth funnel with the thermometer and its adaptor and stir the reaction mixture at 60–5°C under N$_2$ for 24 h.

5. Remove the heating mantle and allow the reaction mixture to cool to room temperature.

6. Vacuum filter the reaction mixture using the larger sintered glass funnel. Rinse the round-bottomed flask with 100 mL of methanol then use this to rinse the residue in the funnel. Repeat this rinsing procedure two more times.

7. Pour c. 500 mL of the combined filtrates into the single-necked 1 L round-bottomed flask and concentrate *in vacuo* (c. 30 mmHg) using a rotary evaporator (bath temp. 50–70°C), cooling the receiver in a dry ice/acetone

bath. Pour the resulting cloudy aqueous mixture into the separating funnel. Repeat this procedure four times with the remainder of the filtrate.
8. Extract the concentrated aqueous mixture with dichloromethane (3 × 100 mL). Wash the combined dichloromethane extracts with distilled water (2 × 50 mL).
9. Place the dichloromethane solution in one of the Erlenmeyer flasks and swirl it with the anhydrous sodium sulfate to remove the water.
10. Remove the sodium sulfate by filtration through the filter funnel and fluted filter paper, rinsing it with dichloromethane (2 × 50 mL).
11. Remove the solvent from the combined solutions by rotary evaporation and dry the crude product overnight *in vacuo* (0.1–0.5 mmHg) to yield a gold solid (52–3 g, 95–8%).
12. Dissolve the crude product in 250 mL of boiling ethyl acetate in the second Erlenmeyer flask and slowly add sufficient hot hexane (50–70 mL) to turn the boiling solution cloudy. Allow the flask to cool slowly to room temperature, seal it with a cork, and store it in a freezer at 0 °C for several hours.
13. Collect the product by vacuum filtration, washing the crystals with ice-cold hexane (3 × 70 mL) and dry them *in vacuo* (0.1–0.5 mmHg) for several hours, yielding 41.3–45.7 g (80–4%) of golden plates, m.p. 138–41 °C (lit.[18] 144–5 °C).

3. Carbon functionalization of octahydroacridine

Octahydroacridine units must be oxidatively functionalized at carbon atoms 4 and 5 in order to build up the fused-ring backbone of torand **1**. The key pyridine-forming reactions (steps 6 and 11 in Scheme 6.1) both involve the condensation of a ketone with an α,β-unsaturated ketone. Unsymmetrically functionalized octahydroacridine derivatives are required so that each unit can be fused to a new pyridine ring first at one end and then at the other. The benzylidene groups serve as latent carbonyl groups that can be unmasked by ozonolysis. Step 2 introduces both a C–O bond at C4 and a benzylidene group at C5 in a convenient, one-pot reaction sequence. Scheme 6.3 shows the intermediates involved in this sequence converting **6** to **7**.

Oxidation at C4 is achieved by the 'Katada' or 'Boekelheide' rearrangement of *N*-oxide **6** to 4-acetoxy-1,2,3,4,5,6,7,8-octahydroacridine **18**. In the first step of this reaction, which is commonly used for selective oxidation of alkylated pyridines, the *N*-oxide reacts with hot acetic anhydride to form an *N*-acetoxypyridinium salt **16**. Loss of a proton at C4 gives the 'anhydro' form **17**, which apparently rearranges to acetate **18** by an ionic pathway.[19] Benzaldehyde is then added to the reaction mixture and condensation at C5 gives 4-acetoxy-5-benzylidene-1,2,3,4,5,6,7,8-octahydroacridine **19**. In the second stage of Protocol 2 the ester group is hydrolyzed with aqueous HBr and the

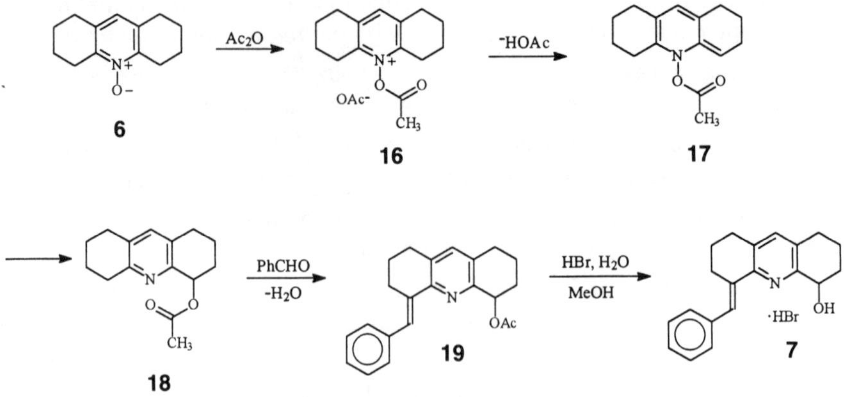

Scheme 6.3

product alcohol is isolated by crystallization of the HBr salt **7**. The isolation procedure minimizes contamination of the product by the major side product, dibenzylidene derivative **13** (*cf.* Scheme 6.1). This side product is apparently formed by condensation of benzaldehyde with **5**, which is either present as an impurity in intermediate **6** or is produced by deoxygenation of **6** in the reaction mixture. In protocol 3 the crystallized HBr salt **7** is converted to the free base **8** by reaction with sodium hydroxide.

Protocol 2.
Synthesis of 5-benzylidene-1,2,3,4,5,6,7,8-octahydroacridin-4-ol hydrobromide (Structure 7, Scheme 6.4)

Caution! Carry out all procedures in a well-ventilated hood, and wear disposable vinyl or latex gloves and chemical-resistant safety goggles.

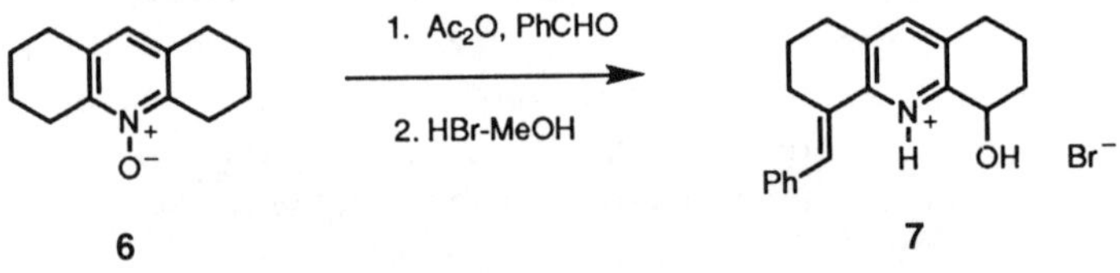

Scheme 6.4

Equipment

- Variable transformer
- Magnetic stirrer and magnetic stirrer bar
- Heating mantle
- Water-cooled condenser
- Nitrogen inlet adaptor
- Source of dry nitrogen
- Thermometer and adaptor
- Three-necked, round-bottomed flask (1 L)
- Two ground glass stoppers
- One piece vacuum distillation apparatus (or head, condenser, and collection adaptor)
- Single-necked, round-bottomed flasks (500 mL and 1 L)
- Glass filter funnel
- Fluted filter paper
- Teflon tape

6: Torands

- Buchner funnel (9 cm)
- Filter paper circles (9 cm)
- Vacuum filter flasks (500 mL and 1 L)
- Separating funnel (1 L)
- Erlenmeyer flasks (500 mL and 3 × 1 L)
- Rotary evaporator
- Insulated cylindrical glass vessel (19 cm id × 14 cm h)

Materials

- 1,2,3,4,5,6,7,8-Octahydroacridine N-oxide, 40.0 g, 0.197 mol
- Acetic anhydride, 300 mL, 324 g, 3.17 mol   **corrosive, lachrymator**
- Benzaldehyde, 100 mL, 105 g, 0.99 mol   **highly toxic, cancer-suspect agent**
- Aqueous 48% HBr solution, 120 mL   **highly toxic, corrosive**
- Methanol, 1.2 L   **highly toxic, flammable liquid**
- Silicone high vacuum grease
- Crushed ice
- Sodium chloride, for ice-salt bath
- Acetone, 1 L   **flammable liquid, irritant**
- Dichloromethane, 2 L   **toxic, irritant**
- Sodium sulfate, anhydrous, c. 125 g   **irritant, hygroscopic**
- Dry ice, 500 g

1. Set up the three-necked, round-bottomed flask with stirrer bar, heating mantle, magnetic stirrer, nitrogen inlet adaptor, reflux condenser, thermometer and adaptor.

2. Flush the apparatus with N_2. Add the 1,2,3,4,5,6,7,8-octahydroacridine N-oxide and acetic anhydride. Place a ground glass stopper in the third neck of the flask and stir the reaction mixture under N_2 at 110–15 °C for 1 h.

3. Cool the reaction mixture to room temperature, add the benzaldehyde, and stir the resulting mixture under N_2 at 140–5 °C for 16 h.

4. Replace the nitrogen inlet adaptor and reflux condenser with the vacuum distillation apparatus and a 500 mL round-bottomed receiving flask.[a] Replace the thermometer and adaptor with the second ground glass stopper. Seal all ground glass joints with silicone grease, then wrap all joints with Teflon tape. Cool the receiving flask in an ice-salt bath (c. −10 °C). Attach the vacuum line to the apparatus and carefully evacuate the system. Heat the solution *in vacuo* (0.1–0.5 mmHg) until 310–20 mL of the solvent has been collected in the receiving flask.

5. Allow the distillation flask to cool to room temperature, replace the vacuum distillation apparatus with the reflux condenser and nitrogen inlet adapter, and replace one stopper with the thermometer and adaptor.

6. Add 300 mL of the methanol and the 48% aqueous HBr solution to the residue. Stir the reaction mixture under N_2 at 65 °C for 24 h. After several hours a yellow precipitate forms.

7. Cool the mixture to room temperature and collect the solid by vacuum filtration. Wash the reaction vessel with 50 mL of methanol and use this to wash the solid in the Buchner funnel. Repeat with another 50 mL portion of methanol. Dry the yellow solid *in vacuo* (0.1–0.5 mmHg) for several hours, yielding a first crop of 39.4–42.0 g (52–7%).

Protocol 2. *Continued*

8. Pour 260 mL of the filtrates into a 1 L round-bottomed flask, add 300 mL of distilled water, and remove the methanol *in vacuo* (30–40 mmHg) by rotary evaporation, using a water bath temperature of 50–65 °C, cooling the receiver in a dry ice/acetone bath. Pour the aqueous mixture into a 1 L separating funnel.

9. Repeat step 8 with the remaining methanol solution. Dissolve any sticky, brown solid left in the round-bottomed flask in 200 mL of dichloromethane, and add this solution to the aqueous mixture. Shake the phases together and separate. Extract the aqueous mixture with dichloromethane (3 × 200 mL).

10. Place the combined dichloromethane extracts in a 1 L Erlenmeyer flask and mix with anhydrous sodium sulfate to remove water.

11. Remove the sodium sulfate by filtration through the filter funnel and fluted filter paper, rinsing with 4 × 250 mL of dichloromethane.

12. Pour 250 mL of the dichloromethane solution into a single-necked 500 mL round-bottomed flask and remove the solvent *in vacuo* (30–40 mmHg) using a rotary evaporator (bath temperature 25–40 °C). Repeat this procedure with portions of the remaining solution until all of the residue has been collected. Dry the resulting sticky brown solid for several hours *in vacuo* (0.1–0.5 mmHg).

13. Mix the dried solid thoroughly with 200 mL of acetone and filter the mixture using a Buchner funnel and vacuum filter flask. Wash this second crop with acetone (3 × 50 mL) and dry it *in vacuo* (0.1–0.5 mmHg) for several hours. The two crops combined give a yield of 55.1–55.9 g (74–5%).

14. In a 1 L Erlenmeyer flask dissolve the two crops of crude product in 500 mL of boiling methanol and reduce the volume of the solution to about 350–375 mL by boiling. Transfer the solution to a warmed 500 mL Erlenmeyer flask and reduce the volume until the solution begins to look faintly cloudy (*c.* 300 mL). Allow the contents to cool slowly to room temperature. Cover the flask and store it at room temperature for about 3 h.[b]

15. Collect the crystals using a Buchner funnel and vacuum filter flask and wash them with ice-cold methanol (3 × 50 mL).

16. Reduce the volume of the mother liquor by boiling until it begins to look cloudy (*c.* 200 mL). Allow the flask to cool slowly to room temperature. Cover and store it at room temperature for about 2 h.

17. Collect the crystals using a Buchner funnel and vacuum filter flask and wash them with ice-cold methanol (3 × 20 mL).

18. Combine the two crops of purified product and dry them *in vacuo* (0.1–0.5 mmHg) for several hours, yielding 47.1–50.2 g (63–8%) of bright yellow crystals as needles, m.p. 233–5 °C.

6: Torands

[a] Mark the 320 mL level on the flask prior to assembling the vacuum distillation apparatus. This is conveniently accomplished by adding 320 mL of distilled water to the flask, marking the water level with a felt pen, pouring out the water and drying the flask with the aid of acetone.
[b] If the flask is allowed to stand too long or is cooled below room temperature side product 13 will crystallize, reducing the purity of the product.

Protocol 3.
Synthesis of 5-benzylidene-1,2,3,4,5,6,7,8-octahydroacridin-4-ol (Structure 8, Scheme 6.5)

Caution! Carry out all procedures in a well-ventilated hood, and wear disposable vinyl or latex gloves and chemical-resistant safety goggles.

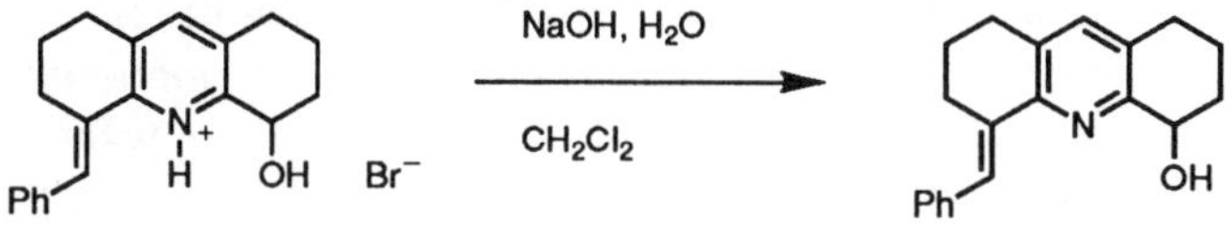

Scheme 6.5

Equipment

- Magnetic stirrer and magnetic stirrer bar
- Nitrogen inlet adaptor
- Source of dry nitrogen
- Single-necked, round-bottomed flasks (250 mL and 1 L)
- Separating funnel (1 L)
- Erlenmeyer flasks (250 mL and 1 L)
- Spatula
- Sintered glass funnel (60 mL, med. porosity)
- Vacuum filter flask (250 mL)
- Glass filter funnel
- Fluted filter paper
- Rotary evaporator

Materials

- 5-Benzylidene-1,2,3,4,5,6,7,8-octahydroacridin-4-ol hydrobromide, 40.0 g, 0.107 mol
- Dichloromethane, 650 mL — toxic, irritant
- 1 M sodium hydroxide solution, 215 mL — corrosive, toxic
- Saturated sodium chloride solution, 215 mL
- Hexane, 200 mL — flammable liquid, irritant
- Sodium sulfate, anhydrous, c. 85 g — irritant, hygroscopic

1. Set up the 1 L round-bottomed flask with stirrer bar, magnetic stirrer, and nitrogen inlet adaptor.

2. Flush the flask with N_2. Add the 5-benzylidene-1,2,3,4,5,6,7,8-octahydroacridin-4-ol hydrobromide and 215 mL of dichloromethane. Replace the nitrogen inlet adaptor and stir the resulting mixture under N_2 at room temperature for 30 min.

3. Add the 1 M sodium hydroxide solution. Replace the nitrogen inlet adaptor and stir the mixture vigorously under N_2 at room temperature for 2 h.

Protocol 3. *Continued*

4. Pour the reaction mixture into the 1 L separating funnel. Rinse the reaction flask with dichloromethane (2 × 15 mL) and add the rinsings to the separating funnel. Add 215 mL of saturated sodium chloride solution to the separating funnel, and shake the solutions together and separate the layers.
5. Extract the aqueous layer with dichloromethane (5 × 60 mL). Combine the dichloromethane solutions in the 1 L Erlenmeyer flask and dry with anhydrous sodium sulfate.
6. Remove the sodium sulfate by filtration through the fluted filter paper, rinsing it with 100 mL of dichloromethane.
7. Remove the solvent *in vacuo* (30–40 mmHg) by rotary evaporation (bath temperature 25–50 °C). Keep the resulting oil warm.
8. Heat 100 mL of hexane to boiling in a 250 mL Erlenmeyer flask. Working quickly, carefully pour the hot hexane into the warm round-bottomed flask containing the oil. Swirl briefly to dissolve the oil in the hexane, then pour the solution back into the Erlenmeyer flask and heat it to boiling.
9. Reduce the volume of the hexane solution to 50 mL and allow it to cool to room temperature, causing the product to separate as an oil on the bottom of the flask.
10. Vigorously mix the oil and supernatant solution (triturate) with a spatula, causing the product to solidify. Triturate until there is no oil remaining. Allow the solid to settle to the bottom of the flask, then collect it *in vacuo* (30–40 mmHg) using the sintered glass funnel and filter flask. Wash the solid with ice-cold hexane (3 × 25 mL) to give a first crop of 27.0–28.2 g (86–90 %).
11. Rinse the Buchner funnel with *c.* 10 mL of hexane and reduce the volume of the combined filtrates to *c.* 40 mL by boiling. Seed the solution with a few grains from the first crop and allow it to cool slowly to room temperature. A second crop will precipitate from the filtrate after several hours at room temperature.
12. Collect the second crop of solid using the Buchner funnel and filter flask, washing it with ice-cold hexane (3 × 5 mL). Dry the combined crops *in vacuo* (0.1–0.5 mmHg) for several hours, yielding 28.1–30.0 g (90–5%) of white solid, m.p. 77–9 °C.

---

4. Oxidation of alcohol 8 to ketone 9

Protocol 4 describes the oxidation of alcohol **8** by dimethylsulfoxide (DMSO) and acetic anhydride at room temperature.[20] This 'Albright–Goldman' method is more economical and simpler to perform on a large scale than other

6: Torands

methods used to oxidize 9-butyloctahydroacridine analogues.[1,14] In the latter series the most significant impurities detected were methylthiomethyl ether and acetate derivatives of the alcohol (cf. **19**, Scheme 6.3). Any side products from the oxidation reaction are largely eliminated during the crystallization step in Protocol 4.

Protocol 4.
Synthesis of 5-benzylidene-2,3,5,6,7,8-hexahydroacridin-4(1H)-one (Structure 9, Scheme 6.6)

Caution! Carry out all procedures in a well-ventilated hood, and wear disposable vinyl or latex gloves and chemical-resistant safety goggles.

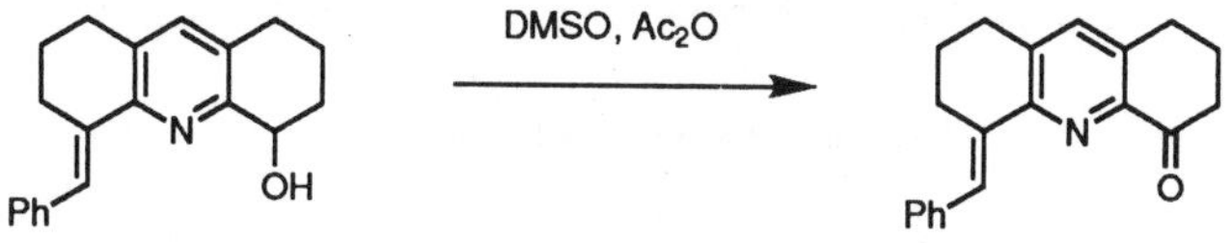

Scheme 6.6

Equipment

- Variable transformer
- Magnetic stirrer and magnetic stirrer bar
- Heating mantle
- Nitrogen inlet adaptor
- Source of dry nitrogen
- Thermometer and adaptor

- Three-necked, round-bottomed flask (500 mL)
- Three ground glass stoppers
- Erlenmeyer flask (500 mL)
- Sintered glass funnel (300 mL, med. porosity)
- Vacuum filter flask (500 mL)

Materials

- 5-Benzylidene-1,2,3,4,5,6,7,8-octahydroacridin-4-ol, 20.0 g, 68.6 mmol
- Dimethyl sulfoxide, anhydrous, 158 mL, 2.2 mol irritant, hygroscopic
- Acetic anhydride, anhydrous, 111 mL, 1.2 mol corrosive, lachrymator
- Ethyl acetate, 550 mL flammable liquid, irritant

1. Set up the magnetic stirrer, round-bottomed flask with stirrer bar, heating mantle, thermometer with adaptor, and nitrogen inlet adaptor.

2. Flush the flask with N_2. Add the 5-benzylidene-1,2,3,4,5,6,7,8-octahydroacridin-4-ol and the anhydrous DMSO. Place the ground glass stopper in the third neck of the flask and stir the mixture at room temperature under N_2 until all of the solid has dissolved.

3. Add the acetic anhydride. Replace the ground glass stopper and stir the resulting solution under N_2 at 39–40 °C for 6 h. Allow the reaction mixture to cool slowly to room temperature, stirring under N_2.

4. Remove the thermometer and the nitrogen inlet adaptor, stopper the other two necks of the flask, and store it at c. 4 °C for several hours.

Protocol 4. *Continued*

5. Collect the solid by vacuum filtration using the sintered glass funnel, wash it with distilled water (3 × 100 mL) and dry it *in vacuo* (0.1–0.5 mmHg) for several hours to give 14.9–16.7 g (75–83 %) of crude product as a pale-yellow solid.

6. Dissolve the crude product in 400 mL of boiling ethyl acetate in the Erlenmeyer flask.[a]

7. Reduce the volume of the solvent by boiling until a fine suspension of solid appears in the solution (c. 200 mL). Cool the mixture slowly to room temperature, stopper the flask, and store it at 0 °C for several hours.

8. Collect the solid by vacuum filtration. Wash it with ice-cold ethyl acetate (3 × 50 mL) and dry it *in vacuo* (0.1–0.5 mmHg) for several hours, yielding 12.7–13.7 g (64–9%) of pale-yellow powder, m.p. 161–4 °C.

[a] If the boiling solution is cloudy, remove undissoved material by hot filtration through fluted filter paper.

5. Formation of a pyridine ring fused to two octahydroacridine units

In steps 5 and 6 of Scheme 6.1 the torand backbone is built up from two molecules of ketone **9** via Mannich salt **10**. Together these reactions constitute the most efficient method known so far for the synthesis of fused pyridines from cyclic ketones. The key intermediates involved in the transformation of **9** to **11** are shown in Scheme 6.7. In Protocol 5, ketone **9** reacts with *N,N*-dimethyl(methylene)ammonium chloride in acetonitrile.[21] The HCl salt of the resulting Mannich compound **10** precipitates from this solvent and the relatively pure product is isolated from the reaction mixture by filtration.

Protocol 6 involves heating Mannich compound **10** in DMSO, apparently causing elimination of dimethylamine and formation of enone **20** as a reactive intermediate. The corresponding enone has been synthesized in the 9-butyl series,[6,22] but comparable yields are obtained in pyridine cyclization reactions involving either the preformed enone or the Mannich HCl salt.[7] Prior to addition of **10**, ketone **9** is heated with ammonium chloride in DMSO to promote the formation of imine **21**. Isomerization of this imine to the enamine tautomer **22**, Michael addition of this nucleophile to enone **20**, and elimination of water account for the formation of product **11**. Like many polycyclic terpyridyl analogues, this product is sparingly soluble in DMSO and is readily isolated by filtration of the reaction mixture.

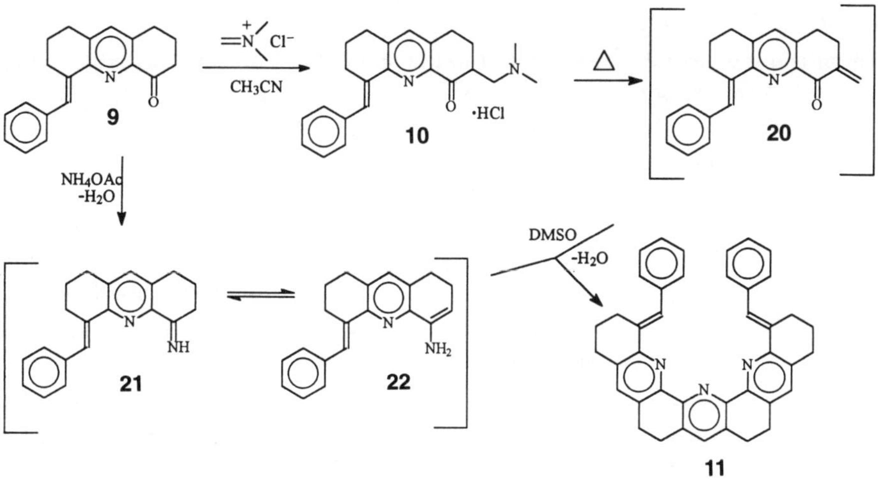

6: Torands

Scheme 6.7

Protocol 5.
Synthesis of 5-benzylidene-3-(*N*,*N*-dimethylaminomethyl)-2,3,5,6,7,8-hexahydroacridin-4(1*H*)-one hydrochloride (Structure 10, Scheme 6.8)

Caution! Carry out all procedures in a well-ventilated hood, and wear disposable vinyl or latex gloves and chemical-resistant safety goggles.

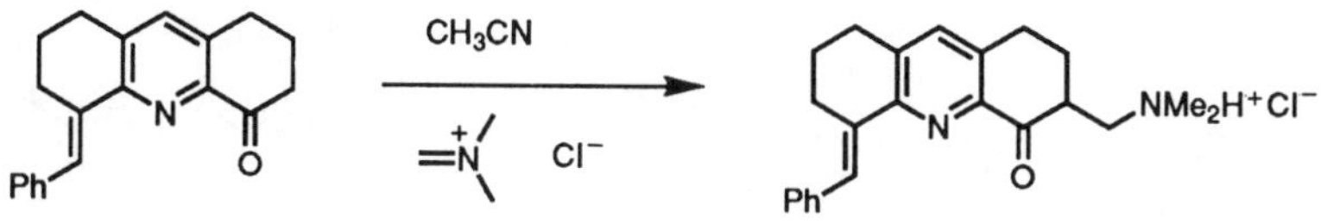

Scheme 6.8

Equipment

- Magnetic stirrer and magnetic stirrer bar
- Nitrogen inlet adaptor
- Source of dry nitrogen
- Single-necked, round-bottomed flask (500 mL)
- Sintered glass funnel (300 mL, med. porosity)
- Vacuum filter flask (500 mL)

Materials

- 5-Benzylidene-2,3,5,6,7,8-hexahydroacridin-4(1*H*)-one, 15.0 g, 51.9 mmol
- *N*,*N*-Dimethyl(methylene)ammonium chloride, 6.31 g, 67.4 mmol — **moisture sensitive, irritant**
- Acetonitrile, 300 mL — **flammable liquid, toxic**
- Ethyl acetate, 150 mL — **flammable liquid, irritant**

1. Dry the round-bottomed flask and stirrer bar in an electric oven (110°C, 2 h),

Protocol 5. *Continued*

attach the gas inlet adaptor, clamp the apparatus over the magnetic stirrer, and allow it to cool under N$_2$.

2. Add the 5-benzylidene-2,3,5,6,7,8-hexahydroacridin-4(1H)-one,N,N-dimethyl-(methylene)ammonium chloride, and acetonitrile to the flask. Stir the reaction mixture at room temperature under N$_2$ for 36 h.
3. Collect the solid product using the sintered glass funnel and vacuum filter flask. Rinse the reaction flask with 50 mL of cold ethyl acetate and wash the solid in the filter funnel. Repeat this washing procedure two more times.
4. Dry the product *in vacuo* (0.1–0.5 mmHg) for several hours to yield 14.1–17.0 g (71–86%) of white solid, m.p. 170–2 °C.

Protocol 6.
Synthesis of 1,15-dibenzylidene-1,2,3,4,6,7,9,10,12,13,14,15-dodecahydroacridino[4,3-*b*]benzo[*j*][1,10]phenanthroline (Structure 11, Scheme 6.9)

Caution! Carry out all procedures in a well-ventilated hood, and wear disposable vinyl or latex gloves and chemical-resistant safety goggles.

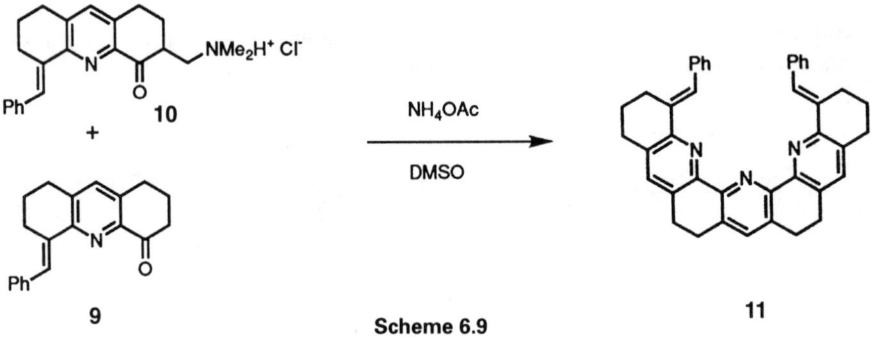

Scheme 6.9

Equipment

- Variable transformer
- Magnetic stirrer and magnetic stirrer bar
- Heating mantle
- Water-cooled condenser
- Nitrogen inlet adaptor
- Source of dry nitrogen
- Thermometer and adaptor
- Three-necked, round-bottomed flask (250 mL)
- Ground glass stopper
- Sintered glass funnel (60 mL, fine porosity)
- Vacuum filter flask (250 mL)
- Erlenmeyer flask (250 mL)

Materials

- 5-Benzylidene-2,3,5,6,7,8-hexahydroacridin-4(1H)-one, 5.0 g, 17 mmol
- 5-Benzylidene-3-(N,N-dimethylaminomethyl)- 2,3,5,6,7,8-hexahydroacridin-4(1H)-one hydrochloride, 6.6 g, 17 mmol
- Ammonium acetate, 3.4 g, 44.1 mmol — **irritant, hygroscopic**
- Dimethyl sulfoxide, anhydrous, 120 mL — **irritant, hygroscopic**

- Ethyl acetate, 75 mL   flammable liquid, irritant
- Pyridine, 250 mL   flammable liquid, toxic
- Acetone, 150 mL   flammable liquid, irritant

1. Dry the three-necked flask and magnetic stirrer bar in an electric oven (110 °C, 2 h) and assemble it with the heating mantle, magnetic stirrer, thermometer and adaptor, reflux condenser, and nitrogen inlet adaptor and allow the apparatus to cool under N_2.
2. Add the 5-benzylidene-2,3,5,6,7,8-hexahydroacridin-4(1H)-one, the ammonium acetate, and the DMSO at room temperature. Stir and heat the mixture under N_2 until the solution temperature is 80 °C.
3. Add the 5-benzylidene-3-(N,N-dimethylaminomethyl)-2,3,5,6,7,8-hexahydroacridin-4(1H)-one hydrochloride to the resulting solution using 2–4 mL of DMSO to rinse the weighing vessel.
4. Stir the reaction mixture under N_2 at 110–15 °C for 20 min. Allow the mixture to cool slowly to room temperature under N_2 with stirring.
5. Collect the product using the sintered glass funnel and vacuum filter flask. Wash the reaction flask with 25 mL of ice-cold ethyl acetate and wash the product. Repeat the washing procedure with ethyl acetate (2 × 25 mL).
6. Dry the product *in vacuo* (0.1–0.5 mmHg) to yield 7.9 g (81%) of bright yellow solid.
7. Dissolve the crude product in boiling pyridine. Concentrate the solution by boiling until it becomes cloudy (c. 100 mL), then allow it to cool to room temperature slowly. Stopper the flask and store it at 20 °C for several hours.
8. Collect the product by vacuum filtration and wash it with ice-cold acetone (3 × 50 mL).[a] Dry the product *in vacuo* (0.1–0.5 mmHg, 100 °C) for 2 d to yield 5.8 g (60%) as bright yellow needles, m.p. 300–2 °C.

[a] Upon standing, more solid will precipitate from the pyridine/acetone solution. An additional 0.2–0.3 g can be collected, increasing the yield to 63%.

---

6. Unmasking carbonyl groups in 12 by ozonolysis of the benzylidene groups in 11

Having served their purpose as masked carbonyl functionalities in the preceding 4 steps of the synthesis, the benzylidene groups in fused terpyridyl **11** are cleaved by ozonolysis in Protocol 7. As shown in Scheme 6.10, the reaction solvent is a mixture of methanol and dichloromethane. Reactant **11** is insoluble in methanol, so dichloromethane is used as a cosolvent to enhance its concentration in the cold (−78 °C) reaction mixture. Methanol is a convenient cosolvent for conducting ozonolysis reactions safely because it cleaves ozonides *in situ* to carbonyl compounds and peroxyhemiacetals or peroxyhemiketals.[23] In this case it is likely that diketone **12** and the methyl hydroperoxyacetal of benzaldehyde are formed (*cf.* Scheme 6.10). The latter side product is reduced to benzaldehyde after dimethyl sulfide is added to the reaction mixture.

Ketone **13** is apparently sensitive to overoxidation and the reaction mixture must be cooled to −78°C before ozone is bubbled through it. Ozone addition must be stopped immediately after the blue colour of the ozone-saturated solution is detected. Another potential side reaction is the condensation of benzaldehyde with diketone **12**; this reaction can be induced by overheating a concentrated solution of the crude product mixture. Accordingly, the residue from rotary evaporation obtained in step 10 of Protocol 7 should not be heated above 45–50°C before benzaldehyde is removed completely by exposure to high vacuum and trituration of the residue with diethyl ether.

Protocol 7.
Synthesis of 3,4,7,9,10,12,13,14-octahydroacridino[4,3-*b*]benzo[*j*] [1,10] phenanthroline-1,15(2*H*,6*H*)-dione (Structure 12, Scheme 6.10)

Caution! Carry out all procedures in a well-ventilated hood, and wear disposable vinyl or latex gloves and chemical-resistant safety goggles.

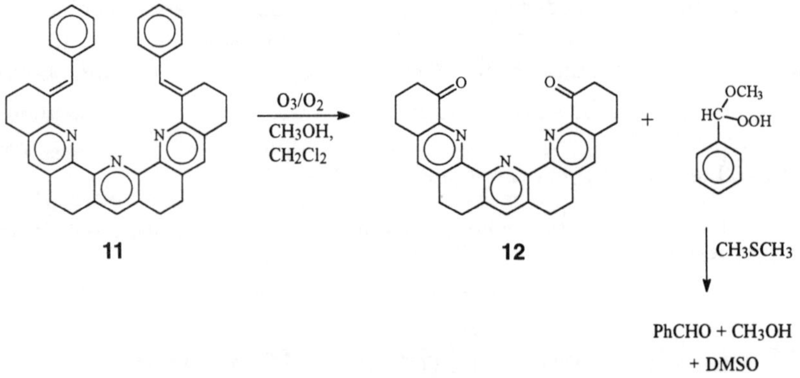

Scheme 6.10

Equipment

- Magnetic stirrer and magnetic stirrer bar
- Nitrogen inlet adaptor
- Source of dry nitrogen
- Source of dry oxygen
- Single-necked, round-bottomed flask (500 mL)
- Ozone generator
- Two Pasteur pipettes
- Insulated cylindrical glass vessel (19 cm in diameter × 14 cm h)
- Low-temperature thermometer
- Rubber septum
- Disposable syringe needle
- Sintered glass funnel (300 mL, fine porosity)
- Vacuum filter flask (500 mL)
- Spatula
- Erlenmeyer flask (500 mL)
- Rotary evaporator

6: Torands

Materials

- 1,15-Dibenzylidene-1,2,3,4,6,7,9,10,12,13,14,15-dodecahydroacridino-[4,3-b]benzo[j][1,10]phenanthroline, 1.6 g, 2.8 mmol
- Dichloromethane, 290 mL — toxic, irritant
- Methanol, 70 mL — highly toxic, flammable liquid
- Dimethyl sulfide, 0.6 mL — flammable liquid, stench
- Diethyl ether, 400 mL — flammable liquid, toxic
- Crushed dry ice, 500 g
- Acetone, 1 L — flammable liquid, irritant

1. Assemble the round-bottomed flask with magnetic stirrer bar and nitrogen inlet adaptor. Flush the flask with N_2.

2. Add the 1,15-dibenzylidene-1,2,3,4,6,7,9,10,12,13,14,15-dodecahydroacridino[4,3-b]benzo[j][1,10]phenanthroline and 210 mL of dichloromethane, and stir the mixture under N_2 at room temperature for 15 min.

3. Add the methanol and stir the resulting mixture under N_2 at room temperature for several min.

4. Clamp the flask in the bath vessel and cool with dry ice and acetone.[a]

5. Bubble N_2 into the solution through a Pasteur pipette for 10 min.

6. Remove the N_2 pipette and bubble O_3 into the solution through a second pipette until the solution turns dark blue, requiring about 5 min.[b]

7. Remove the O_3 pipette and bubble N_2 through the solution again until it is pale yellow, requiring about 20 min.

8. Add the dimethyl sulfide.

9. Cover the reaction flask with a rubber septum and pierce the septum with a syringe needle. Allow the flask and its contents to warm to room temperature.

10. Remove the solvents *in vacuo* (30–40 mmHg) using a rotary evaporator and dry the residue *in vacuo* (0.1–0.5 mmHg) for several hours.

11. Triturate the solid with ether (4 × 25 mL) and allow it to dry in air for 1 h.

12. Dissolve the solid in 80 mL of dichloromethane and pour the solution into the Erlenmeyer flask. Add 300 mL of ether, cover the flask and store it at 20 °C for several hours.

13. Collect the precipitate by vacuum filtration. Dry the product *in vacuo* (0.1–0.5 mmHg) to yield a pale-gold powder, 1.03 g (85 %), m.p. 270 °C (dec).

[a] The temperature of the reaction mixture must be lower than −70 °C and should be checked by means of the low-temperature thermometer.
[b] The reaction time depends on the concentration of O_3/O_2 and the rate of gas flow. The reaction must be carefully watched to prevent overoxidation of the product. Remove the O_3 inlet pipette as soon as the blue colour appears.

7. Functionalization of the third octahydroacridine unit

Diketone **12** must be condensed with a third, doubly functionalized octahydroacridine unit in the last step of the torand synthesis (*cf.* Scheme 6.1). The following protocols (8–10) describe the three-step synthesis of torand precursor **15** from octahydroacridine **5**. The reagents involved in these three steps are shown in Scheme 6.11. According to Protocol 8, octahydroacridine is condensed with benzaldehyde in the presence of acetic anhydride,[24] as in the second stage of Protocol 2 (*cf.* Scheme 6.3). The crystalline product **13** precipitates from the reaction mixture in high yield and purity. Ozonolysis of **13** (Protocol 9) is conducted by the method described in Protocol 7 for conversion of **11** to diketone **12** (*cf.* Scheme 6.10) and the same precautions apply. The product diketone **14**[25] requires no further purification after removal of benzaldehyde by trituration with diethyl ether. The third octahydroacridine unit is then readied for torand cyclization in Protocol 10 by condensing diketone **14** with Brederick's reagent, *t*-butoxybis(dimethylamino)methane,[26] which is commercially available. The bis[β-(dimethylamino)]enone product **15** is easily purified by precipitation from ether/dichloromethane.

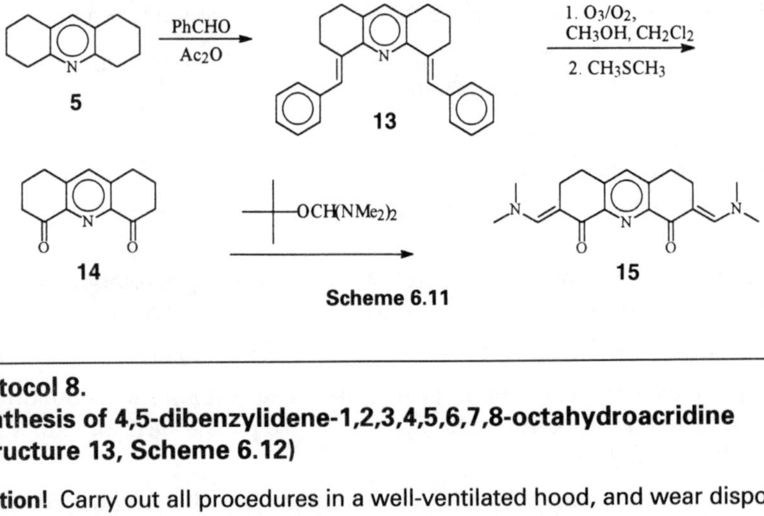

Scheme 6.11

Protocol 8.
Synthesis of 4,5-dibenzylidene-1,2,3,4,5,6,7,8-octahydroacridine (Structure 13, Scheme 6.12)

Caution! Carry out all procedures in a well-ventilated hood, and wear disposable vinyl or latex gloves and chemical-resistant safety goggles.

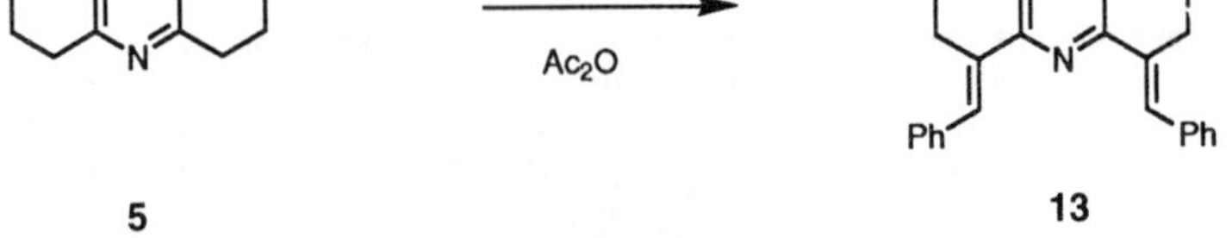

Scheme 6.12

6: Torands

Equipment

- Variable transformer
- Magnetic stirrer and magnetic stirrer bar
- Heating mantle
- Water-cooled condenser
- Nitrogen inlet adaptor
- Source of dry nitrogen
- Thermometer and adaptor
- Three-necked, round-bottomed flask (500 mL)
- Three ground glass stoppers
- Sintered glass funnel (300 mL, med. porosity)
- Vacuum filter flask (500 mL)

Materials

- 1,2,3,4,5,6,7,8-octahydroacridine 10.0 g, 53.5 mmol,   irritant
- Benzaldehyde, 92 mL (97 g), 0.91 mol   highly toxic, cancer-suspect agent
- Acetic anhydride, anhydrous, 84 mL 0.89 mol,   corrosive, lachrymator
- Diethyl ether, 85 mL   flammable liquid, toxic
- Ethanol, 35 mL   highly toxic, flammable liquid
- Methanol, 35 mL   highly toxic, flammable liquid

1. Assemble the round-bottomed flask with stirrer bar, magnetic stirrer, heating mantle, thermometer and adaptor, reflux condenser, and nitrogen inlet adaptor. Flush the apparatus with nitrogen.
2. Add the 1,2,3,4,5,6,7,8-octahydroacridine, benzaldehyde, and acetic anhydride.
3. Replace the stopper and stir the reaction mixture at 150 °C under nitrogen for 24 h.
4. Allow the reaction flask to cool to room temperature, seal it with the stoppers, then store it at 0 °C for several hours.
5. Collect the product by vacuum filtration. Wash it first with ice-cold ether and then with 65 mL of ice-cold 1 : 1 (v/v) ethanol/methanol.
6. Dry the product *in vacuo* for several hours to give 17.3 g (89%) of pale-gold needles, m.p. 183–5 °C (lit.[24] 184–6 °C).

Protocol 9.
Synthesis of 2,3,7,8-tetrahydroacridine-4,5(1*H*,6*H*)-dione (Structure 14, Scheme 6.13)

Caution! Carry out all procedures in a well-ventilated hood, and wear disposable vinyl or latex gloves and chemical-resistant safety goggles.

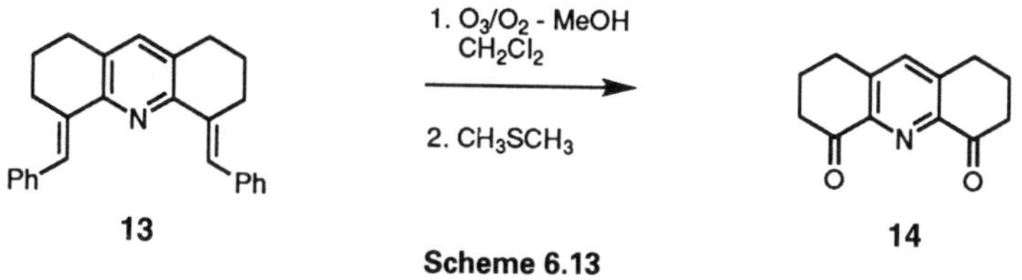

Scheme 6.13

Thomas W. Bell and Julia L. Tidswell

Protocol 9. *Continued*

Equipment

- Magnetic stirrer and magnetic stirrer bar
- Nitrogen inlet adaptor
- Source of dry nitrogen
- Source of dry oxygen
- Single-necked, round-bottomed flask (500 mL)
- Ozone generator
- Two Pasteur pipettes
- Rubber septum
- Disposable syringe needle
- Spatula
- Insulated glass vessel (19 cm in diameter × 14 cm h)
- Rotary evaporator

Materials

- 4,5-Dibenzylidene-1,2,3,4,5,6,7,8-octahydroacridine, 10.0 g, 27.5 mmol
- Dichloromethane, 200 mL   **toxic, irritant**
- Methanol, 50 mL   **highly toxic, flammable liquid**
- Diethyl ether, 200 mL   **flammable liquid, toxic**
- Dimethyl sulfide, 6 mL   **flammable liquid, stench**
- Crushed dry ice, 500 g
- Acetone, 1 L   **flammable liquid, irritant**

1. Assemble the magnetic stirrer, round-bottomed flask with stirrer bar, and nitrogen inlet adaptor. Flush the flask with nitrogen.
2. Add the 4,5-dibenzylidene-1,2,3,4,5,6,7,8-octahydroacridine and the dichloromethane. Stir the mixture under N_2 at room temperature for about 10 min to obtain a clear solution.
3. Add the methanol and stir the resulting solution under N_2 at room temperature for several minutes.
4. Clamp the flask in the bath vessel and cool with dry ice and acetone.[a]
5. Bubble N_2 into the solution through the Pasteur pipette for 10 min.
6. Remove the N_2 pipette and bubble O_3 into the solution through the second Pasteur pipette until the solution turns dark blue, requiring about 30 min.[b]
7. Remove the O_3 pipette and bubble N_2 through the solution again until it turns pale yellow, requiring 1–2 h.
8. Add the dimethyl sulfide.
9. Cover the reaction flask with a rubber septum and pierce the septum with a syringe needle. Allow the solution to warm to room temperature.
10. Remove the solvents *in vacuo* (30–40 mmHg) using a rotary evaporator.
11. Triturate the resulting oil with ether (4 × 50 mL) to yield a pale-yellow solid. Dry the product *in vacuo* (0.1–0.5 mmHg) for several hours to yield 4.3 g (72%) of a yellow powder, m.p. 152–3 °C (lit.[25] 150–1 °C).

[a] The temperature of the reaction mixture must be lower than −70 °C and should be checked by means of the low-temperature thermometer.
[b] The reaction time depends on the concentration of O_3/O_2 and the rate of gas flow. The reaction must be carefully watched to prevent overoxidation of the product. Remove the O_3 inlet pipette as soon as the blue colour appears.

Protocol 10.
Synthesis of 3,6-bis(dimethylaminomethylene)-2,3,7,8-tetrahydroacridine-4,5 (1H,6H)-dione (Structure 15, Scheme 6.14)

Caution! Carry out all procedures in a well-ventilated hood, and wear disposable vinyl or latex gloves and chemical-resistant safety goggles.

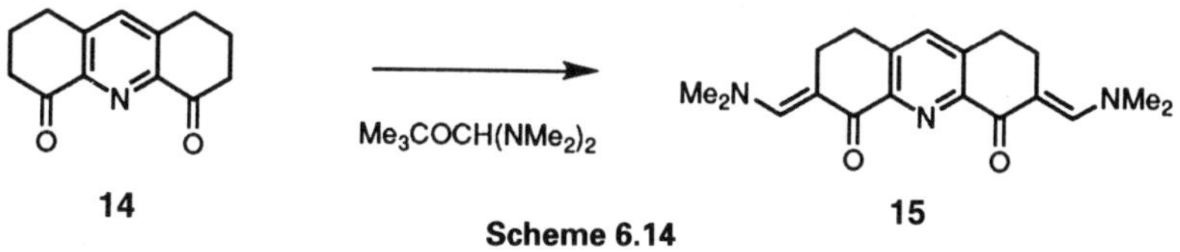

Scheme 6.14

Equipment

- Variable transformer
- Magnetic stirrer and magnetic stirrer bar
- Heating mantle
- Water-cooled condenser
- Nitrogen inlet adaptor
- Source of dry nitrogen
- Ground glass stopper

- Thermometer and adaptor
- Three-necked, round-bottomed flask (250 mL)
- Sintered glass funnel (30 mL, fine porosity)
- Vacuum filter flask (500 mL)
- Spatula
- Erlenmeyer flask (250 mL)

Materials

- 2,3,7,8-Tetrahydroacridine-4,5(1H,6H)-dione, 2.0 g, 9.3 mmol
- Bredereck's reagent, 10.0 mL, 8.4 g, 24 mmol
- Diethyl ether, 500 mL **flammable liquid, toxic**
- Dichloromethane, 30 mL **toxic, irritant**

1. Assemble the magnetic stirrer, round-bottomed flask with stirrer bar, heating mantle, thermometer with adaptor, condenser, and nitrogen inlet adaptor. Flush the apparatus with N_2.

2. Add the 2,3,7,8-tetrahydoacridine-4,5(1H,6H)-dione and the Bredereck's reagent.

3. Stir the reaction mixture under N_2 at 85–90°C for 1 h.

4. Allow the dark reaction mixture to cool to room temperature with stirring under N_2. Add 40 mL of ether and stir the resulting mixture under N_2 at room temperature for 2 h.

5. Allow the precipitated solid to settle to the bottom of the flask, decant the supernatant solution, and triturate the residue with ether (3 × 30 mL).

6. Collect the product by vacuum filtration, wash it with ether (2 × 40 mL), and dry it *in vacuo* (0.1–0.5 mmHg) for several hours to yield 3.0 g (97%) of crude product as a green solid, m.p. 220°C (dec).

7. Dissolve the crude product in the dichloromethane. Add 120 mL of ether and mix thoroughly.

Protocol 10. *Continued*

8. Collect the precipitated product by vacuum filtration, wash it with ether (2 × 40 mL) and dry it *in vacuo* (0.1–0.5 mmHg) for several hours to yield 2.8 g (73%) of product as a golden-green solid, m.p. 240°C (dec).

8. Torand cyclization

Protocol 11 describes the final step in the synthesis of torand **1**, macrocyclization of diketone **12** with bis[(β-dimethylamino)enone] **15** and conversion of the intermediate bis(pyrylium) macrocycle **23** to torand **1**, as shown in Scheme 6.15. The macrocyclization reaction was modelled after the synthesis of dodecahydro-18,21-dioxoniakekulene, involving reaction of a diketone with a bis(β-chlorovinyl) aldehyde in the presence of trifluoromethanesulfonic acid.[27] The synthesis of torand **1** differs from previous approaches to heterokekulenes[8,27] in that the two pyrylium rings in the intermediate **23** are not directly opposed with respect to the central 18-membered ring. In Protocol 11 the bis(pyrylium) salt is not isolated;[8,27] ammonium acetate is added to the reaction mixture and torand **1** forms at a higher temperature than required in the macrocyclization step. Like the tributyl analogue **2**,[6,7] torand **1** is conveniently isolated as its monotriflate salt. Hexaazakekulene derivatives are powerful complexing agents for metal cations[6,7] and aqueous solutions used in Protocol 11 should be prepared with water that has been carefully purified by deionization and distillation to exclude traces of metals.

Protocol 11.
Synthesis of 1,2,4,5,7,8,10,11,13,14,16,17-dodecahydro-19, 20,21,22,23,24-hexaazakekulene (Structure 1, Scheme 6.15)

Caution! Carry out all procedures in a well-ventilated hood, and wear disposable vinyl or latex gloves and chemical-resistant safety goggles.

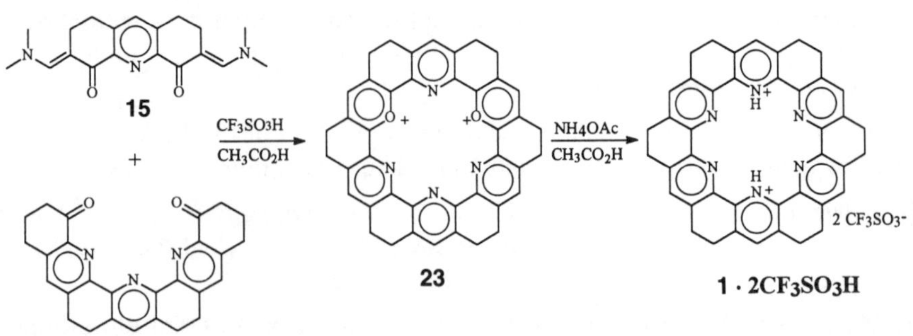

Scheme 6.15

6: Torands

Equipment

- Variable transformer
- Magnetic stirrer and stirrer bar
- Silicone oil bath with nichrome heating wire
- Power cord with clips
- Thermometer
- Pressure tube with threaded Teflon stopper
- Volumetric flask (100 mL)
- Separating funnel (125 mL)
- Erlenmeyer flasks (25 and 125 mL)
- Single-necked, round-bottomed flask (250 mL)
- Sintered glass funnel (60 mL, fine porosity)
- Vacuum filter flask (125 mL)

Materials

- 3,4,7,9,10,12,13,14-Octahydroacridino[4,3-b] benzo[j][1,10]phenanthroline-1,15(2H,6H)-dione, 0.38 g, 0.9 mmol
- 3,6-Bis(dimethylaminomethylene)-2,3,7,8-tetrahydroacridine 4,5(1H,6H)dione, 0.29 g, 0.89 mmol
- Acetic acid, 2.9 mL — corrosive
- Trifluoromethanesulfonic acid, 1.2 mL, 13.6 mmol — corrosive, hygroscopic
- Ammonium acetate, 3.5 g, 45 mmol — irritant, hygroscopic
- 1 M trifluoromethanesulfonic acid solution in deionized, distilled water, 80 mL
- Dichloromethane, 100 mL — toxic, irritant
- Ethanol, 50 mL — highly toxic, flammable liquid
- Acetone, 10 mL — flammable liquid, irritant
- Glycine, 3.0 g, 40 mmol
- Deionized, distilled water
- Source of dry nitrogen

1. Set up the magnetic stirrer and oil bath and use the power cord to attach the heating element to the variable transformer. Clamp the thermometer so that it will be as close to the pressure tube as possible.

2. Combine the 3,4,7,9,10,12,13,14-octahydroacridino[4,3-b]benzo[j][1,10]phenanthroline-1,15(2H,6H)-dione, the 3,6-bis(dimethylaminomethylene)-2,3,7,8-tetrahydroacridine-4,5(1H,6H)-dione, and the acetic acid in the pressure tube, add the stirrer bar and close the tube. Swirl the contents in the tube until all solids are dissolved. Add 1.2 mL (13 mmol) of trifluoromethanesulfonic acid, and seal the tube.

3. Immerse the pressure tube in the oil bath at least up to the level of the contents. Heat the bath to 100°C and stir the mixture in the tube for 1.5 h. Remove the tube from the oil bath and allow it to cool for 15 min.

4. Carefully open the tube and add the ammonium acetate.

5. Reseal the tube and immerse it in the oil bath up to the level of the contents. Heat the bath to 165–70°C and stir the mixture in the tube for 6.5 h.

6. Allow the tube to cool to room temperature, remove the stirrer bar and add the contents of the tube to the separating funnel.

7. Dilute the reaction mixture with 35 mL of dichloromethane. Wash this solution with 1 M aq trifluoromethanesulfonic acid solution (4 × 15 mL). Extract the aqueous mixture with dichloromethane (3 × 20 mL). Wash the combined organic solutions with 0.2 M glycine/glycine triflate buffer (3 × 15 mL)[a] and remove the solvent *in vacuo* (30–40 mmHg) by rotary evaporation.

Protocol 11. *Continued*

8. Dissolve the resulting sticky solid in 5 mL of acetone and transfer this solution to the 25 mL Erlenmeyer flask. Rinse the round-bottomed flask with acetone (3 × 1.5 mL) and ethanol (3 × 1.5 mL) and add the rinsings to the Erlenmeyer flask. Add 5 mL of ethanol to the flask and stir. Reduce the volume of solvents by blowing a stream of N_2 over the surface for 30 min. Cover the flask with aluminium foil to protect it from dust and let it stand at room temperature for 1 d.

9. Stopper the flask and store it at 0 °C for 1 d. Collect the product by vacuum filtration, wash it with cold ethanol (3 × 10 mL), and dry it *in vacuo* (0.1–0.5 mmHg, 115 °C, 1 d) yielding 0.13 g (15%) of a golden microcrystalline solid.[b]

[a] The buffer is prepared in a volumetric flask by dissolving 3.00 g (40 mmol) of glycine in c. 20 mL of deionized, distilled water, adding 20.0 mL of 1 M trifluoromethanesulfonic acid and diluting to the 100 mL mark with deionized, distilled water.
[b] The product gives the following spectoscopic and microanalytical data: ^{1}H NMR: (DMSO-d_6, 300 MHz), δ3.01 (s, 24 H), 7.88 (s, 6 H); FTIR (KBr, cm^{-1}): 3442 (m, br), 1397 (m), 1257 (s), 1222 (m), 1152 (m), 1030 (s), 637 (s). Analysis calculated for $C_{42}H_{30}N \cdot CF_3SO_3H \cdot 2.75\ H_2O$(%): C, 63.11; H, 4.50; N, 10.27; S, 3.92; found (%): C, 62.73; H 4.14; N, 10.48; S, 4.32.

References

1. Bell, T. W.; Firestone, A. *J. Am. Chem. Soc.* **1986**, *108*, 8109–8111.
2. Bell, T. W.; Firestone, A.; Ludwig, R. *J. Chem Soc., Chem. Commun.* **1989**, 1902–1904.
3. Bell, T. W.; Liu, J. *Angew. Chem.* **1990**, *102*, 931–933; *Angew. Chem., Int. Ed. Engl.* **1990**, *29*, 923–925.
4. Bell. T. W.; Cragg, P. J.; Drew, M. G. B.; Firestone, A.; Kwok, D.-I. A. *Angew. Chem.* **1992**, *104*, 319; *Angew. Chem., Int. Ed. Engl.* **1992**, *31*, 345–347.
5. Bell, T. W.; Cragg, P. J.; Drew, M. G. B.; Firestone, A.; Kwok, D.-I. A. *Angew. Chem.* **1992**, *104*, 321; *Angew. Chem., Int. Ed. Engl.* **1992**, *31*, 348–350.
6. Bell, T. W. In *Crown Compounds, Toward Future Applications*; Cooper, S. R., ed.; VCH Publishers: New York, **1992**, pp. 305–318.
7. Bell. T. W.; Cragg, P. J.; Drew, M. G. B.; Firestone, A.; Kwok, D.-I. A.; Liu, J.; Ludwig, R. T.; Papoulis, A. T. *Pure Appl. Chem.* **1993**, *65*, 361–366.
8. Ransohoff, J. E. B.; Staab, H. A. *Tetrahedron Lett.* **1985**, *26*, 6179–6182.
9. Newkome, G. R.; Lee, H.-W. *J. Am. Chem. Soc.* **1983**, *105*, 5956–5957.
10. Bell, T. W.; Beckles, D. L.; Cragg, P. J.; Liu, J.; Maioriello, J.; Papoulis, A. T.; Santora, V. J. In *Fluorescent Chemosensors of Ion and Molecule Recognition;* Czarnik, A., ed., ACS Symposium Series No. 538, ACS Books: Washington (D.C.), **1993**, pp. 85–103.
11. Bell, T. W.; Rothenberger, S. D. *Tetrahedron Lett.* **1987**, *28*, 4817–4820.
12. Bell, T. W.; Cho, Y.-M.; Firestone, A.; Healy, K.; Liu, J.; Ludwig, R. T.; Rothenberger, S. D. *Org. Synth., Coll. Vol.* **1993**, *8*, 87–93.
13. Bell, T. W.; Liu, J. *J. Am. Chem. Soc.* **1988**, *110*, 3673–3674.
14. Bell, T. W.; Firestone, A.; Liu, J.; Ludwig, R. T.; Rothenberger, S. D. In *Inclusion*

Phenomena and Molecular Recognition; Atwood, J. L., ed., Plenum: New York, **1990**, pp. 49–56.
15. Bell, T. W.; Santora, V. J. *J. Am. Chem. Soc.* **1992**, *114*, 8300–8302.
16. Bell, T. W.; Heiss, A. M.; Jousselin, H.; Ludwig, R. T. In *Supramolecular Stereochemistry*; Siegel, J. S., ed., Kluwer: Dordrecht, The Netherlands, **1995**, 161–168.
17. Ogata, Y.; Tabushi, I. *Bull. Chem. Soc. Jpn.* **1958**, *31*, 969–973.
18. Risaliti, A.; De Martino, U. *Ann. Chim. (Rome)* **1963**, *53(6)*, 819–827.
19. Cohen, T.; Deets, G. L. *J. Am. Chem. Soc.* **1972**, *94*, 932–938.
20. Albright, J. D.; Goldman, L. *J. Am. Chem. Soc.* **1967**, *89*, 2416–2423.
21. Risch, N.; Esser, A. *Synthesis* **1988**, 337–339.
22. Bell, T. W.; Jousselin, H. *J. Am. Chem. Soc.* **1991**, *113*, 6283–6284.
23. Pappas, J. J.; Keaveney, W. P.; Gancher, E.; Berger, M. *Tetrahedron Lett.* **1966**, 4273–4278.
24. Tilichenko, M. N.; Vysotskii, V. I. *J. Gen. Chem. USSR (Engl. Transl.)* **1962**, 81–82.
25. Thummel, R. P.; Jahng, Y. *J. Org. Chem.* **1985**, 2407–2412.
26. Bredereck, H.; Simchen, G.; Rebsdat, S.; Kantlehner, W.; Horn, P.; Wahl, R.; Hoffman, H.; Grieshaber, P. *Chem. Ber.* **1968**, *101*, 41–50.
27. Katritzky, A. R.; Marson, C. M. *J. Am. Chem. Soc.* **1983**, *105*, 3279–3283.

Acknowledgement

The authors gratefully acknowledge the important contributions of Julia Munsch, Wilbur B. Knowall, Jr, and Stacy Shlachtman to this body of work, which was supported by the U.S. National Institutes of Health (PHS grant GM 32937).

7

Calixarenes

ARTURO ARDUINI and ALESSANDRO CASNATI

1. Introduction

The term calix[n]arenes indicates a class of phenolic metacyclophanes derived from the condensation of phenols and aldehydes. The name was coined by Gutsche and derives from the Latin 'calix' because of the vase-like structure that these macrocycles assume when all the aromatic rings are oriented in the same direction.[1] The bracketed number indicates the number of aromatic rings and hence defines the size of the macrocycle. To identify the phenol from which the calixarene is derived, the *para* substituent is designated by name. Thus the cyclic tetramer derived from *p-t*-butylphenol and formaldehyde is named *p-t*-butylcalix[4]arene, or with a more systematic but still simplified nomenclature proposed by Gutsche and used in this chapter 5,11,17,23-Tetrakis(1,1-dimethylethyl)-25,26,27,28-tetrahydroxycalix[4]arene, **1** (Scheme 7.1). The systematic name reported by Chemical Abstracts is pentacyclo[19.3.1.1^{3,7}.1^{9,13}.1^{15,19}]octacosa-1(25),3,5,7(28),9,11,13(27),15,17,19(26),21,23-dodecaene-25,26,27,28-tetrol-5,11,17,23-tetrakis(1,1-dimethylethyl).

Although nowadays the term calixarene tends to be used for all [1$_n$]-metacyclophanes, this chapter will deal with [1$_n$]-metacyclophanes bearing phenolic OH groups in the intraannular (*endo*, e.g. calix[n]arenes) or extraannular (*exo*, e.g. resorc[4]arenes) position.

Calixarenes are ditopic molecular structures on which functional groups and binding sites can be organized and oriented in space for the selective binding of ions and neutral molecules. In these molecules two distinct regions are present: the lower rim, where the phenolic oxygens are situated and the upper rim defined by the *para* positions of the aromatic nuclei. The increasing interest in this class of compounds in the field of supramolecular or host–guest chemistry arises from the possibility of functionalization of both rims in a regio- and stereocontrolled way. Much of this work, together with the complexing properties of calixarene-based hosts, has already been reported in other books[2,3] or review articles.[4]

2. Synthesis of calixarenes

Two general routes are available for the synthesis of calixarenes: the base- and acid-promoted one-step synthesis and the convergent stepwise synthesis (fragment condensation).[2,3]

Both the acid-catalysed and the base-promoted one-step syntheses provide the most powerful procedures to obtain symmetrically substituted calixarenes having only one type of phenolic unit in the cyclic array.

Convergent stepwise syntheses (fragment condensation) are usually based on the condensation of two synthons from which calixarenes can be isolated in 10–25% yield (Fig. 7.1). This strategy is complementary to the one-step procedure and also offers the advantage of obtaining calixarenes with different phenol rings incorporated in the molecular skeleton[5] or having substituents at the methylene bridges.[6]

2.1 Base-promoted one-step synthesis

In general, the base-catalysed condensation of a *p*-substituted phenol and formaldehyde yields a complex mixture of linear and cyclic oligomers and the outcome of the reaction is difficult to control.[7] There are, however, a few cases in which the reaction can be controlled: when the starting phenol bears a bulky substituent at the *p*-position (e.g. *t*-butyl or *t*-octyl) the condensation reaction can be directed to cyclic products in excellent yields. The wide-ranging investigations of Gutsche and his co-workers resulted in both reproducible and high-yield procedures for the synthesis of *p-t*-butylcalixarenes containing an even number of phenolic rings. From these studies it has been concluded that the octamer is the product of kinetic control, the tetramer the product of thermodynamic control, while the hexamer is favoured under

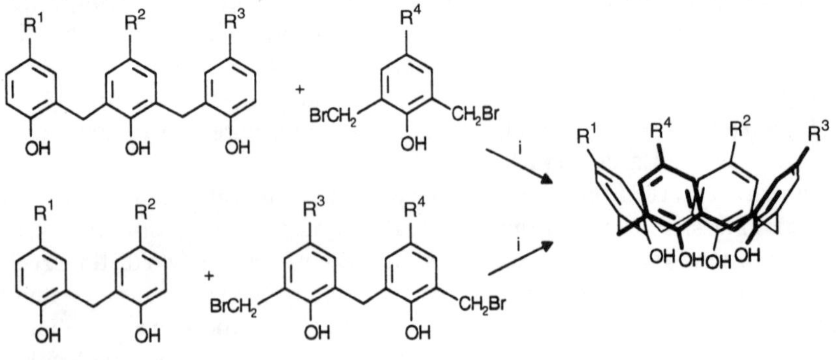

R^1=alkyl, C_6H_5; R^2=alkyl, NO_2, COOEt, N=N-C_6H_5; R^3=alkyl, C_6H_5, Cl, COOEt

R^4=alkyl, C_6H_5, Cl, COOEt, CH_2COOEt. i, $TiCl_4$

Fig. 7.1 Convergent syntheses of calix[4]arenes: the 3 + 1 and 2 + 2 approach.

7: Calixarenes

kinetic control in a metal-templated reaction. In fact, whereas the use of alkali metal cations with large ionic radii (e.g. K⁺ or Rb⁺) favours the formation of the hexamer, higher temperatures give mainly the tetramer and both the octamer and the hexamer may be converted to the tetramer under basic conditions.

Calixarenes containing an odd number of phenolic rings are less accessible. *p-t*-Butylcalix[5]arene was originally isolated by Ninagawa and Matsuda[8] and more recently Steward and Gutsche[9] have reported an improved procedure from which the cyclic pentamer could be obtained in about 15% yield. Calix[7]arene was obtained in about 29% yield by Nakamoto and Ishida.[10] The larger members of the series, calix[9]- and calix[10]arene, have been isolated only in trace amounts.[11] *p-t*-Butylcalix[4]arene **1** is the most extensively studied member of the calixarene series and can be easily synthesized by a base (NaOH) catalysed reaction. The so-called 'one-flask' procedure has been optimized by Gutsche *et al.*[12]

Protocol 1.
Synthesis of 5,11,17,23-tetrakis(1,1-dimethylethyl)-25,26,27,28-tetrahydroxycalix[4]arene (Structure 1, Scheme 7.1)

Caution! Carry out all procedures in a well-ventilated hood, and wear disposable vinyl or latex gloves and chemical-resistant safety goggles.[a]

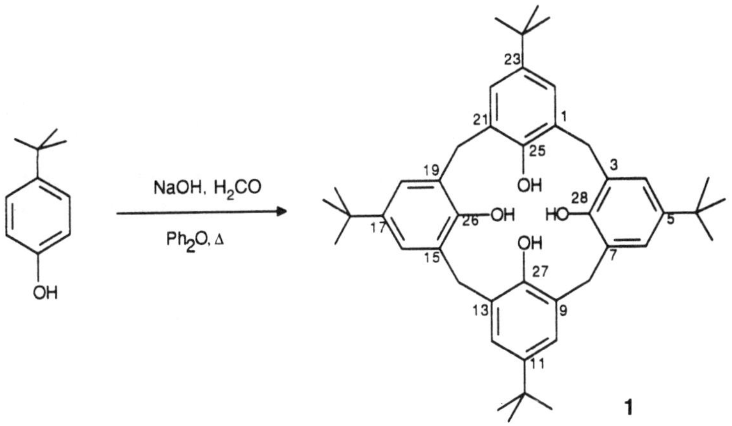

Scheme 7.1

Equipment[a]

- Four-necked, round-bottomed flask (3 L)
- Mechanical stirrer
- Laboratory jack
- Nitrogen inlet adaptor
- Source of dry nitrogen
- Heating mantle equipped with a variable transformer[b]
- Thermometer
- Allihn condenser
- Still head
- Buchner funnel
- Filter flask
- Beaker

147

Protocol 1. Continued

Materials

• 4-t-Butylphenol (FW 150.22), 250 g, 1.66 mol	irritant
• Formaldehyde 37% wt. water solution (FW 30.03), 155 mL, 2.07 mol	toxic, irritant
• Sodium hydroxide pellets (FW 39.99), 1.2 g, 0.03 mol	corrosive, highly irritant
• Phenyl ether, 2 L	harmful by inhalation, irritant
• Toluene, 100 mL + 3 L for recrystallization	flammable
• Ethyl acetate, 1.5 L + 200 mL for washings	flammable
• Acetic acid, 200 mL	corrosive, irritant

1. Position the heating mantle on a laboratory jack and place the 3 L flask on the heating mantle.
2. Clamp the flask to a support and equip the flask with a mechanical stirrer, a nitrogen inlet and a still head connected to a tube dipped into a beaker of water. Close the fourth neck with a blowed glass stopper.
3. Purge the flask with nitrogen and maintain a gentle flow throughout the reaction period.
4. Remove the glass stopper and with the aid of a funnel load the reaction vessel with 4-t-butylphenol (250 g), formaldehyde solution (155 mL), and sodium hydroxide (1.2 g) dissolved in 3 mL of water. Close the neck and start to stir.
5. Set the temperature to 120°C and start to heat. Within a few minutes the heterogeneous mixture turns homogeneous and pale yellow. Allow the vapour to exit the flask through the still head and be trapped by the water. As the reaction mixture becomes more viscous turn the heating off, but continue to stir until a thick mass has formed. At this point stop the stirring.[c]
6. Remove the heating mantle, allow the reaction to cool to room temperature and add the phenyl ether (2 L)[d] and toluene (100 mL)[e] to the solid mass.
7. Reposition the heating mantle, set the temperature to 260°C and heat. Allow the vapour to exit the reaction mixture until the internal temperature has reached 180–90°C.[f]
8. Replace the still head with the Allihn condenser and reflux the heterogeneous light-brown mixture for 4 h; during this period the mixture becomes homogeneous and dark brown.
9. Remove from the heat and allow the mixture to cool to 60°C. Add ethyl acetate (1.5 L), cool to room temperature and continue to stir for an additional hour.
10. Filter the pale brown crystals by suction filtration using a Buchner funnel.
11. Transfer the solid to a 500 mL beaker and triturate with 200 mL of acetic acid.

7: Calixarenes

12. Filter the suspension on a Buchner funnel and wash the crystals with small portions of ethyl acetate 4 × 50 mL.
13. Dissolve the crystals by adding in small portions to 3 L of toluene heated at 100 °C in a conical flask and heat the mixture under reflux until complete dissolution occurs.
14. Allow the solution to cool down to room temperature and separate the glistening crystals by suction filtration on a Buchner funnel. After drying, 175 g (55%) of product 1 are obtained as a 1:1 complex with toluene[g] (m.p. 342–4 °C).

[a] The synthesis of calix[4]arene requires effective temperature control, therefore all glassware must be examined carefully.
[b] It should reach 260 °C.
[c] The viscosity of the mass increases until it becomes a foamy mass which blocks the stirring paddle.
[d] Phenyl ether is a persistent stinking chemical, and disposable gloves must be worn.
[e] Toluene facilitates the azeotropic removal of water.
[f] Occasionally during this phase an abundant formation of foam takes place; if it starts to overflow, add a few mL of ethanol (dropwise) and continue to heat.
[g] The presence of toluene usually does not interfere during subsequent functionalization reactions. Toluene can be removed by heating at 110 °C for 24–48 h under vacuum (0.1 mmHg).

When the cyclization reaction is carried out under basic conditions with KOH in refluxing xylene, the hexamer is isolated in nearly quantitative yield (Scheme 7.2).[13]

Protocol 2.
Synthesis of 5,11,17,23,29,35-hexakis(1,1-dimethylethyl)-37, 38,39,40,41,42-hexahydroxycalix[6]arene (Structure 2, Scheme 7.2)

Caution! Carry out all procedures in a well-ventilated hood, and wear disposable vinyl or latex gloves and chemical-resistant safety goggles.

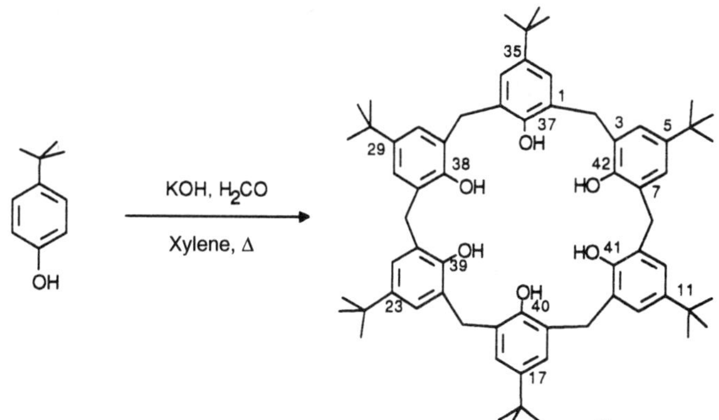

Scheme 7.2

Protocol 2. Continued

Equipment

- Mechanical stirrer
- Heating mantle
- Three-necked, round-bottomed flask (2 L)
- Single-necked, round-bottomed flask (1 L)
- Allihn condenser
- Gas inlet adapter
- Source of dry nitrogen
- Separating funnel
- Dean–Stark trap
- Buchner funnel

Materials

- *p-t*-Butylphenol (FW 150.22), 100 g, 0.66 mol — **irritant**
- Formaldehyde, 37% wt. in water (FW 30.03), 135 mL, 1.8 mol — **toxic, irritant**
- Potassium hydroxide pellets (FW 56.11), 15.0 g, 0.267 mol — **corrosive, highly irritant**
- Xylene (mixture of isomers) 1.25 L — **flammable**
- Chloroform, 1.7 L — **highly toxic**
- Acetone, 1 L — **flammable**
- 1 M HCl, 800 mL — **corrosive, irritant**
- Magnesium sulfate, anhydrous, for drying

1. Assemble the three-necked flask, the condenser, the gas inlet adaptor, the mechanical stirrer and a Dean–Stark trap on to the heating mantle.

2. Purge the flask with nitrogen and maintain a gentle gas flow throughout the reaction period in order to facilitate the removal of water.

3. Load the reaction flask with 100 g of 4-*t*-butylphenol, 135 mL of formaldehyde solution and 15.0 g of potassium hydroxide pellets. Start stirring and heat to 80–100°C.

4. Continue to stir and heat until the reaction mixture becomes a foamy yellow solid and stirring is no longer possible (usually 1–2 h). Then stop stirring.

5. Immediately add 1 L of xylene in order to avoid overheating and decomposition. As the solid dissolves, start to stir again.

6. Continue the heating at 100–10°C until all the water has been collected in the trap.[a] Only at this point increase the temperature of the heating mantle until the solution begins to reflux. Maintain a gentle reflux for 3 h.

7. Allow the reaction mixture to cool to room temperature, and then filter the white precipitate on a Buchner funnel and wash it with 250 mL of xylene.

8. Place the solid in a conical flask, add 1.5 L of chloroform[b] and 0.8 L of 1 M HCl and stir[c] the biphasic mixture for 20 min.

9. Transfer the mixture into a separating funnel, separate the lower chloroform layer and extract the water with another portion of chloroform (200 mL). Combine the organic extracts, wash them with water (2 × 200 mL) and dry over magnesium sulfate.

10. Filter off the magnesium sulfate, collect the chloroform solution into a round bottomed flask and concentrate it to *c.* 1 L.

7: Calixarenes

11. Add 1 L of hot acetone and allow this solution to cool. A white precipitate is formed which is filtered on a Buchner funnel to give p-t-butylcalix[6]arene **2** (m.p. 372–4°C,[d] 90–5 g, 88–90% yield).[e]

[a] Incomplete removal of water at this temperature results in the formation of a consistent amount of cyclic octamer, which is very difficult to separate from the desired calix[6]arene.
[b] The p-t-butylcalix[6]arene is not completely soluble at this concentration.
[c] After stirring the solution turns yellow to light orange.
[d] Determined in a sealed capillary.
[e] Due to very high melting point and low solubility, an accurate TLC analysis of the product is important to verify its purity. Eluants and retention factors for p-t-butylcalix[4]-, [6]- and [8]-arene are, respectively: petroleum ether/dichloromethane 1/1: 0.60, 0.70, 0.80; tetrachloroethylene: 0.32, 0.12, 0.08 (after one run); 0.67, 0.32, 0.26 (after three runs).

When the reaction is carried out under basic conditions with NaOH and paraformaldehyde in refluxing xylene, the cyclic octamer is isolated in good yield (Scheme 7.3).[14]

Protocol 3.
Synthesis of 5,11,17,23,29,35,41,47-octakis(1,1-dimethylethyl)-49,50, 51,52,53,54,55,56-octahydroxycalix[8]arene (Structure 3, Scheme 7.3)

Caution! Carry out all procedures in a well-ventilated hood, and wear disposable vinyl or latex gloves and chemical-resistant safety goggles.

Equipment

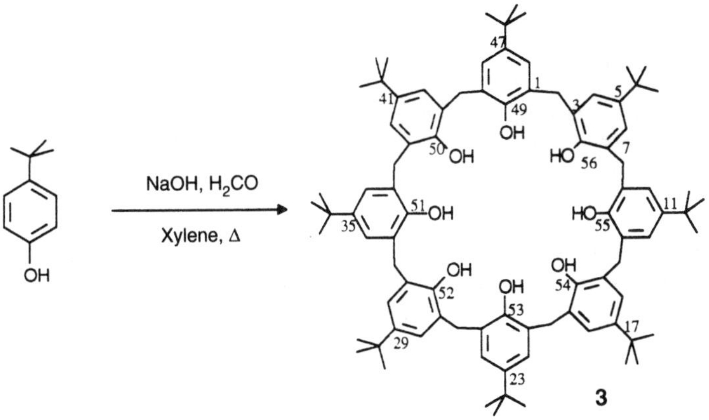

Scheme 7.3

Materials
- Mechanical stirrer
- Heating mantle
- Three-necked, round-bottomed flask (2 L)
- Single-necked, round-bottomed flask (1 L)
- Buchner funnel
- Gas inlet adaptor
- Source of dry nitrogen
- Separating funnel
- Allihn condenser
- Mortar and pestle

Protocol 3. Continued

Materials

• p-t-Butylphenol (FW 150.22), 100 g, 0.66 mol	irritant
• Paraformaldehyde (FW 30.03), 35 g, 1.16 mol	toxic, irritant
• 10 M Sodium hydroxide, 2 mL, 0.02 mol	corrosive, highly irritant
• Xylene (mixture of isomers), 600 mL	flammable
• Chloroform, 1.2 L	highly toxic
• Toluene, 400 mL	flammable
• Diethyl ether, 400 mL	highly flammable
• Acetone, 400 mL	flammable

1. Position the three-necked flask on the heating mantle and equip it with the condenser, the gas inlet adaptor and the mechanical stirrer.
2. Purge the reaction vessel with nitrogen and load it with 100 g of p-t-butylphenol, 35 g of paraformaldehyde, 2 mL of a 10 M sodium hydroxide solution and 600 mL of xylene.
3. Heat to reflux for 4 h,[a] and then allow to cool to room temperature.
4. Collect the precipitate on a Buchner funnel and wash it, in sequence, respectively with 400 mL portions of toluene, ethyl ether, acetone and water.
5. Break the solid up with a mortar and pestle and transfer it into a flask and leave it under reduced pressure for 24 h.
6. Heat the product with 1.2 L of chloroform and after cooling to room temperature, filter the solid on a Buchner funnel and collect product 3 (m.p. 418–20°C,[b] 67–70 g, 62–5%).

[a] After 30 min the reaction mixture becomes homogeneous and after 1 h a white precipitate starts to form.
[b] Measured in a sealed capillary; variations of this melting point to lower values are due to the presence of included solvents.

2.2 Acid-catalysed one-step synthesis

The acid-catalysed reaction of resorcinol with aldehydes leads to cyclic tetramers which are grouped in the class of resorc[4]arenes (*exo*-calixarenes).[3] This reaction is quite general with respect to the aldehyde and to the starting resorcinol, which can bear several substituents in the 2-position.[15] Niederl and Högberg showed in a typical example, the synthetic procedure by which an ethanol solution of resorcinol, acetaldehyde and 37% hydrochloric acid maintained at 80°C for 16 h, afforded 75% of the all *cis* C-methylresorc[4]arene. A shorter reaction time leads to the 'kinetic' stereoisomer (*trans,cis,trans,cis*).[16] This procedure is general for many aldehydes (Scheme 7.4), but the use of formaldehyde is limited to some special cases.[17]

7: *Calixarenes*

Protocol 4.
Synthesis of 2,c-8,c-14,c-20-tetraundecyl-4,6,10,12,16,18,22,24-octahydroxyresorc[4]arene (all *cis* isomer) (Structure 4, Scheme 7.4)

Caution! Carry out all procedures in a well-ventilated hood, and wear disposable vinyl or latex gloves and chemical-resistant safety goggles.

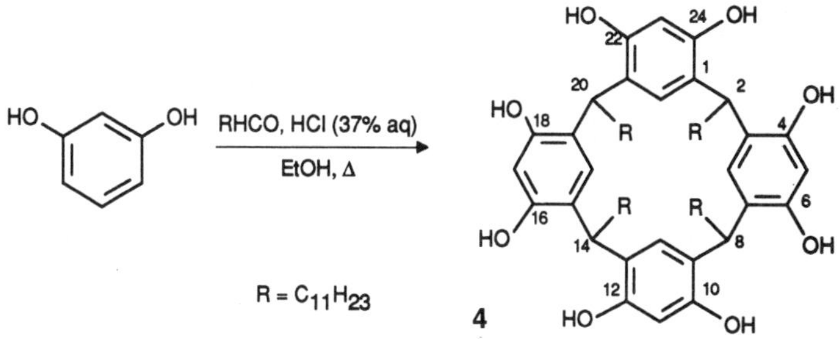

Scheme 7.4

This procedure is representative of the preparation of several C-alkylresorc[4]arenes.[18]

Equipment

- Single-necked, round-bottomed flask (250 mL)
- Hot plate stirrer
- Teflon-coated magnetic stirring bar
- Oil bath
- Ice bath
- Allihn condenser
- Dropping funnel (50 mL)
- Buchner funnel
- Filter flask

Materials

- Resorcinol (FW 110.11),[a] 18.06 g, 0.164 mol — irritant
- Dodecyl aldehyde (FW 184.32),[b] 30.23 g, 0.164 mol — irritant
- 37% HCl, 20 mL — corrosive, highly irritant
- 95% ethyl alcohol, 60 mL — flammable
- Methanol, c. 200 mL — flammable, poisonous
- pH Indicator paper

1. Clamp the reaction flask to the hot plate stirrer and equip it with the magnetic stirrer bar.
2. Load the vessel with 60 mL of ethanol, 20 mL of concentrated HCl and resorcinol and stir until complete dissolution.
3. Dip the reaction flask into an ice bath and cool the homogeneous mixture to 0–5 °C.
4. Through a dropping funnel add the dodecyl aldehyde slowly over 1 h (0.6 mL/min) maintaining the temperature below 5 °C; at the end of the addition a heterogeneous mixture should be obtained.

Protocol 4. *Continued*

5. Remove the cooling bath, drain the external surface of the reaction flask from water and dip it into the oil bath.
6. Remove the dropping funnel, position the Allihn condenser and start to heat slowly; the precipitate should rapidly dissolve.
7. Reflux the reaction for 8 h and then cool to room temperature.
8. Filter the yellow/orange precipitate over a Buchner funnel and wash it with water up to neutrality.[c]
9. Recrystallize the solid from hot methanol (*c.* 200 mL), filter and dry the colourless crystals[d] of compound **4** (30.0 g, yield 75%, m.p. 300–1 °C).

[a] A.C.S. reagent grade.
[b] Freshly distilled.
[c] Test the pH of the filtrate with pH indicator papers.
[d] Store the pure compound in the dark under nitrogen.

3. Functionalization at the lower rim (phenolic OH groups)

The functionalization of calixarenes at the lower rim can be either partial (selective) or complete and can be carried out with a large variety of reagents.[4]

3.1 Selective alkylation at the lower rim

The partial and selective functionalization allows the design and synthesis of a large variety of receptors for ions and neutral molecules bearing mixed functionalities. Especially in calix[4]arenes, the significant differences in the pK_a values of the phenolic OH groups allow their stepwise deprotonation and hence their selective functionalization.

3.1.1 Preparation of 1,3-dialkoxycalix[4]arenes

Distal or 1,3-dialkylation at the lower rim of calix[4]arenes can be easily and nearly quantitatively achieved.[19] The reaction is tolerant of different R groups and has been widely used in calixarene chemistry. It uses a stoichiometric amount of a mild base such as potassium carbonate and alkyl halide or tosylate (Scheme 7.5).

Protocol 5.
Synthesis of 5,11,17,23-tetrakis(1,1-dimethylethyl)-25,27-dihydroxy-26, 28-dimethoxycalix[4]arene (Structure 5, Scheme 7.5)

Caution! Carry out all procedures in a well-ventilated hood, and wear disposable vinyl and latex gloves and chemical-resistant safety goggles. Alkylating reagents are highly toxic.

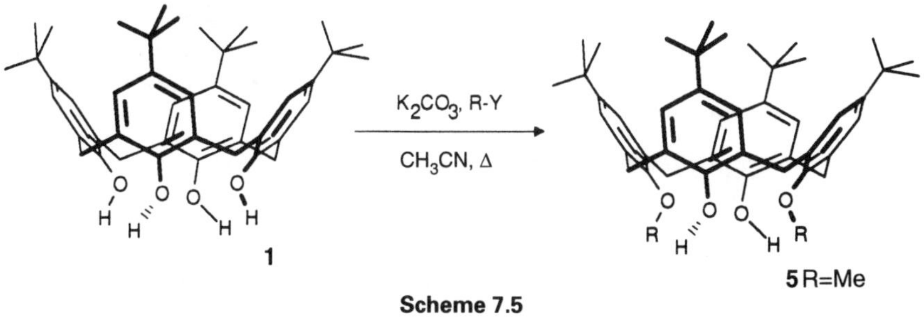

Scheme 7.5

The following procedure is representative of the di-alkylation of several *para*-substituted calix[4]arenes.

Equipment

- Hot plate stirrer
- Oil bath
- Two-necked, round-bottomed flask (500 mL)
- Calcium chloride drying tube
- Source of dry nitrogen
- Gas inlet adaptor
- Teflon-coated magnetic stirring bar
- Separating funnel (500 mL)
- Allihn condenser

Materials

- *p-t*-Butylcalix[4]arene·toluene[a] 1 (FW 741.0), 10.0 g, 13.5 mmol
- Dry acetonitrile,[b] 200 mL — flammable, irritant
- Anhydrous potassium carbonate (FW 138.2), 2.05 g, 14.85 mmol
- Methyl iodide (FW 141.9), 1.70 mL, 27.0 mmol — highly toxic, harmful by inhalation
- Dichloromethane, 210 mL
- Sodium sulfate, anhydrous, for drying
- Methanol, 40 mL — flammable, poisonous
- 10% (v/v) HCl — corrosive

1. Clean all the glassware and dry it for 1 h in a 120°C electric oven.
2. Support the flask on a stirrer with a clamp and assemble the condenser on the main neck. Let a small nitrogen flux pass through the glassware until it returns to room temperature. Then insert a calcium chloride drying tube on top of the condenser.[c]
3. Load the reaction flask with *p-t*-butylcalix[4]arene (10.0 g, 13.5 mmol), acetonitrile (200 mL), potassium carbonate (2.05 g, 14.85 mmol) and methyl iodide (1.70 mL, 27.0 mmol).

Protocol 5. *Continued*

4. Dip the bottom of the flask into the oil bath and heat and stir the mixture under reflux for 48 h.

5. The progress of the reaction is monitored by TLC. From the flask take a sample of 0.5 mL of the solution, remove the solvent on the rotary evaporator, add 3 mL of dichloromethane and 5 mL of 10% HCl. TLC (SiO_2:[d] hexane/ethyl acetate 4/1) of the organic phase in comparison with a sample of compound **1** should reveal, upon irradiation (UV lamp, 254 nm), the complete disappearance of the starting calixarene and the presence of a new spot, which colours red upon spraying with a 50% (w/w) aqueous solution of iron trichloride and heating.

6. After the completion of the reaction, distil off the acetonitrile under reduced pressure on a rotary evaporator. Take up the residue in 200 mL of 10% HCl[e] and 200 mL of dichloromethane and transfer to a separating funnel.[f]

7. Separate the organic phase and wash it twice with distilled water (2 × 200 mL).

8. Separate the organic phase and dry it over anhydrous sodium sulfate. Filter it through a fluted filter paper. Concentrate the filtrate under reduced pressure using a rotary evaporator to a volume of approximatively 15 mL.

9. Add 40 mL of methanol and after evaporation of the dichloromethane collect over a Buchner funnel white crystals of 1,3-dimethoxy-*p*-*t*-butylcalix[4]arene (**5**: R = CH_3) (m.p. 255–6 °C, 8.4 g, 96%).

[a] If *p*-*t*-butylcalix[4]arene **1** is used as the 'toluene-free' ligand, 8.76 g of macrocycle are needed.
[b] A.C.S. reagent grade acetonitrile was dried by standing over molecular sieves (3 Å) for at least 3 h.
[c] *Caution!* Because of the stoichiometric amount of alkylating agent used, the reaction should be carried out in a flask protected from moisture by a $CaCl_2$ drying tube.
[d] Merck, silica gel 60 F_{254}.
[e] The addition of 10% HCl may result in a violent evolution of CO_2.
[f] Due to the possibility of CO_2 evolution vent the separating funnel frequently during extractions.

3.1.2 Preparation of 1,3,5-trialkyloxy-*p*-*t*-butylcalix[6]arenes

So far only the *p*-*t*-butylcalix[6]arene **2** has been symmetrically alkylated in the 1,3,5-positions, using K_2CO_3 as a base and relatively few alkylating agents.[20,21] The reaction is not very selective and a mixture of mono- and polyalkylated products are obtained from which 1,3,5-trialkoxy-*p*-*t*-butylcalix[6]arenes may be isolated by flash chromatography. The yields are modest and range between 15 and 27% (Scheme 7.6).

7: Calixarenes

Protocol 6.
Synthesis of 5,11,17,23,29,35-hexakis(1,1-dimethylethyl)-37,39,41-trihydroxy-38,40,42-trimethoxycalix[6]arene (Structure 6 (R = CH$_3$), Scheme 7.6)

Caution! Carry out all procedures in a well-ventilated hood, and wear disposable vinyl or latex gloves and chemical-resistant safety goggles. Alkylating reagents are highly toxic.

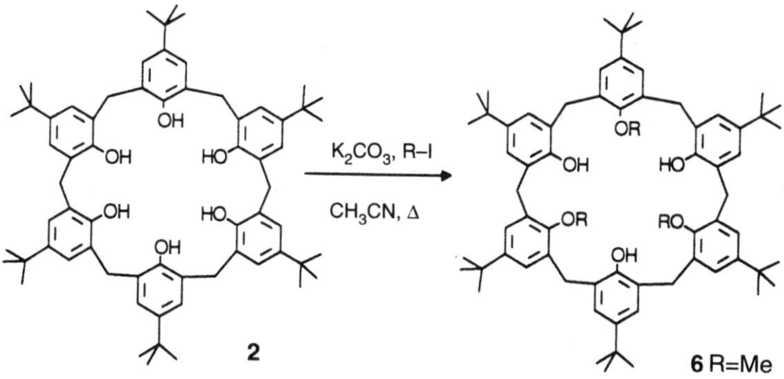

Scheme 7.6

Equipment

- Hot plate stirrer
- Oil bath
- Two-necked, round-bottomed flask (250 mL)
- Calcium chloride drying tube
- Teflon-coated magnetic stirring bar
- Separating funnel (500 mL)
- Allihn condenser
- Gas inlet adaptor
- Dry nitrogen source
- Column for chromatography
- Single-necked, round bottomed flask (100 mL)

Materials

- p-t-Butylcalix[6]arene, **2** (FW 973.3), 3.0 g, 3.1 mmol
- Dry acetone,[a] 150 mL flammable
- Anhydrous potassium carbonate (FW 138.2), 1.28 g, 9.3 mmol
- Methyl iodide (FW 141.9), 1.70 mL, 12.4 mmol highly toxic
- Dichloromethane, for extraction, c. 70 mL harmful by inhalation
- 2 M HCl, 70 mL corrosive
- Sodium sulfate, anhydrous, for drying
- Methanol, for crystallization, c. 20 mL flammable, poisonous
- Silica gel (32–63 μm, 60 Å)
- Hexane, c. 1.5 L flammable, irritant
- Tetrahydrofuran, c. 1.5 L flammable, irritant

1. Clean all glassware and dry it for 1 h in a 120 °C electric oven.
2. Support the hot flask on a stirrer using a clamp and insert the condenser on the main neck. Let a small nitrogen flux pass through the glassware, connecting the top of the condenser with the dry nitrogen source.[b]

Protocol 6. *Continued*

3. Put the calixarene 2 into the flask and add dry acetone (150 mL), potassium carbonate (1.28 g) and methyl iodide (1.70 mL).

4. Dip the bottom of the assembled flask into the oil bath, turn the stirring and the heating on and boil the mixture under reflux for 18 h under a nitrogen atmosphere.

5. Remove the acetone under reduced pressure using a rotary evaporator, quench the residue[c] with 2 M HCl (70 mL) and add dichloromethane (70 mL).

6. Transfer the two phases into a separating funnel,[d] separate the lower organic phase and wash it twice with distilled water (2 × 100 mL).

7. Collect the organic phase and dry it over anhydrous sodium sulfate. Filter it through a filter paper into a round-bottomed flask and remove the solvent on a rotary evaporator.

8. Apply the residue to a silica gel column (5.0 cm × 40 cm filled with 300 g of silica) and elute with hexane:THF[e] = 9:1 collecting the first product coming out (R_f = 0.40) to yield a white powder.

9. Recrystallize the product by slow evaporation from CH_2Cl_2/CH_3OH (1:5) to give white crystals of product **6** (R = CH_3) (m.p. 273–4 °C,[20] 0.86 g, 27%).

[a] A.C.S. reagent grade acetone was dried by standing over molecular sieves (3 Å) for at least 3 h.
[b] *Caution!* Because of the stoichiometric amount of alkylating agent used, the reaction should be carried out under a nitrogen atmosphere.
[c] The addition of 10% HCl may result in a brisk evolution of CO_2.
[d] Due to the possibility of CO_2 evolution vent the separating funnel frequently during extractions.
[e] THF was distilled in order to remove BHT (2,6-di-*t*-butyl-*p*-cresol) which is used as a stabilizing agent. (Caution! Peroxidation may occur.)

3.2 Exhaustive alkylation at the lower rim

The introduction of four alkyl groups, bulkier than ethyl, at the lower rim of calix[4]arenes locks the macrocycle in one of the four possible stereoisomeric configurations: cone, partial cone, 1,3-alternate, 1,2-alternate (Fig. 7.2).

The stereochemical outcome of the reaction is highly dependent on the nature of the base, countercation, solvent and temperature.[22] Due to the larger dimension of the annulus, in general the complete functionalization of calix[6]- and calix[8]arenes at the lower rim leads to conformationally mobile derivatives.

3.2.1 Preparation of tetraalkoxycalix[4]arenes in the cone structure

The conditions required to obtain only the cone isomer involve the use of NaH in DMF at room temperature (Scheme 7.7).

7: *Calixarenes*

Cone

Partial cone

1,2-Alternate

1,3-Alternate

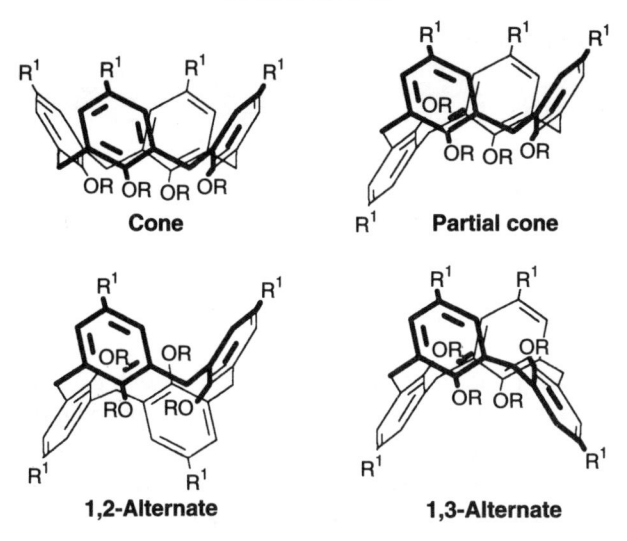

Fig. 7.2 The four possible isomers of tetraalkoxycalix[4]arenes.

Protocol 7.
Synthesis of 25,26,27,28-tetrapropyloxycalix[4]arene (Structure 8 (cone isomer) (R = *n*-Pr), Scheme 7.7)

Caution! Carry out all procedures in a well-ventilated hood, and wear disposable vinyl or latex gloves and chemical-resistant safety goggles. Alkylating reagents are highly toxic.

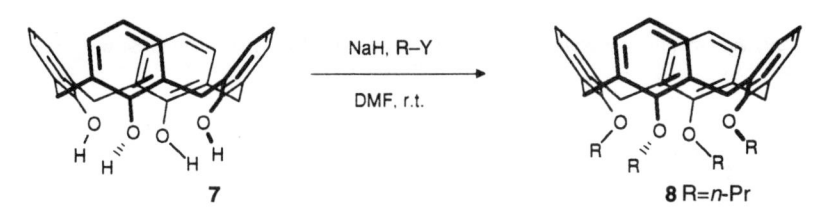

Scheme 7.7

The following procedure is representative of the synthesis of several calix[4]arenes and alkyl halides.

Equipment

- Hot plate stirrer
- Oil bath
- Double-necked, round-bottomed flask (250 mL)
- Teflon-coated magnetic stirring bar
- Single-necked, round-bottomed flask (1 L)
- Gas inlet adaptor
- Source of dry nitrogen
- Separating funnel (500 mL)
- Buchner funnel

Protocol 7. Continued

Materials
- Calix[4]arene, 7 (FW 424.5), 4.0 g, 9.4 mmol
- Dry N,N-dimethylformamide, 80 mL — toxic
- Sodium hydride (FW 24, 60% in oil), 3.9 g, 98.0 mmol — flammable, harmful
- 1-Iodopropane (FW 169.92), 9.2 mL, 94.4 mmol — highly toxic
- Petroleum ether (40–60°), 30 mL — highly flammable
- Methanol, 40 mL — flammable, poisonous
- 2 M HCl, 15 mL — corrosive
- Brine
- Magnesium sulfate, anhydrous, for drying

1. Support the dry round-bottomed flask on a stirrer with a clamp and connect one neck with the dry nitrogen source using a gas inlet adaptor.
2. Put the sodium hydride (3.9 g) in the flask containing the magnetic stirring bar and add 30 mL of 40–60° petroleum ether. Stir this slurry for 1 min, then leave the solid to deposit and with a pipette remove all the solvent.[a]
3. Repeat step 2 and then allow nitrogen to gently bubble through the flask until most of the solvent evaporates.
4. Add DMF, 80 mL, the calix[4]arene (4.0 g) (*caution!* due to evolution of hydrogen, add this compound slowly) and 1-iodopropane (9.2 mL).
5. After 12 h, slowly add 150 mL of 2 M HCl (*caution!*), filter the precipitate on a Buchner funnel and wash it with 150 mL of distilled water.
6. Transfer this solid into a 100 mL flask, add methanol (30 mL) and gently heat at 60°C.
7. Place the flask in a refrigerator and after 2–3 h filter the resulting solid on a Buchner funnel and recover product **8** (m.p. 191–3°C, 5.40 g, 96% yield).

[a] *Caution!* The removed petroleum ether can contain some NaH, which should be quenched with methanol.

3.2.2 Preparation of tetraalkoxycalix[4]arenes in the 1,3-alternate structure

When alkylation is carried out using an excess of Cs_2CO_3 as a base in DMF at or above 80°C or in refluxing acetonitrile, the 1,3-alternate isomer is mainly formed (Scheme 7.8).[23]

7: *Calixarenes*

Protocol 8.
Synthesis of 25,26,27,28-tetrapropyloxycalix[4]arene (Structure 9 (1,3-alternate isomer) (R = *n*-Pr), Scheme 7.8)

Caution! Carry out all procedures in a well-ventilated hood, and wear disposable vinyl or latex gloves and chemical-resistant safety goggles. Alkylating reagents are highly toxic.

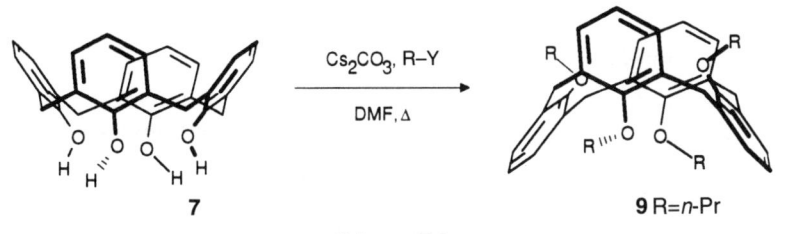

Scheme 7.8

7 → 9 R=*n*-Pr (Cs$_2$CO$_3$, R–Y, DMF, Δ)

The following procedure is representative of the alkylation of several calix[4]arenes.

Equipment
- Hot plate stirrer
- Oil bath
- Two-necked, round-bottomed flask (50 mL)
- Single-necked, round-bottomed flask (100 mL)
- Teflon-coated magnetic stirring bar
- Gas inlet adaptor
- Source of dry nitrogen
- Separating funnel (500 mL)
- Buchner funnel
- Allihn condenser

Materials
- Calix[4]arene 7 (FW 424.5), 0.5 g, 1.18 mmol
- Dry *N,N*-dimethylformamide,[a] 20 mL — toxic
- Anhydrous caesium carbonate (FW 325.8), 5.80 g, 17.7 mmol
- *n*-Propyl-*p*-toluenesulfonate (FW 214.27), 7.58 g, 35.4 mmol — toxic
- Dichloromethane, for extraction, 200 mL — harmful by inhalation
- Methanol, 10 mL — flammable, poisonous
- Brine
- 1 M HCl 100 mL — corrosive
- Potassium iodide, 1.0 g
- Triethylamine, 1 mL — flammable, irritant
- Acetonitrile, 30 mL — flammable, irritant

1. Dry all the glassware in an oven. Assemble the two-necked flask, the condenser, gas inlet adaptor, stirring bar and stopper and allow them to cool to room temperature under N$_2$.

2. Pour 20 mL of dry *N,N*-dimethylformamide, the calix[4]arene (0.5 g) and caesium carbonate (5.80 g) in the assembled flask and heat at 80°C for 30 min.

Protocol 8. Continued

3. Add propyl *p*-toluenesulfonate (7.58 g) and stir at 80°C for 7 h.
4. After cooling slowly pour the reaction mixture into the separating funnel and add 200 mL of water.
5. Extract with dichloromethane (3 × 50 mL), combine the lower organic layers and wash them with 1 M HCl (50 mL) and brine (3 × 50 mL).
6. Remove dichloromethane under reduced pressure in the single-necked round-bottomed flask.
7. Suspend the residue in acetonitrile (30 mL), and add triethylamine (1 mL) and potassium iodide (about 1 g) and heat under reflux for 1 h.
8. After removal of the solvent take up the residue in dichloromethane (50 mL) and pour it into a separating funnel with 1 M HCl (50 mL). Separate the organic layer and wash it with water (2 × 50 mL).
9. Remove dichloromethane under reduced pressure, triturate the residue with cold methanol (*c.* 20 mL) and filter the precipitate on a Buchner funnel, collecting a mixture of tetrapropyloxy isomers (90% of which is the 1,3-alternate isomer).
10. Recrystallize the 1,3-alternate compound **9** from CH_2Cl_2/CH_3OH (m.p. 249–51°C,[23] 0.31 g, 45%).

3.2.3 Preparation of hexaalkoxycalix[6]arenes

The exhaustive alkylation of calix[6]arenes at the lower rim to introduce different groups (alkyl, CH_2COX, CH_2CONX_2, CH_2X, CH_2Ph, *etc.*) can be achieved (yields 50–95%) using several reaction conditions (NaH in THF/DMF, $BaO/Ba(OH)_2$ in DMF, K_2CO_3 in acetone or acetonitrile).[24] The use of K_2CO_3 in dry acetonitrile is an efficient method which, however, requires more reactive electrophiles (Scheme 7.9).

7: *Calixarenes*

Protocol 9.
Synthesis of 5,11,17,23,29,35-hexakis(1,1-dimethylethyl)-37, 38,39,40,41,42-hexakis[(N,N-diethylaminocarbonyl) methoxy]calix[4]arene (Structure 10 [R = CH$_2$CON(C$_2$H$_5$)$_2$], Scheme 7.9)

Caution! Carry out all procedures in a well-ventilated hood, and wear disposable vinyl or latex gloves and chemical-resistant goggles. Alkylating agents are highly toxic.

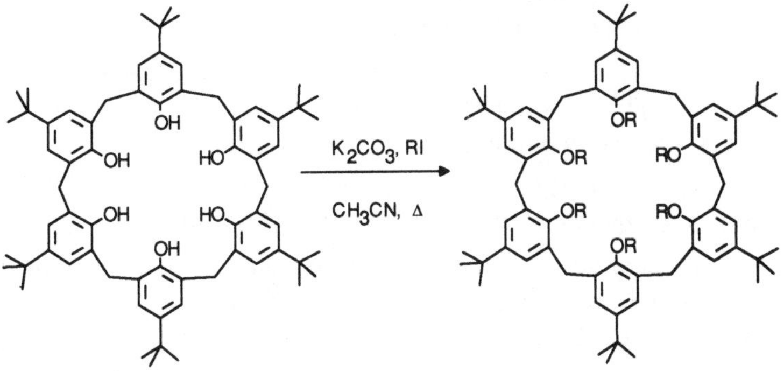

Scheme 7.9

The following procedure is representative of the alkylation of several calix[6]arenes.

Equipment

- Hot plate stirrer
- Oil bath
- Two-necked, round-bottomed flask (100 mL)
- Calcium chloride drying tube
- Buchner funnel
- Allihn condenser
- Separating funnel (250 mL)
- Teflon-coated magnetic stirring bar
- Single-necked, round-bottomed flask (100 mL)

Materials

- p-t-Butylcalix[6]arene 2 (FW 973.3), 1.0 g, 1.03 mmol
- Dry acetonitrile,[a] 50 mL — flammable, irritant
- Anhydrous potassium carbonate (FW 138.2), 1.71 g, 12.3 mmol
- 2-Chloro-N,N-diethylacetamide (FW 149.5), 1.84 g, 12.3 mmol — toxic
- Potassium iodide[b] (FW 166.01), 2.04 g, 12.3 mmol — irritant
- Dichloromethane, for extraction, 50 mL — harmful by inhalation
- 2 M HCl, 50 mL — corrosive
- Magnesium sulfate, anhydrous, for drying
- Methanol for crystallization, 20 mL — flammable, poisonous

1. Clean all the glassware and dry it for 1 h in a 120°C electric oven.
2. Support the flask on a stirrer using a clamp and assemble the condenser. Place the calcium chloride drying tube on the top of the condenser.

Protocol 9. Continued

3. Add acetonitrile (50 mL) and 2-chloro-*N,N*-diethylacetamide (1.84 g) to the flask and start stirring and heating at reflux.
4. Add potassium iodide (2.04 g) to the reaction mixture and wait for 45 min.[b]
5. Add *p-t*-butylcalix[6]arene **2** (1.0 g) and continue reflux for 18–24 h.
6. Remove acetonitrile under vacuum, quench the residue (*caution!*)[c] with 2 M HCl (50 mL) and dichloromethane (50 mL).
7. Transfer the two phases into a separating funnel,[c] separate the organic phase and wash twice with distilled water (2 × 70 mL).
8. Collect the organic phase and dry it over anhydrous magnesium sulfate. Filter it through a filter paper into a round-bottomed flask and remove the solvent at a rotary evaporator.
9. Leave the flask with the oily residue under high vacuum (<0.1 mmHg, 3 h) in order to remove most of the unreacted alkyl halide.
10. Add methanol (*c.* 20 mL) and, after cooling, filter the white precipitate on a Buchner funnel and collect the product **10** [R = $CH_2CON(C_2H_5)_2$] (m.p. 238–40°C, 0.90 g, 53%).[20]

[a] A.C.S. reagent grade acetonitrile was dried by standing over molecular sieves (3 Å) for at least 3 h.
[b] The use of potassium iodide in step 4 is necessary only when alkyl chlorides are employed.
[c] Due to the possibilty of CO_2 evolution vent the separating funnel frequently during extractions.

4. Functionalization at the upper rim (aromatic nuclei)

4.1 Preparation of *p-H*calix[*n*]arenes

The most readily available calixarenes are compounds derived from *p-t*-butylphenol, **1–3**. In order to introduce functional groups at the upper rim of the calix, the *t*-butyl group has to be removed, and this can be performed in 70–90% yield by reacting the *p-t*-butylcalixarenes with an excess of anhydrous aluminium trichloride in toluene in the presence of phenol at room temperature (Scheme 7.10). The unsubstituted *p-H*calixarenes thus obtained are versatile intermediates which give access to hosts having a large variety of properties.

7: *Calixarenes*

Protocol 10.
Synthesis of 25,26,27,28-tetrahydroxycalix[4]arene (Structure 7, Scheme 7.10)

Caution! Carry out all procedures in a well-ventilated hood, and wear disposable vinyl or latex gloves and chemical-resistant goggles. HCL is produced in this reaction.

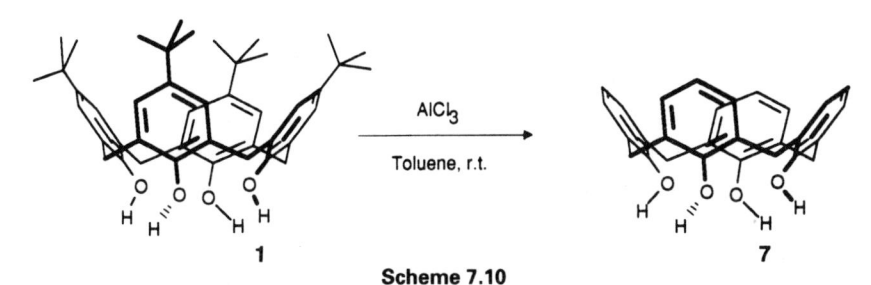

Scheme 7.10

The following de-*t*-butylation procedure is representative of all the calix[*n*]arene series.

Equipment

- Mechanical stirrer
- Gas inlet adaptor
- Dry nitrogen source
- Three-necked, round-bottomed flask (250 mL)
- Beaker (500 mL)

- Vacuum adaptor
- Separating funnel, 1 L
- Buchner funnel
- Single-necked, round-bottomed flask (500 ml)

Materials

- *p-t*-Butylcalix[4]arene (1)·toluene (FW 741.0), 10 g, 13.5 mmol
- Toluene, 100 mL   **flammable**
- Anhydrous aluminium trichloride (FW 133.3), 10.0 g, 75.02 mmol   **corrosive**
- Phenol (FW 94.11), 1.75 g, 18.60 mmol   **highly toxic**
- Dichloromethane, for extraction, 500 mL   **harmful by inhalation**
- Sodium sulfate, anhydrous, for drying
- Calcium chloride, for trap
- Diethyl ether, for precipitation, 150 mL   **highly flammable**
- 1 M HCl, 300 mL   **corrosive**
- Ice
- Methanol, for crystallization, 50 mL   **flammable, poisonous**
- Chloroform, for crystallization, 50 mL   **harmful by inhalation**

1. Dry all glassware in an electric oven at 120°C for 1 h.
2. Clamp the hot flask to a support and allow to cool to room temperature in a nitrogen atmosphere and equip it with a mechanical stirrer.
3. Fill the vacuum adaptor with calcium chloride and position it to a neck of the flask; connect it to the water aspirator (gentle flow) to trap HCl produced during the reaction.

Protocol 10. *Continued*

4. Load the flask with *p-t*-butylcalix[4]arene **1** (10 g), toluene (100 mL) and phenol (1.75 g) and stir for 10 min.

5. Add, with vigorous stirring, to the colourless heterogeneous mixture, aluminium trichloride (10 g) at once and continue to stir the mixture, which rapidly turns deep red. Within 30 min a viscous deep-red sticky phase separates on the walls of the flask.

6. Stir at room temperature for 2–3 h while monitoring reaction progress by TLC.[a]

7. After completion of the reaction pour the reaction mixture into a 500 mL beaker containing 200 g of crushed ice (*caution!* HCl develops).

8. Rinse the reaction vessel with 100 mL of dichloromethane and *c.* 100 g of crushed ice and add this washing to the beaker content.

9. Stir the beaker content with a glass rod, transfer it to a separating funnel and add 400 mL of dichloromethane.

10. Swirl the funnel and separate the organic phase.

11. Wash the organic phase three times with 100 mL portions of 1 M HCl and then twice with water (2 × 100 mL).

12. Separate the organic phase and dry it over anhydrous sodium sulfate (30 min).

13. Filter off the sodium sulfate, collecting the filtrate in a 500 mL round-bottomed flask and distil the solvent under reduced pressure using a rotary evaporator.

14. Add 50 mL of diethyl ether to the oily orange residue, for the precipitation of the calix[4]arene **7**. Allow the heterogeneous mixture to stand at −15°C for 1 h.

15. Collect the solid by suction filtration on a Buchner funnel.

16. Pour the pale-yellow solid into a 100 mL beaker and triturate it with 100 mL of diethyl ether. Allow to stand at −15°C for 1 h and filter on a Buchner funnel. The resulting white powder is the calix[4]arene **7**, m.p. 315–18°C (crystallized from methanol/chloroform) (80%).

[a] Draw a sample of *c.* 0.5 mL from the reaction flask taking care to collect also some viscous phase, add 2 mL of 1 M HCl and 2 mL of dichloromethane and stir vigorously up to complete homogeneity of both phases. Run a TLC of the organic phase using hexane/ethyl acetate 4/1. Upon irradiation with a UV lamp, it should reveal the complete disappearance of the starting material together with the appearance of other spots due to phenols. Spray the TLC plate with a 50% (w/w) aqueous solution of iron(III) chloride hexahydrate and heat the plate with a flameless heat gun. Compound **7** creates a distinctive brown colour.

7: *Calixarenes*

4.2 Preparation of *p*-sulfonatocalix[4]arenes

The introduction of –SO₃H groups at the upper rim of calixarenes can be easily obtained by sulfonation with concentrated H₂SO₄ (Scheme 7.11).[25] This method represents a rapid and efficient way of preparing water-soluble calixarenes.

Recently, other methods[26] for the introduction of –SO₂X (X = OH, Cl) groups at the upper rim of calix[4]arenes ethers which are locked in the cone conformation have also been reported.

Protocol 11.
Synthesis of 25,26,27,28-tetrahydroxy-5,11,17,23-tetrasulfonatocalix[4]arene, sodium salt (Structure 11, Scheme 7.11)

Caution! Carry out all procedures in a well-ventilated hood, and wear disposable vinyl or latex gloves and chemical-resistant safety goggles.

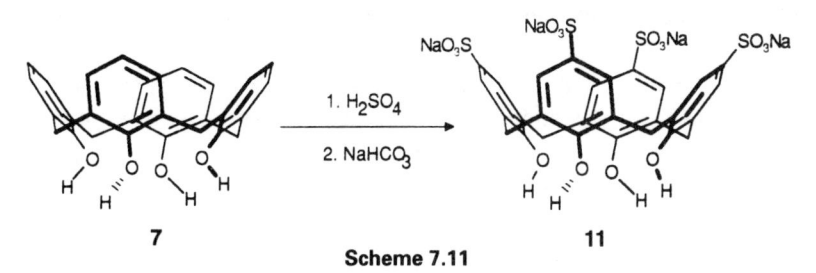

Scheme 7.11

This procedure is also representative of the sulfonation of several calixarene ethers.

Equipment

- Hot plate stirrer
- Oil bath
- Single-necked, round-bottomed flasks (25 mL and 500 mL)
- Slow speed (blue ribbon) filter paper
- Buchner funnel with fibrous glass frit (low porosity)
- Teflon-coated magnetic stirring bar
- Erlenmeyer flask, (100 mL)
- pH indicator paper

Materials

- Calix[4]arene 7 (FW 424.5), 3.0 g, 7.07 mmol
- Sulfuric acid (FW 98, *d* = 1.84, 96%), 10 mL, 180 mmol   corrosive
- Barium carbonate (FW 197.35)
- Sodium carbonate (FW 105.99)
- 0.1 M HNO₃   corrosive
- Methanol, 80 mL   flammable, poisonous
- diluted NaOH aqueous solution

1. Support the 25 mL flask on the magnetic stirrer, add the calix[4]arene 7 (3.0 g, 7.07 mmol), concentrated sulfuric acid (96%, 10 mL) and the magnetic stirring bar.

Protocol 11. *Continued*

2. Dip the flask in the oil bath and heat (60°C) under stirring for 4–5 h.[a]
3. Cool the flask in a refrigerator and filter the white precipitate through the glass frit under vacuum until most of the sulfuric acid has been eliminated.
4. Dissolve the precipitate with water in an Erlenmeyer flask and slowly add, under stirring and gently heat, barium carbonate (BaCO$_3$) and water until the solution is neutral.
5. Filter off the resulting barium sulfate, first through a normal and then through a slow speed (blue ribbon) filter paper, carefully washing the precipitates with hot water.
6. Collect all the aqueous filtrates in a round-bottomed flask (500 mL) and evaporate to dryness under reduced pressure.
7. Take up the residue in water (c. 15–20 mL) and adjust the pH to 8 with Na$_2$CO$_3$.
8. Filter off the resulting carbonates and add methanol (c. 40 mL) to the solution.
9. Vacuum filter on a Buchner funnel the white precipitate formed, which consists of 4–5 g of p-sulfonatocalix[4]arene **11** impure with carbonates.
10. Acidify this product at pH ~3 with HNO$_3$ (0.1 M) and boil this solution in an Erlenmeyer flask (100 mL).
11. Carefully neutralize the solution with a dilute aqueous solution of NaOH and concentrate it under reduced pressure, until a precipitate starts to form.
12. Slowly add methanol (~40 mL) and after cooling, filter the purified product **11** on a Buchner funnel (m.p. >300°C, 3.1 g, 45%).

[a] The reaction is complete when a drop of the reaction mixture poured in 1–2 mL of water gives no water-insoluble material.

4.3 Preparation of *p*-iodocalix[4]arenes

Bromination or iodination of calixarenes can be conveniently performed by direct substitution with the appropriate halogen. The halogenated macrocycles can be used for further functionalization. Because of the higher reactivity of Ar–I bonds, iodination is preferred. (Scheme 7.12).[27]

Protocol 12.
Synthesis of 25,26,27,28-tetrakis(2-ethoxyethoxy)-5,11,17,23-tetraiodocalix[4]arene (Structure 12 (R = C₂H₄OC₂H₅), Scheme 7.12)

Caution! Carry out all procedures in a well-ventilated hood, and wear disposable vinyl or latex gloves and chemical-resistant safety goggles.

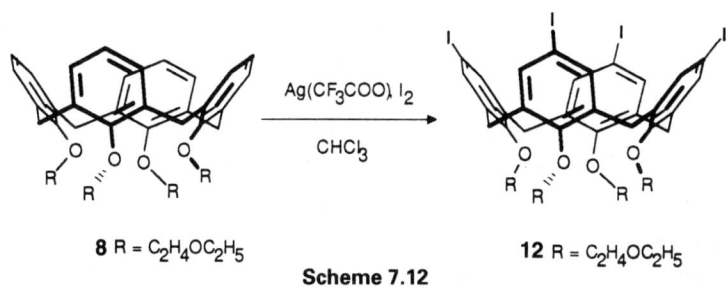

Scheme 7.12

8 R = C₂H₄OC₂H₅ → 12 R = C₂H₄OC₂H₅
Ag(CF₃COO), I₂ / CHCl₃

The starting calixarene 8[a] is obtained (72% yield) following Protocol 7 using 1-bromo-2-ethoxyethane as alkylating reagent.

Equipment

- Two-necked, round-bottomed flask (2 L)
- Teflon-coated magnetic stirring bar
- Hot plate stirrer
- Allihn condenser
- Erlenmeyer flask (3 L)
- Oil bath

Materials

- 25,26,27,28-Tetrakis(2-ethoxyethoxy)calix[4]arene 8 (R = C₂H₄OC₂H₅) in the cone structure (FW 712.92), 10.0 g, 14.0 mmol
- Chloroform,[b] 1.5 L   **highly toxic**
- Silver trifluoroacetate[c] (FW 220.88), 13.60 g, 61 mmol
- Iodine (FW 253.81), 17.76 g, 70 mmol   **highly irritant**
- Sodium metabisulfite 20% (w/w), aqueous solution, 600 mL
- Sodium sulfate, anhydrous, for drying
- Dichloromethane, 40 mL   **harmful by inhalation**
- Methanol, for recrystallization, 250 mL   **flammable, poisonous**

1. Clamp the reaction flask to a support and equip it with an Allihn condenser and the magnetic stirring bar.

2. Through the second neck introduce 1.5 L of chloroform, 10 g of calixarene 8 (R = C₂H₄OC₂H₅) and silver trifluoroacetate (13.60 g) and then close the neck with a ground-glass stopper.

3. Dip the flask in the oil bath and reflux the reaction mixture with vigorous stirring for 1 h.

4. Decrease the temperature to 50°C and add iodine in c. 5–6 g portions, continuing to stir. After the last addition of iodine stir for an additional hour.[d]

Protocol 12. Continued

5. Turn the heating off and allow the heterogeneous violet mixture to cool to room temperature.
6. Filter off the yellow silver iodide precipitate with a fluted filter paper, collecting the filtrate in a 3 L Erlenmeyer flask.
7. Transfer the violet solution to a separating funnel and wash twice with 300 mL portions of a 20% (w/w) sodium metabisulfite solution, swirling the funnel vigorously to reduce the excess of iodine until there is a complete bleaching of the organic phase.
8. Separate the organic phase and dry it over anhydrous sodium sulfate. Filter with a fluted filter paper in a round-bottomed flask.
9. Evaporate the filtrate using a rotary evaporator.
10. Dissolve the residue in the minimum quantity of dichloromethane (30–40 mL) and add 250 mL of methanol.
11. Heat the flask with a flameless heat gun to gentle reflux.
12. Allow to cool to room temperature and collect the white powder of compound **12** (R = $C_2H_4OC_2H_5$) by suction filtration using a Buchner funnel (m.p. 167–9°C, 15.3 g, 90%).

[a] m.p. 115–16°C.[28]
[b] A.C.S. reagent grade stabilized with 0.75% ethanol.
[c] Silver trifluoroacetate (98%) and iodine (99.8%) were purchased from Aldrich Chemical Co. Inc.
[d] The resulting heterogeneous mixture should be deep violet; if not, add more iodine until there is a persistent colour.

4.4 Preparation of *p*-nitrocalix[4]arenes

The introduction of nitro groups at the upper rim of calix[4]arenes is a quick and useful method for the preparation of functionalized calix[4]arenes. This reaction is quite general and can be conveniently carried out on a wide variety of calix[4]arenes substituted at the lower rim. When the starting calixarene is partially substituted at the lower rim, the substitution takes place at the more reactive phenol rings.

Particularly convenient is the reaction carried out on tetraalkoxytetra-*t*-butylcalix[4]arenes in which concomitant removal of the *t*-butyl groups and introduction of four nitro groups occurs (*ipso* nitration) in 37–67% yield (Scheme 7.13).[29]

7: *Calixarenes*

Protocol 13.
Synthesis of 25,26,27,28-tetrakis(2-ethoxyethoxy)-5,11,17,23-tetranitrocalix[4]arene (Structure 14 (R = $C_2H_4OC_2H_5$), Scheme 7.13)

Caution! Carry out all procedures in a well-ventilated hood, and wear chemical-resistant safety goggles. Thick vinyl gloves are recommended when using 100% nitric acid.

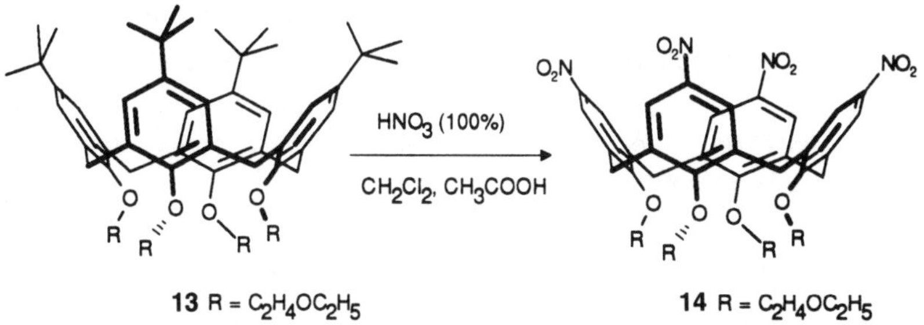

Scheme 7.13

The following procedure is representative of several tetraalkoxy calix[4]arenes. The starting calixarene **13**[a] is obtained (80% yield) following Protocol 7 using *p-t*-butylcalix[4]arene **1** and 1-bromo-2-ethoxyethane as alkylating reagent.

Equipment
- Magnetic stirrer
- Beaker (400 mL)
- Three-necked, round-bottomed flask (100 mL)
- Ice bath
- Calcium chloride drying tube
- Conical flask, 500 mL
- Teflon-coated magnetic stirring bar
- Separating funnel, 500 mL
- Nitrogen inlet adaptor
- Source of dry nitrogen

Materials
- 25,26,27,28-Tetrakis(ethoxyethoxy)-5,11,17,23-tetrakis(1,1-dimethylethyl)calix[4]arene, **13**(R = $C_2H_4OC_2H_5$) in the cone structure (FW 936), 2.81 g, 3.0 mmol
- Dry dichloromethane,[b] 30 mL   **harmful by inhalation**
- Glacial acetic acid, 30 mL   **corrosive, irritant**
- 100% HNO_3 (FW 63.01), 10 mL, 240 mmol   **corrosive**
- Dichloromethane, for extraction, 100 mL   **harmful by inhalation**
- Sodium sulfate, anhydrous, for drying
- Methanol, for recrystallization 100 mL   **flammable, poisonous**

1. Equip the reaction flask with the magnetic stirring bar, the calcium chloride drying tube and the nitrogen inlet adaptor and purge the reaction vessel with a gentle flux of dry nitrogen.

2. Add to the flask 30 mL of dry dichloromethane, the calixarene **13** (R = $C_2H_4OC_2H_5$, (2.13 g) and 30 mL of glacial acetic acid and cool the resulting homogeneous mixture to 0°C.

Protocol 13. *Continued*

3. With a glass syringe, add 10 mL of 100% nitric acid and stir at 0°C until the dark-red colour disappears (about 1 h).
4. Pour the reaction mixture into a beaker containing 200 mL of water (*caution!*).
5. Separate the organic layer and extract the aqueous phase twice with 50 mL portions of dichloromethane.
6. Dry the combined organic phases with anhydrous sodium sulfate.
7. Filter the sodium sulfate off and evaporate the solvent to dryness.
8. Recrystallize the residue with methanol (*c.* 100 mL) and filter the solid by suction filtration. After drying, 2.0 g of tetranitro compound **14** (R = $C_2H_4OC_2H_5$) is isolated (m.p. 174–6 °C, 76%).

[a] m.p. 175–6 °C.[30]
[b] Dichloromethane should be distilled and dried by standing overnight over 3 Å molecular sieves.

References

1. Gutsche, C. D.; Muthukrishnan, R. J. *J. Org. Chem.* **1978**, *43*, 4905–4906.
2. Gutsche, C. D. *Calixarenes*; Monographs in Supramolecular Chemistry, Vol. 1, Stoddart, J. F., ed., The Royal Society of Chemistry: Cambridge, **1989**.
3. Vicens, J.; Böhmer, V. (eds), *Calixarenes: A Versatile Class of Macrocyclic Compounds*; Kluwer Academic Publisher: Dordrecht, **1991**.
4. Gutsche, C. D. *Acc. Chem. Res.* **1983**, *16*, 161–170. Gutsche, C. D. *Top. Curr. Chem.* **1984**, *123*, 1–47. Van Loon, J. D.; Verboom, W.; Reinhoudt, D. N. *Org. Prep. & Procedure Int.* **1992**, *24*, 437–462. Shinkai, S. *Tetrahedron* **1993**, *40*, 8933–8968; Böhmer, V. *Angew. Chem. Int. Ed. Engl.* **1995**, *34*, 713–745.
5. Böhmer, V.; Chhim, P.; Kämmerer, H. *Makromol. Chem.* **1979**, *180*, 2503–2506. Böhmer, V.; Merkel, L.; Kunz, U. *J. Chem. Soc., Chem. Commun.* **1987**, 896–897.
6. Tabatabai, M.; Vogt, W.; Böhmer, V. *Tetrahedron Lett.* **1990**, *31*, 3295–3298. Sartori, G.; Maggi, R.; Bigi, F.; Arduini, A.; Pastorio, A.; Porta, C. *J. Chem. Soc., Perkin Trans. 1*, **1994**, 1657–1658. Grütten, C.; Böhmer, V.; Vogt, W.; Thondorf, I.; Biali, S. E.; Grynszpan, F. *Tetrahedron Lett.* **1994**, *35*, 6267–6270.
7. Knop, A.; Pilato, L. A. *Phenolic Resins*; Springer Verlag: Berlin, **1985**.
8. Ninagawa, A.; Matsuda, H. *Makromol. Chem., Rapid Commun.* **1982**, *3*, 65–67.
9. Steward, D. R.; Gutsche, C. D. *Org. Prep. & Procedure Int.* **1993**, *25*, 137–139.
10. Nakamoto, Y.; Ishida, S. *Makromol. Chem., Rapid Commun.* **1982**, *3*, 705–707.
11. Gutsche, C. D.; Alam, I. unpublished results reported in Ref. 2.
12. Gutsche, C. D.; Iqbal, M. *Org. Synth., Coll. Vol.* **1993**, *8*, 75–77.
13. Gutsche, C. D.; Dhawan, B.; Leonis, M.; Stewart, D. *Org. Synth., Coll. Vol.* **1993**, *8*, 77–79.
14. Munch, S. H.; Gutsche, C. D. *Org. Synth., Coll. Vol.* **1993**, *8*, 80–81.
15. Cometti, J.; Dalcanale, E.; Du Vosel, A.; Levelut, A. M. *J. Chem. Soc., Chem. Commun.* **1990**, 163–165.

7: Calixarenes

16. Högberg, S. A. G. *J. Org. Chem.* **1980**, *45*, 4498–4500.
17. Konishi, H.; Iwasaki, Y.; Okano, T.; Kiji, *J. Chem. Lett.* **1989**, 1815–1816. Konishi, H.; Morikawa, O. *J. Chem. Soc., Chem. Commun.* **1993**, 34–35.
18. Abis, L.; Dalcanale, E.; Du Vosel, A.; Spera, S. *J. Org. Chem.* **1988**, *53*, 5475–5479.
19. van Loon, J. D.; Arduini, A.; Coppi, L.; Verboom, W.; Ungaro, R.; Pochini, A.; Harkema, S.; Reinhoudt, D. N. *J. Org. Chem.* **1990**, *55*, 5639–5646.
20. Casnati, A.; Minari, P.; Pochini, A.; Ungaro, R.; Nijenhuis, W. F.; de Jong, F.; Reinhoudt, D. N. *J. Isr. Chem. Soc.* **1992**, *32*, 79–87.
21. Casnati, A.; Minari, P.; Pochini, A.; Ungaro, R. *J. Chem. Soc., Chem. Commun.* **1991**, 1413–1414. Neri, P.; Consoli, G. M. L.; Cunsolo, F.; Piattelli, M. *Tetrahedron Lett.* **1994**, *35*, 2795–2798.
22. Groenen, L. C.; Ruël; B. H. M.; Casnati, A.; Timmerman, P.; Verboom, W.; Harkema, S.; Pochini, A.; Ungaro, R.; Reinhoudt, D. N. *Tetrahedron Lett.* **1991**, *32*, 2675–2678.
23. Verboom, W.; Datta, S.; Asfari, Z.; Harkema, S.; Reinhoudt, D. N. *J. Org. Chem.* **1992**, *57*, 5394–5398.
24. Andreetti, G. D.; Calestani, G.; Ugozzoli, F.; Arduini, A.; Ghidini, E.; Pochini, A.; Ungaro, R. *J. Incl. Phenom.* **1987**, *5*, 123–126. Arnaud-Neu, F.; Collins, E.; Deasy, M.; Ferguson, G.; Harris, S. J.; Kaitner, B.; Lough, A. J.; McKervey, M. A.; Marques, E.; Ruhl, B. L.; Schwing-Weill, M. J.; Seward, E. M. *J. Am. Chem. Soc.* **1989**, *111*, 8681–8691.
25. Shinkai S.; Araki, K.; Tsubaki, T.; Arimura, T.; Manabe, O. *J. Chem. Soc., Perkin Trans 1* **1987**, 2297–2299. Arena, G.; Cali, R.; Lombardo, G. G.; Casnati, A.; Rizzarelli, E.; Sciotto, D.; Ungaro, R. *Supramol. Chem.* **1992**, *1*, 19–24. Scharff, J. P.; Mahjoubi, M.; Perrin, R. *Nouv. J. Chim.* **1991**, *15*, 883–887.
26. Casnati, A.; Ting, Y.; Berti, D.; Fabbi, M.; Pochini, A.; Ungaro, R.; Sciotto, D.; Lombardo, G. G. *Tetrahedron* **1993**, *49*, 9815–9822. Morzherin, Y.; Rudkevich, D. M.; Verboom, W.; Reinhoudt, D. N. *J. Org. Chem.* **1993**, *58*, 7602–7605.
27. Arduini, A.; Pochini, A.; Sicuri, A. R.; Secchi, A.; Ungaro, R. *Gazz. Chim. Ital.* **1994**, *124*, 129–132.
28. Arduini A.; Casnati, A.; Fabbi, M.; Minari, P.; Pochini, A.; Sicuri, A. R.; Ungaro, R. *Supramol. Chem.* **1993**, *1*, 235–246.
29. Verboom, W.; Durie, A.; Egberink, R. J. M.; Asfari, Z.; Reinhoudt. D. N. *J. Org. Chem.* **1992**, *57*, 1313–1316.
30. Chang, S. K.; Cho, I. *J. Chem. Soc., Perkin Trans. 1* **1986**, 211–214.

8

Spherands, hemispherands and calixspherands

WILLEM VERBOOM and DAVID N. REINHOUDT

1. Introduction

In 1979 Cram[1] reported the synthesis of a novel class of macrocyclic host molecules composed of anisyl units that have rigid preorganized cavities. These so-called spherands, the prototype of which is **1**, form very stable complexes with small alkaline cations (Li$^+$, Na$^+$). The high thermodynamic stabilities ($K_a > 10^{13}$ L mol^{-1}) are due to the rigidity of the molecular framework which creates a high electron density directed in the cavity, generated by the oxygen lone pairs of anisyl moieties.[2] Upon complexation, the electron–electron repulsion of neighbouring oxygen lone pairs is relieved. This is a fundamentally different concept when compared with the complexation of cations by more flexible macrocyclic ligands[3], which require reorganization of the ligand prior to or during complexation, although in the complexation of cations by crown ethers, relief of oxygen–oxygen repulsions may also contribute. The concept of preorganization[4] has been further extended by Cram and his group, for example by variation of the ring size and by substitution of one or more anisyl units by methoxycyclohexyl, urea or phenanthroline[5] subunits.

This preorganization principle not only affects the thermodynamic stability of the complex but also has a pronounced effect on the kinetic stability of the alkali complexes of the spherands. The anisyl groups in the spherands provide the preorganized ligating donor sites, and they shield the cavity from solvent molecules. As a result, the rates of complexation by the spherands are decreased and the rates of decomplexation are decreased even more when compared with related flexible crown ethers.

In addition to the spherands, Cram synthesized compounds in which at least half of the binding sites may be considered to be preorganized, the so-called hemispherands.[6] In the parent molecule **2**, the molecular cavity is composed of a rigid *m*-teranisyl moiety in which the oxygen binding sites are conformationally organized prior to complexation, and a flexible polyether

8: Spherands, hemispherands and calixspherands

chain. The synthesis and binding properties of several hemispherands composed of three or four anisyl units or anisyl and cyclic urea units has been reported.[2,4] Optimal binding for different alkali cations has been achieved by variation of the cavity size from 18 to 23 ring atoms through an enlargement of the conformationally mobile part of the hemispherands.[7] Although in some cases thermodynamically stable complexes are formed, they are not kinetically stable.

Reinhoudt[8-10] has reported the synthesis of spherand-type molecules that have a larger cavity and resemble the spherands in their high kinetic stability. The calixspherands, e.g. 3, are calix[4]arenes (see Chapter 7) which are bridged at the lower rim with a rigid terphenyl moiety. Space-filling molecular models indicate that the size of the cavity in 3 is comparable with the size of the potassium ion. The calixspherand is the first ligand to form kinetically stable complexes with the larger alkali cations potassium and rubidium.

2. Spherands

The host molecule 1, which is the prototype of the spherands, was prepared for the first time by Cram[11] from substituted mono-, di- and trimeric anisyl units. These reactions involve halogen–lithium exchange and subsequently the oxidative coupling of the aryllithiums with iron(III) acetylacetonate [Fe(acac)$_3$] under high-dilution conditions. The highest yield (28%) of spherand 1 was obtained by dimerization of 1,1′:3′,1″-terphenyl 4; in the other reactions both the cyclic hexamer and the octamer were obtained. Reinhoudt has repeated the synthesis of spherand 1 by lithiation of 4 with s-butyllithium in THF at –78 °C followed by oxidation of the diaryllithium compound in benzene. Yields up to 18% (as 1·LiCl) have been obtained (Scheme 8.1). The yields are extremely sensitive to minor impurities, reaction conditions and experimental skill! For instance, use of n-butyllithium gives rise to substantial amounts of alkylated products and this decreases the yield of spherand 1 to about 5%. The methyl group in the *para* positions of spherand 1 serves as a blocking group to prevent undesired substitution reactions at

Table 8.1 Alkyllithium reagent used and yields of spherand–lithium chloride complexes.

Starting compound	Reagent	Product	Yield (%)*
5a	n-BuLi	6a·LiCl	12
5b	t-BuLi	6b·LiCl	17
5c	s-BuLi	6c·LiCl	12

* Averaged values of at least six experiments.

that position during the synthesis of the starting materials. Analogously the synthesis of spherands **6a–c**·LiCl has been carried out in several experiments using either *n*-, *s*- or *t*-butyllithium.[12] The organometallic reagents giving the highest yield of the spherands **6a–c**·LiCl are given in Table 8.1. The described strategy can also be applied for the preparation of spherands containing 1,3-aryl or 2-methoxycyclohexyl units, for example.[2,4]

Protocol 1.
Synthesis of 31,32,33,34,35,36-hexamethoxy-4,9,14,19,24,29-hexamethylheptacyclo[25.3.1.1^{2,6}.1^{7,11}.1^{12,16}.1^{17,21}.1^{22,26}] hexatriaconta-1(31),2,4,6(36),7,9,11(35),12,14,16(34),17,19,21(33),22,24,26(32),27,29-octadecaene·LiCl (Structure 1·LiCl, Scheme 8.1)

Caution! Carry out all procedures in a well-ventilated hood, and wear disposable vinyl or latex gloves and chemical-resistant safety goggles.

The following procedure for the synthesis of spherand **1** by dimerization of 1,1′:3′,1″-terphenyl **5** is representative for this type of spherand ring closure reaction.

Equipment

- Triple-necked, round-bottomed flasks (2 × 250 mL, 5 L)
- Single-necked, round-bottomed flasks (250 mL, 500 mL, 2 L)
- Magnetic stirrer bars
- Addition funnel (100 mL)
- Alcohol thermometer
- Needle
- Reflux condenser
- Argon balloon
- CO$_2$(s)–acetone bath for cooling
- Oil bath
- Magnetic hot plate stirrer
- Rotary evaporator
- Filter funnel
- Fluted filter papers
- 2 septa
- Separating funnel (1 L)

Materials

- 3,3″-Dibromo-2,2′,2″-trimethoxy-5,5′,5″-trimethyl[1,1′:3′,1″-terphenyl] **4**[11] (FW 519), 4.9 g, 9.4 mmol
- Dry THF,[a] 100 mL   flammable
- Fe(acac)$_3$ (FW 353.2), 15 g, 42.6 mmol
- Benzene, 1.7 L   flammable, highly toxic
- *s*-Butyllithium, 1.3 M solution in cyclohexane, 22 mL, 28.6 mmol   pyrophoric, moisture sensitive
- FeCl$_3$·6H$_2$O (FW 270.3), 7.0 g, 25 mmol   hygroscopic, corrosive
- 2 M hydrochloric acid, 600 mL   corrosive, irritant
- Water, 100 mL
- Diethyl ether,[b] 100 mL   flammable
- Dichloromethane, 300 mL   toxic
- 0.2 M LiCl solution saturated with EDTA, 1200 mL
- Deionized water, 600 mL
- Toluene, 150 mL   flammable

All reactions should be carried out in oven-dried glassware (electric oven, 110 °C, 2 h) under a nitrogen or argon atmosphere.

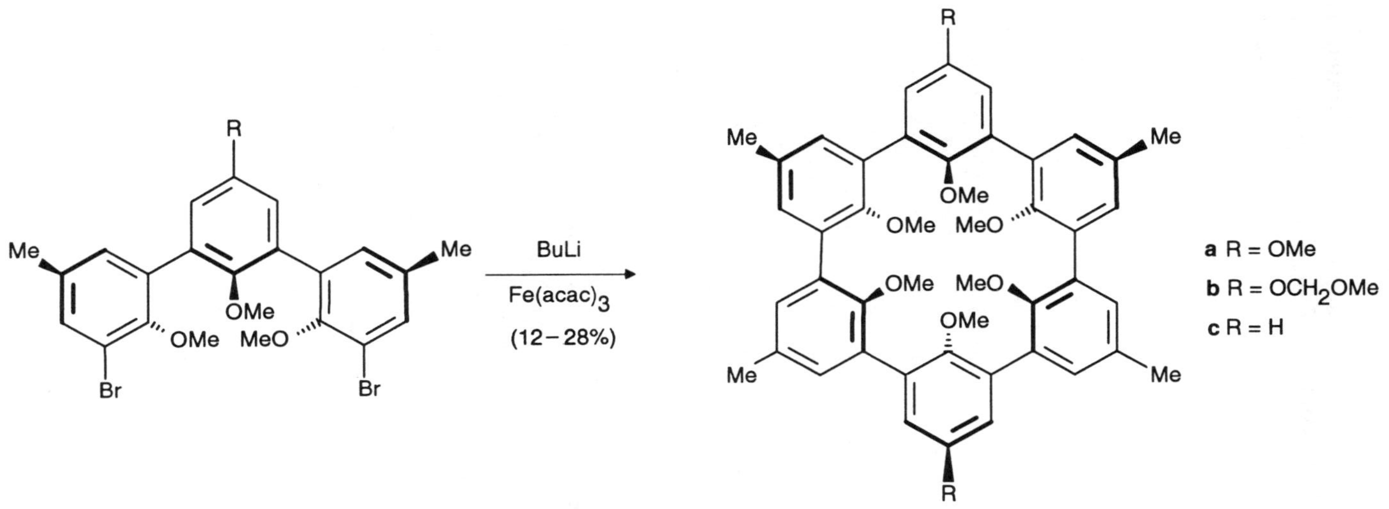

Scheme 8.1

Protocol 1. *Continued*

1. Assemble a 250 mL triple-necked flask, stirrer bar, thermometer, dropping funnel and argon ballon in a solid CO_2–acetone bath (–78°C).

2. Put dibromoterphenyl **4** (4.9 g) in the 250 mL triple-necked flask. Add 100 mL of dry THF and stir (the reaction mixture) at –78°C under argon.

3. Weigh 15 g of dry Fe(acac)$_3$ in a 5 L triple-necked round-bottomed flask equipped with a reflux condenser. Add 1.7 L of dry benzene and reflux while vigorously stirring under argon.

4. Add 22 mL of a 1.3 M solution of *s*-butyllithium in cyclohexane dropwise to the terphenyl solution at –78°C, and stir at that temperature for 10 min. Replace the dropping funnel by a septum.

5. Transfer the cold reaction mixture via a steel cannula through a septum into the vigorously refluxing solution using a small stream of argon. Reflux the resulting mixture for 45 min. The reaction mixture is a suspension of a heavy red precipitate.

6. Allow the mixture to cool. Dissolve 7.0 g of $FeCl_3 \cdot 6H_2O$ in 600 mL of 2 M hydrochloric acid. Add this solution to the reaction mixture and stir for 10 h at 25°C.

7. Remove the organic solvents under reduced pressure by means of a rotary evaporator. Collect the suspension through a filter funnel and a fluted filter paper, and wash with 100 mL of water. Transfer the solid to a 250 mL round-bottomed flask and dry under vacuum at 25°C.

8. Add 100 mL of diethyl ether and concentrate the suspension to 75 mL by boiling under reflux. Filter the suspension through a filter funnel and a fluted filter paper, and dissolve the yellow solid in 300 mL of dichloromethane in a 2 L round-bottomed flask.

9. Stir this solution vigorously successively with two 600 mL portions of 0.2 M LiCl solution saturated with EDTA and finally with 600 mL of deionized water. Separate the layers after each treatment with a separating funnel.

10. Transfer the organic layer to a 500 mL round-bottomed flask and remove the solvent under reduced pressure on a rotary evaporator to a 100 mL volume. Add 100 mL of toluene, and evaporate under reduced pressure on a rotary evaporator until crystals appear (*c.* 100 mL volume).

11. Allow the mixture to cool to 25°C and filter through a filter funnel and a fluted filter paper. Wash the product with 50 mL of cold toluene and dry to yield compound **1**·LiCl as a white solid (1.0 g, 28% yield, m.p. >400°C).

[a] Distil THF from benzophenone ketyl under an inert atmosphere (nitrogen or argon) and use immediately.
[b] Distil diethyl ether from lithium aluminium hydride under an inert atmosphere (nitrogen or argon) and use immediately.

3. Hemispherands

Several strategies have been reported for the preparation of hemispherands. The first example involved the attachment of the poly(ethylene glycol) chain to an appropriately functionalized terphenyl unit. This approach will be illustrated by the synthesis of hemispherand **2**.[6] The starting teranisyl **7** can be easily prepared by oxidation of *p*-cresol with $FeCl_3 \cdot 6H_2O$ followed by methylation with dimethyl sulfate.[6] Dilithiation of **7** with *n*-butyllithium in the presence of TMEDA in diethyl ether followed by treatment with carbon dioxide provides diacid **8a** in 93% yield. Reduction of this diacid with borane–THF gives diol **8b** (96%), which upon subsequent reaction with phosphorus tribromide affords dibromide **8c** in 97% yield. An alternative approach for the preparation of diol **8b** involves treatment of the dilithio compound with excess DMF followed by reduction with sodium borohydride.[13–15] Reaction of dibromide **8c** with diethylene glycol in the presence of sodium hydride as a base under high dilution conditions gives hemispherand **2** in 64% yield (Scheme 8.2).[16] It can also be prepared without high dilution in a yield of 17% and from diol **8b** and diethylene glycol ditoluenesulfonate without high dilution in 28% yield.[6] In principle two diastereoisomeric products can be produced, since the central hole is much too small in any conceivable conformation to allow two methoxyl groups attached to adjacent aryls to pass one another. The ^1H NMR spectrum of **2** proves the *meso* structure; the inner and outer $ArCH_3$ and $ArOCH_3$ possess different chemical shifts, and the $ArCH_2O$ groups exhibit the anticipated pair of AB quartets.

The same procedure can be applied for the preparation of hemispherands with modified terphenyl units (other substituents) and hemispherands in which the central aromatic ring is a pyridine ring.[14,15]

Protocol 2.
Synthesis of 2,6-bis(3-carboxy-2-methoxy-5-methylphenyl)-4-methylanisole (Structure 8a[6], Scheme 8.2)

Caution! Carry out all procedures in a well-ventilated hood, and wear disposable vinyl or latex gloves and chemical-resistant safety goggles.

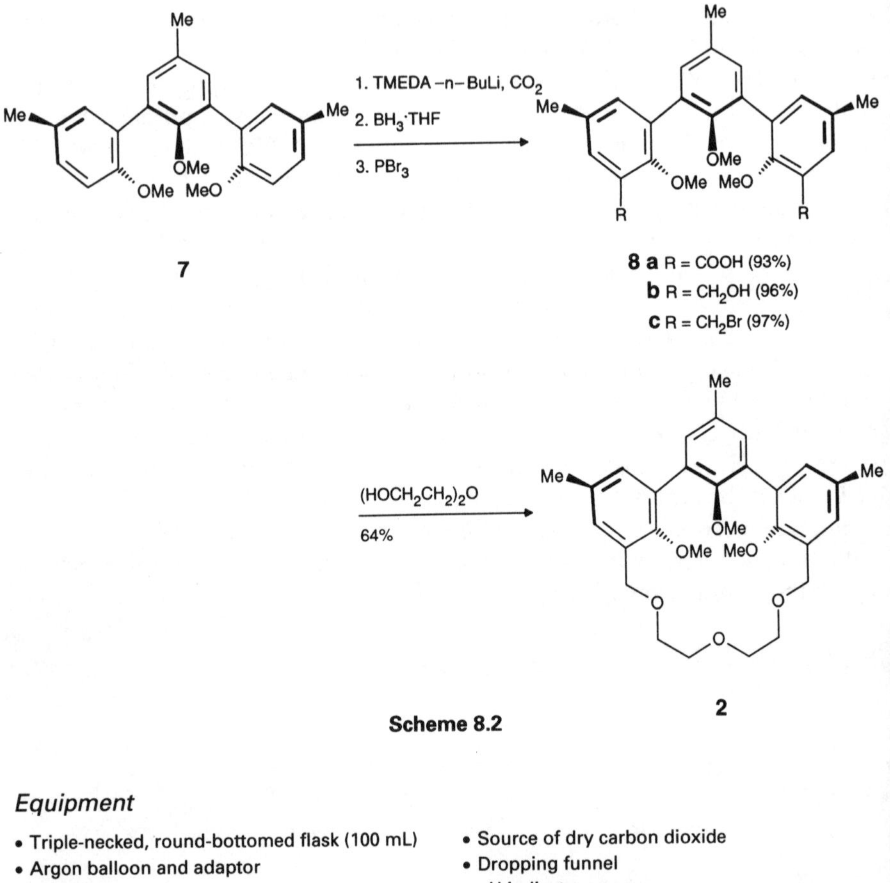

8 a R = COOH (93%)
b R = CH$_2$OH (96%)
c R = CH$_2$Br (97%)

Scheme 8.2

Equipment

- Triple-necked, round-bottomed flask (100 mL)
- Argon balloon and adaptor
- Stopper
- Magnetic hot plate stirrer
- Magnetic stirrer bar
- 2 septa
- Electric heat gun
- Teflon syringe and needle (50 mL)
- All-glass syringe with a needle-lock Luer (2.5 mL) and needle
- All-glass syringe with a needle-lock Luer (10 mL) and needle
- Gas line adaptor
- Source of dry carbon dioxide
- Dropping funnel
- pH indicator paper
- Separating funnel (500 mL)
- Conical flask (500 mL)
- Filter funnel
- Fluted filter paper
- Single-necked, round-bottomed flask (500 mL)
- Rotary evaporator
- Buchner flask and accessory funnel
- Filter paper
- Vacuum line
- Source of dry argon

8: Spherands, hemispherands and calixspherands

Materials

- 2,6-Bis(2'-methoxy-5'-methylphenyl)-4-methylanisole 7, 2.0 g, 5.52 mmol
- Anhydrous diethyl ether,[a] 50 mL — **flammable**
- Tetramethylethylenediamine (FW 116.2), 2.4 mL, 11.04 mmol — **irritant**
- n-Butyllithium, 1.6 M solution in hexane, 6.9 mL, 11.04 mmol — **moisture sensitive, toxic, flammable**
- Dry THF,[b] 25 mL — **flammable**
- Concentrated hydrochloric acid — **irritant, corrosive**
- Demineralized water, 300 mL
- Ethyl acetate, 300 mL — **flammable**
- Sodium chloride, saturated aqueous solution, 100 mL
- Pentane, 100 mL — **flammable**

1. Dry all glassware in an electric oven (110 °C, 2 h). Assemble the triple-necked flask, argon balloon via the adaptor, septum, stopper, stirrer bar and magnetic stirrer. Heat all glassware with the electric heat gun and allow to cool under an argon atmosphere.

2. Place 2.0 g of teranisyl 7 in the triple-necked flask as quickly as possible to retain the argon atmosphere.

3. Assemble the teflon syringe and the needle and flush the syringe with argon. Fill the syringe via the septum with 50 mL of diethyl ether and add this to the flask containing 7 via the septum. Dissolve 7 in the diethyl ether.

4. Assemble the all-glass syringes and needles while hot and allow the assembled syringes to cool in a desiccator under argon. Flush the syringes with argon.

5. Fill the 2.5 mL syringe with 2.4 mL of tetramethylethylenediamine and add this to the solution via the septum. Fill the 10 mL syringe with 6.9 mL of n-butyllithium in hexane and add this in the same way. Stir the solution for 3 h at room temperature.

6. Attach the gas line adaptor to the flask and bubble CO_2 vigorously through the mixture for 30 min. Add THF (c. 20 mL) to replace the evaporated ether. Stir the resulting mixture for 14 h under an argon atmosphere.

7. Attach the dropping funnel to the flask, fill the funnel with concentrated hydrochloric acid, and acidify the mixture to pH 2. Check the pH with the indicator paper. Add 100 mL of water and 100 mL of ethyl acetate to the mixture and dissolve the solids.

8. Transfer the solution to the separating funnel and separate both layers. Extract the aqueous layer with two portions of 100 mL of ethyl acetate. Combine the organic layers and wash with two portions of 100 mL of water and one portion of 100 mL saturated NaCl solution. Separate both layers.

9. Transfer the organic solution to the 500 mL conical flask, add some $MgSO_4$ (c. 5 g) and a stirrer bar. Dry the solution by stirring the suspension on a

Protocol 2. Continued

magnetic stirrer for about an hour. Filter the suspension through the filter funnel containing a fluted filter paper into the single-necked round-bottomed flask.

10. Remove the solvent on a rotary evaporator. Add 100 mL of pentane to the residue to dissolve any monoacid, and filter the suspension by using the Buchner flask and funnel with a filter paper. Allow the residue to dry under vacuum to give **8a** (2.3 g, 93% yield, m.p. 225–8 °C).

[a] For the preparation of dry diethyl ether, see Protocol 1.
[b] For the preparation of dry THF, see Protocol 1.

Protocol 3.
Synthesis of 2,6-bis(3-hydroxymethyl-2-methoxy-5-methylphenyl)-4-methylanisole (Structure 8b[6], Scheme 8.2)

Caution! Carry out all procedures in a well-ventilated hood, and wear disposable vinyl or latex gloves and chemical-resistant safety goggles.

Equipment

- Triple-necked, round-bottomed flask (100 mL)
- Argon balloon and adaptor
- Stopper
- Septum
- Magnetic hot plate stirrer
- 2 Magnetic stirrer bars
- Water-cooled reflux condenser
- Oil bath
- Electric heat gun
- All-glass syringe with needle-lock Luer (10 mL) and needle
- Teflon syringe (50 mL) and needle
- Conical flask (50 mL)
- Magnetic stirrer
- Separating funnel (250 mL)
- Conical flask (250 mL)
- Filter funnel
- Fluted filter paper
- Single-necked, round-bottomed flask (250 mL)
- Rotary evaporator
- Vacuum line
- Source of dry argon

Materials

- 2,6-Bis(3-carboxy-2-methoxy-5-methylphenyl)-4-methylanisole **8a**, 1.0 g, 2.20 mmol
- Dry THF,[a] 30 mL **flammable**
- Borane–tetrahydrofuran complex (FW 85.9), 1 M in THF, 6.6 mL, 6.60 mmol **irritant, flammable**
- Demineralized water
- Potassium carbonate **irritant**
- Diethyl ether,[b] 150 mL **flammable**
- Sodium chloride, saturated aqueous solution, 50 mL

1. Dry all glassware in an electric oven (110 °C, 2 h). Assemble the triple-necked flask, argon balloon via the adaptor, septum, stopper, stirrer bar, reflux condenser, oil bath and magnetic hot plate stirrer. Heat all glassware with the electric heat gun and allow to cool under an argon atmosphere.

8: Spherands, hemispherands and calixspherands

2. Place 1.0 g of **8a** in the triple-necked flask as quickly as possible to retain the argon atmosphere.
3. Assemble the teflon syringe and the needle and flush the syringe with argon. Fill the syringe via the septum with 30 mL of THF and add this to the flask containing **8a** via the septum. Dissolve **8a** in the THF.
4. Assemble the all-glass syringe and needle while hot and allow the assembled syringe to cool in a desiccator under argon. Flush the syringe with argon, fill the syringe with the solution of BH_3 in THF, and add this via the septum to the stirring solution of **8a**.
5. Stir the mixture at room temperature under an argon atmosphere for 3 h and boil under reflux for an additional 3 h. Allow the reaction mixture to cool to room temperature.
6. Prepare a saturated solution of K_2CO_3 in water in the 50 mL conical flask.
7. Carefully add water (c. 5 mL) to the reaction mixture to destroy the remaining BH_3, and subsequently add 15 mL of the saturated K_2CO_3 solution. Stir this mixture for 14 h at room temperature, add 50 mL of diethyl ether and stir to allow the solids to dissolve.
8. Transfer the solution to the separating funnel and separate both layers. Wash the aqueous layer with two portions of 50 mL of diethyl ether. Combine the organic layers and wash with two portions of 50 mL of water, and one portion of 50 mL of saturated aqueous NaCl solution.
9. Transfer the ethereal solution to the 250 mL conical flask and add some $MgSO_4$ and a stirrer bar. Dry the solution by stirring the suspension on a magnetic stirrer for about an hour. Filter the suspension through a filter funnel with a fluted filter paper into the single-necked flask.
10. Remove the solvent on a rotary evaporator and allow the product to dry under vacuum to give **8b** as a foam (0.90 g, 96%).

[a] For the preparation of dry THF, see Protocol 1.
[b] For the preparation of dry diethyl ether, see Protocol 1.

Protocol 4.
Synthesis of 2,6-bis(3-bromomethyl-2-methoxy-5-methylphenyl)-4-methylanisole (Structure 8c[6], Scheme 8.2)

Caution! Carry out all procedures in a well-ventilated hood, and wear disposable vinyl or latex gloves and chemical-resistant safety goggles.

Equipment

- Triple-necked, round-bottomed flask (500 mL)
- Argon balloon and adaptor
- Dropping funnel (2 mL)
- Stopper
- Magnetic hot plate stirrer
- Magnetic stirrer bar
- Electric heat gun
- Source of dry argon

Protocol 4. Continued

- Separating funnel (1 L)
- Conical flask (500 mL)
- Filter funnel
- Fluted filter paper
- Single-necked, round-bottomed flask (500 mL)
- Rotary evaporator
- Vacuum line

Materials

- 2,6-Bis(3-hydroxymethyl-2-methoxy-5-methylphenyl)-
  4-methylanisole **8b**, 5.0 g, 11.8 mmol
- Toluene, 300 mL   **flammable**
- Phosphorus tribromide (FW 270.7), 1.15 mL, 11.8 mmol   **toxic, irritant**
    flammable
- Technical grade diethyl ether, 200 mL   **flammable**
- Sodium bicarbonate saturated aqueous solution, 400 mL
- Demineralized water 200 mL
- Saturated aqueous sodium chloride solution, 200 mL

1. Dry all glassware in an electric oven (110°C, 2 h). Assemble the triple-necked flask, dropping funnel, argon balloon via the adaptor, stopper, stirrer bar and magnetic stirrer. Heat the glassware with the electric heat gun and allow to cool to room temperature under an argon atmosphere.

2. Place 5.0 g of **8b** and 300 mL of toluene in the flask and fill the dropping funnel with 1.15 mL of PBr$_3$ as quickly as possible to retain the argon atmosphere. Dissolve **8b** and add PBr$_3$ dropwise to the solution at room temperature. Stir for 14 h.

3. Transfer the reaction mixture to the separating funnel and shake with 200 mL of diethyl ether and 200 mL of a saturated aqueous solution of NaHCO$_3$. Separate both layers and wash the organic layer subsequently with a second portion of 200 mL of the NaHCO$_3$ solution, 200 mL of water and 200 mL of brine. Separate both layers.

4. Transfer the organic layer to the conical flask and add some MgSO$_4$ and a stirrer bar. Dry the solution by stirring for about 1 h on a magnetic stirrer. Filter the suspension through the filter funnel with a fluted filter paper into the single-necked flask.

5. Remove the solvent on a rotary evaporator and dry the product under vacuum to give **8c** as a colourless foam (6.3 g, 97%).

Protocol 5.
Synthesis of 2,3,4,5,6,7,8,9,10-tri[1,3-(2-methoxy-5-methylbenzo)]-12,15,18,-trioxacyclooctadeca-2,5,8-triene (Structure 2[16], Scheme 8.2)

Caution! Carry out all procedures in a well-ventilated hood, and wear disposable vinyl or latex gloves and chemical-resistant safety goggles.

8: Spherands, hemispherands and calixspherands

Equipment

- Triple-necked, round-bottomed flask (500 mL)
- Reflux condenser
- Septum
- Stopper
- Oil bath
- Magnetic hot plate stirrer
- Magnetic stirrer bar
- 3 argon balloons and adaptors
- Single-necked, round-bottomed flasks (2 × 100 mL, 500 mL)
- Electric heat gun
- Source of dry argon
- Perfusor and 2 × 50 mL perfusor syringes and needles
- 2 Teflon tubes and adaptors
- Rotary evaporator
- Refrigerator
- Buchner flask (100 mL) and accessory funnel
- Filter paper
- Vacuum line

Materials

- Sodium hydride (FW 24.0, 60% dispersion in mineral oil), 1.80 g, 45.0 mmol **flammable, moisture sensitive**
- Dry THF,[a] 350 mL **flammable**
- 2,6-Bis(3-bromomethyl-2-methoxy-5-methylphenyl)-4-methylanisole **8c**, 7.05 g, 11.3 mmol
- Diethylene glycol (FW 106.1), 1.19 g, 11.3 mmol **irritant**
- Water, c. 10 mL
- Dichloromethane, 50 mL **toxic**
- Methanol, 50 mL **flammable, toxic**
- Sodium perchlorate (FW 122.4), 1.38 g, 11.3 mmol

1. Dry all glassware in an electric oven (110 °C, 2 h). Assemble the triple-necked 500 mL round-bottomed flask, condenser, argon balloon via adaptor, stopper, septum, stirrer bar, oil bath and hot plate magnetic stirrer. Heat the glassware with the electric heat gun and allow to cool to room temperature under an argon atmosphere.

2. Weigh the sodium hydride (1.8 g) into the flask and add 250 mL of dry THF as quickly as possible to retain the argon atmosphere. Boil the suspension under reflux.

3. Heat the single-necked 100 mL flasks with the electric heat gun and allow them to cool to room temperature under an argon atmosphere using the argon balloons and adaptors. Assemble the perfusor syringes and needles and flush with argon.

4. Place **8c** (7.05 g) in one of the flasks and add 50 mL of dry THF as quickly as possible to retain the argon atmosphere. Fill one of the perfusor syringes with the solution of **8c** in THF.

5. Place diethylene glycol (1.19 g) in the other flask and add 50 mL of dry THF as quickly as possible to retain the argon atmosphere. Dissolve the diethylene glycol in the THF and fill the other perfusor syringe with this solution.

6. Assemble the perfusor, the perfusor syringes and the teflon tubes via the adaptors. Bring the Teflon tubes into the triple-necked flask via the septum and add both solutions to the boiling suspension at a constant rate with stirring over a 10-h period.

7. After additional heating under reflux for 6 h, cool the reaction mixture to

Protocol 5. *Continued*

room temperature, and subsequently add about 10 mL of water to destroy remaining sodium hydride. Transfer the reaction mixture to the single-necked 500 mL flask and remove the solvent on a rotary evaporator.

8. Dissolve the remaining product in 50 mL of dichloromethane and 50 mL of methanol and add 1 equiv. of sodium perchlorate. Concentrate the solution on a rotary evaporator to a volume of about 20 mL. Cool the remaining suspension in a refrigerator for 2 d.

9. Filter the suspension using the Buchner flask and funnel with filter paper and wash the crystals with ice-cold methanol. Allow the product to dry under vacuum (0.1 mmHg, 20°C, 2 h) to give pure **2** (64% yield, m.p. 208–9°C).

a For the preparation of dry THF, see Protocol 1.

In a second approach[17] to hemispherands the central aromatic ring is introduced in the last step of the synthesis via a Suzuki coupling[18] using a zero-valent palladium catalyst. The advantage of this route is that, in the last step, modified central aromatic rings can be introduced. The reaction sequence is summarized in Scheme 8.3. Reaction of the easily accessible 2-bromo-6-(chloromethyl)-4-methylanisole **9** (prepared in two steps from 4-methylanisole)[19] with diethylene glycol in THF in the presence of sodium hydride as a base affords compound **10a** as a colourless oil in 83% yield after bulb to bulb distillation. Transformation of **10a** into the diboronic acid **10b** is carried out by treatment with *n*-butyllithium in THF followed by reaction of the resulting dilithio compound with trimethyl borate and subsequent treatment with 3% HCl. The diboronic acid **10b** cannot be purified, on account of slow decomposition, and has to be used as such in the macrocyclization step. Subjection of crude **10b** to modified Suzuki cross-coupling conditions (toluene/aq Na_2CO_3/2% $Pd(PPh_3)_4$) with aryl dibromides **11a–d** (prepared from commercially available phenols by alkylation with methyl iodide or allyl bromide) gives the hemispherands **12a–d**, after chromatography, in 12–21% yield (calculated on **10a**). Variation of the cation (Li^+, Na^+, K^+, Cs^+) indicates that the Na^+ and K^+ cations probably serve as template ions in the macrocyclization reaction.

Protocol 6.
Synthesis of hemispherands (Structures 12a-d with different R[1] and R[2] groups[17], Scheme 8.3)

Caution! Carry out all procedures in a well-ventilated hood, and wear disposable vinyl or latex gloves and chemical-resistant safety goggles.

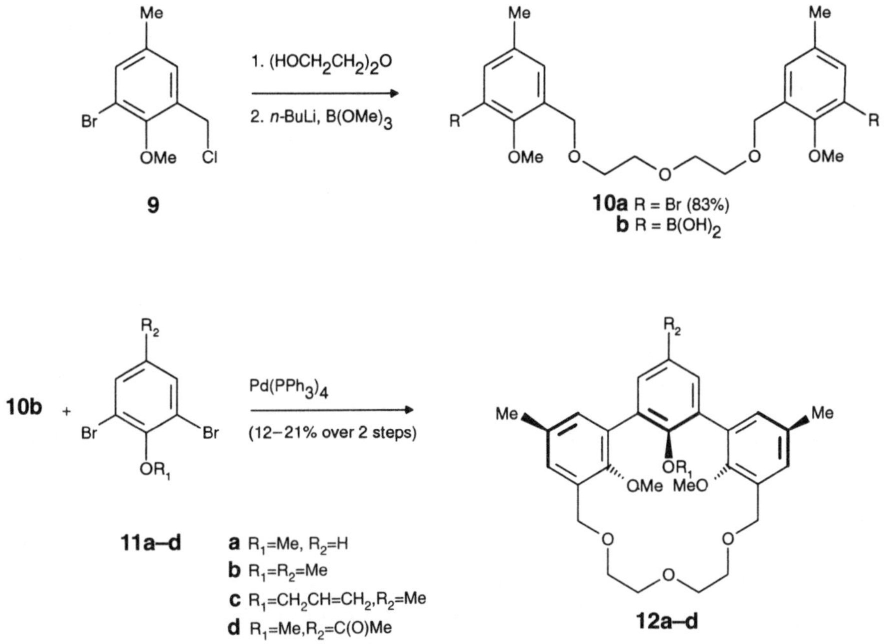

Scheme 8.3

Equipment

- Triple-necked, round-bottomed flasks (2 × 250 mL)
- Single-necked, round-bottomed flasks (3 × 250 mL, 2 × 100 mL)
- 2 Magnetic stirrer bars
- Water-cooled condenser
- Dropping funnel (250 mL)
- Source of dry nitrogen
- Source of dry argon
- Gas inlet
- Magnetic hot plate stirrer
- Separating funnel (250 mL)
- Rotary evaporator
- Bulb to bulb vacuum distillation setup
- Bath for cooling
- Septum
- Syringe
- pH indicator paper
- Column for chromatography
- Filter funnel
- Filter paper

Materials

- Sodium hydride (FW 24.0, 60% dispersion in mineral oil), 0.56 g, 14 mmol — **flammable**
- Hexane for washing the sodium hydride — **flammable, irritant**
- Anhydrous THF,[a] 200 mL — **flammable**
- Diethylene glycol (FW 106.1), 0.50 g, 4.7 mmol — **irritant**

Protocol 6. Continued

- 2-Bromo-6-(chloromethyl)-4-methylanisole 9,[19] 2.35 g, 9.4 mmol
- Methanol, 25 mL (also for chromatography), c. 75 mL — flammable, toxic
- Dry ice
- Acetone for cooling — flammable
- n-Butyllithium (1.6 M in hexane), 1.17 mL, 1.88 mmol — flammable, corrosive, moisture sensitive
- Trimethyl borate (FW 103.9), 1.37 g, 13.2 mmol — flammable, irritant
- Diethyl ether,[b] 160 mL — flammable
- 3% hydrochloric acid — corrosive, irritant
- 2 M sodium hydroxide — irritant
- Concentrated hydrochloric acid — corrosive, irritant
- Tetrakis(triphenylphosphine)palladium(0)(FW 1155.6), 23 mg, 0.02 mmol — light and air sensitive
- 2 M Na_2CO_3 aqueous solution, 50 mL — irritant
- Toluene, 50 mL — flammable
- Aryl dibromide 11a–d, 1 mmol
- 95% ethanol — flammable
- Dichloromethane, 60 mL (also for chromatography), c. 900 mL — toxic
- 1 M hydrochloric acid, 30 mL — corrosive, irritant
- Saturated methanolic KBr solution, 15 mL — irritant
- Silica gel for chromatography — irritant, dust
- KBr for chromatography — irritant

All reactions should be carried out in dry glassware under a nitrogen atmosphere.

1. Assemble the apparatus consisting of a triple-necked 250 mL round-bottomed flask, a magnetic stirrer bar, a condenser, a dropping funnel and a gas inlet on a hot plate stirrer.
2. Carefully weigh the sodium hydride (0.56 g) into the flask and add hexane so that the sodium hydride is below the liquid level. Stir this suspension gently, allow the solid to settle and decant the liquid. Repeat this washing. Add 50 mL of anhydrous THF and the diethylene glycol (0.50 g). Stir the reaction mixture for 1 h at room temperature.
3. Prepare a solution of 2-bromo-6-(chloromethyl)-4-methylanisole 9[19] (2.35 g) in 100 mL of anhydrous THF and transfer this to the dropping funnel. Add this solution dropwise to the reaction mixture.
4. Boil the reaction mixture under reflux for 5 h.
5. Allow the mixture to cool to room temperature and add 10 mL of methanol. Remove the stirrer bar and pour the contents into a 250 mL round-bottomed flask. Remove the solvent on a rotary evaporator.
6. The residue is submitted to bulb to bulb distillation in which the fraction boiling between 160 and 170°C at 0.05 mmHg is collected to give compound 10a (83% yield).
7. Weigh 1.0 g of compound 10a into a 100 mL single-necked round-bottomed flask. Add 50 mL of anhydrous THF. Close the flask with a sep-

8: Spherands, hemispherands and calixspherands

tum pierced with a needle attached to an argon-filled balloon. Cool the mixture to −78 °C using a dry ice/acetone bath.

8. Stir the mixture vigorously and add a cooled (−78 °C) n-butyllithium solution (1.17 mL) by means of a syringe through the septum. Stir for 2 h at −78 °C.

9. Dissolve 1.37 g of trimethyl borate in 10 mL of dry diethyl ether in a 250 mL round-bottomed flask and cool this solution to −78 °C.

10. Connect a Teflon tube between the two stirring solutions and transfer the solution of the dilithio compound of **10a** into the borate ester solution by means of an overpressure of argon at −78 °C.

11. Stir this solution while allowing the temperature to rise slowly to room temperature.

12. Add 50 mL of 3% HCl and stir for 10 h at room temperature.

13. Transfer the contents of the flask to a 250 mL separating funnel and separate the layers. Wash the organic layer five times with 25 mL of 2 M NaOH. Cool the combined aqueous layers to 5 °C and acidify with concentrated HCl to pH = 1. Extract this solution twice with 75 mL of diethyl ether.

14. Transfer the organic layers to a 250 mL single-necked round-bottomed flask and remove the solvent on a rotary evaporator to give crude compound **10b**.

15. Weigh Pd(PPh$_3$)$_4$ (23 mg) into a 250 mL triple-necked round-bottomed flask equipped with a condenser and add 50 mL of a 2 M aqueous solution of Na$_2$CO$_3$ and 50 mL of toluene. Heat this mixture and stir vigorously.

16. Dissolve crude compound **10b** (0.46 g, 1.0 mmol) and aryl dibromide **11a–d** (1 mmol) in 5 mL of 95% ethanol. Add this solution by funnel to that in the triple-necked flask and boil the mixture under reflux for 15 h.

17. Allow the mixture to cool to room temperature and transfer the contents of the flask to a 250 mL separating funnel. Separate the layers and extract the aqueous layer with 3 × 20 mL of dichloromethane. Combine the organic layers and wash them with 3 × 10 mL of 1 M HCl and then with 20 mL of water.

18. Transfer the organic layer to a 100 mL single-necked, round-bottomed flask and remove the solvent on a rotary evaporator.

19. Dissolve the residue in 15 mL of methanol and stir for 1 h with 15 mL of a saturated methanolic KBr solution. Remove the solvent on a rotary evaporator.

20. Apply the residue to a silica gel column (2.5 cm × 20 cm, wet SiO$_2$:KBr, 10:1). Elute the non-ionic impurities by using dichloromethane as the eluent. Change the eluent to dichloromethane:methanol, 93:7 to give hemispherands **12a–d** as their KBr salts. Combine the fractions containing **12a–d**.KBr and wash them with 3 × 10 mL of 3% HCl and 10 mL of water.

Protocol 6. *Continued*

21. Dry the organic solution with MgSO$_4$ and filter it through a filter funnel and a fluted paper.
22. Remove the solvent on a rotary evaporator and recrystallize the residue with 95% ethanol to give the hemispherands as white solids (**12a**: 14% yield, m.p. 191–3°C; **12b**: 15% yield, m.p. 208–10°C; **12c**: 21% yield, m.p. 133–5°C; **12d**: 12% yield, m.p. 151–2.5°C).

a For the preparation of dry THF, see Protocol 1.
b For the preparation of dry diethyl ether, see Protocol 1.

The main problem in the synthesis of hemispherands is the modest yield of the macrocyclization reaction when the rigid *m*-teranisyl unit is incorporated into the starting compound. An alternative method involves the synthesis of a more flexible intermediate macrocycle, which generally proceeds in better yield, followed by the introduction of the rigid triaryl moiety. The introduction of the third aromatic ring involves the condensation reaction of an incorporated dibenzyl ketone moiety with nitromalonodialdehyde.[14] Tautomerization of the resulting 2,6-disubstituted 4-nitrocyclohexa-2,5-dien-1-one provides the aromatization energy, which compensates for the increased O····O repulsion in the resulting hemispherand. Reaction of diol **13** [prepared in several steps from 1,3-bis(3-bromo-2-methoxy-5-methylphenyl)-2-propanone] with diethylene glycol ditoluenesulfonate and sodium hydride as a base in THF gives macrocycle **14a** in 53% yield (Scheme 8.4). The X-ray structure of **14a** shows that the methoxy groups are situated at different faces of the macro-ring and their methyl groups are directed towards the centre of the cavity. The cyclic ketal function is directed outside the ring and the conformation of the ethyleneoxy bridge is all-*gauche*. Deprotection of **14a** in a methanol/hydrochloric acid mixture affords **14b** in 98% yield. Subsequent reaction of **14b** with the sodium salt of nitromalonodialdehyde and sodium hydroxide in an ethanol/water mixture gives hemispherand **5**. The reaction can be forced to completion only when a large excess of nitromalonodialdehyde is used, giving **15** in 89% yield (Scheme 8.4). The free hydroxyl group in **15** can easily be methylated using methyl iodide and potassium carbonate (95% yield).

8: Spherands, hemispherands and calixspherands

Protocol 7.
Synthesis of 23,24-dimethoxy-7,21-dimethyl-11,14,17-trioxatricyclo [17.3.1.1^{5,9}]-tetracosa-1(23),5,7,9(24),19,21-hexaen-3-one (Structure 14b[14], Scheme 8.4)

Caution! Carry out all procedures in a well-ventilated hood, and wear disposable vinyl or latex gloves and chemical-resistant safety goggles.

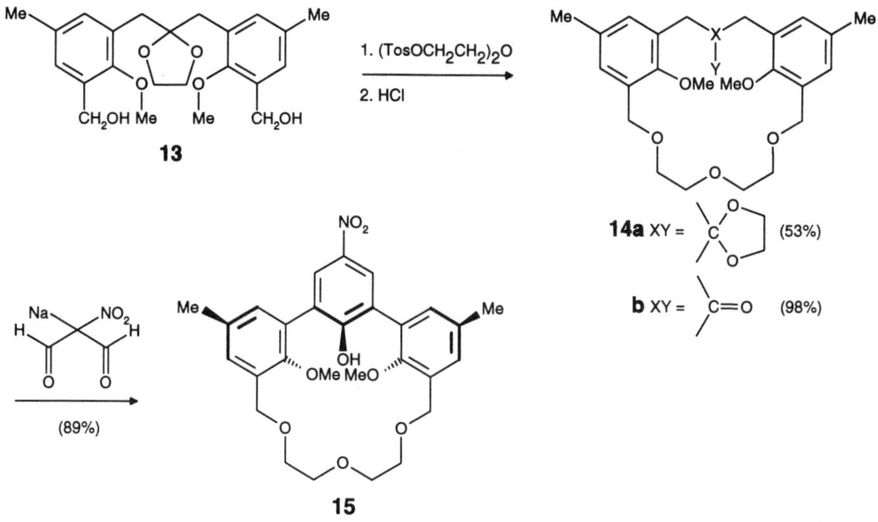

Scheme 8.4

Equipment

- Triple-necked, round-bottomed flask (500 mL)
- Single-necked, round-bottomed flasks (500 mL, 250 mL, 2 × 100 mL, 50 mL)
- Condenser
- Oil bath
- Magnetic hot plate stirrer
- Electric heat gun
- Stopper
- Septum
- Magnetic stirrer bar

- Rotary evaporator
- Separating funnel (250 mL)
- Filter funnel
- Filter papers
- Perfusor and 2 × 50 mL perfusor syringes and needles
- 2 Teflon tubes and adaptors
- Argon balloon and adaptor
- Source of dry argon

Materials

- 1,3-Bis[3-(hydroxymethyl)-2-methoxy- 5-methylphenyl]-2-propanone ethylene acetal 13[14] 0.80 g, 2 mmol
- Sodium hydride (FW 24.0, 80% dispersion in mineral oil), 0.12 g, 4 mmol   **flammable**
- Anhydrous THF,[a] 200 mL   **flammable**
- Diethylene glycol ditoluenesulfonate, 0.91 g, 2.2 mmol
- Chloroform, 150 mL   **highly toxic, cancer-suspect agent**
- Methanol for recrystallization, 100 mL   **flammable, toxic, hygroscopic**
- 4 M hydrochloric acid, 60 mL   **corrosive, irritant**

193

Protocol 7. *Continued*

1. Dry all glassware in an electric oven (110°C, 2 h). Assemble the triple-necked 500 mL round-bottomed flask, reflux condenser, argon balloon via adaptor, stopper, septum, stirrer bar, oil bath, and magnetic stirrer bar. Heat the glassware with the electric heat gun and allow to cool to room temperature under an argon atmosphere.

2. Weigh the sodium hydride (0.12 g) in the flask and add 150 mL of dry THF as quickly as possible. Boil the suspension under reflux.

3. Prepare a solution of **13** (0.80 g) in 50 mL of dry THF in a single-necked 100 mL flask and subsequently fill one of the perfusor syringes with this solution.

4. Prepare a solution of diethylene glycol ditoluenesulfonate (0.91 g) in 50 mL of dry THF in another single-necked 150 mL flask and subsequently fill the other perfusor syringe with this solution.

5. Assemble the perfusor, the perfusor syringes and the Teflon tubes via the adaptors. Connect the Teflon tubes to the triple-necked 500 mL flask via the septum and add both solutions at a constant rate with stirring over a 10-h period.

6. After additional boiling under reflux for 8 h, allow the reaction mixture to cool to room temperature and subsequently add about 5 mL of water to destroy any remaining sodium hydride. Remove the stirrer bar and transfer the reaction mixture to the single-necked 500 mL flask. Remove the solvent on a rotary evaporator.

7. Treat the residue with 50 mL of chloroform and 50 mL of water. Transfer the contents of the flask to a 250 mL separating funnel, separate the layers and extract the aqueous layer with 2 × 50 mL of chloroform. Combine the organic layers and wash with 50 mL of water.

8. Dry the organic layer with $MgSO_4$ and filter it through a filter funnel and a fluted filter paper into a 250 mL round-bottomed flask and remove the solvent on a rotary evaporator.

9. Recrystallize the residue from methanol to give compound **14a** as colourless crystals (53% yield, m.p. 115–16°C).

10. Weigh 0.53 g of this compound into a 50 mL round-bottomed flask. Add 6 mL of 4 M HCl and 6 mL of methanol and stir at room temperature for 16 h.

11. Filter the suspension through a filter funnel and a fluted filter paper.

12. Recrystallize the residue from methanol to give compound **14b** as colourless crystals (98% yield, m.p. 121–2°C).

[a] For the preparation of dry THF, see Protocol 1.

Protocol 8.
Synthesis of 25,26-dimethoxy-9,23-dimethyl-4-nitro-13,16,19-trioxa-tetracyclo-[19.3.1.1^{2,6}.1^{7,11}]heptacosa-1,(25),2,4,6(27),7,9,11(26),21,23-nonaen-27-ol (Structure 15[14], Scheme 8.4)

Caution! Carry out all procedures in a well-ventilated hood, and wear disposable vinyl or latex gloves and chemical-resistant safety goggles.

Equipment

- Single-necked, round-bottomed flask (100 mL)
- Water-cooled condenser
- Separating funnel (100 mL)
- Filter funnel
- Filter papers
- Magnetic stirrer bar
- Oil bath
- Magnetic hot plate stirrer
- Rotary evaporator

Materials

- Compound **14b**, 0.105 g, 0.24 mmol
- Ethanol, 20 mL flammable
- Nitromalonodialdehyde sodium salt, 0.55 g, 4.0 mmol irritant
- Sodium hydroxide (FW 40.0), 0.037 g, 0.92 mmol corrosive
- 2 M hydrochloric acid corrosive, irritant
- Chloroform, 30 mL highly toxic, cancer-suspect agent
- Acetic acid for recrystallization corrosive, irritant

The reaction should be carried out in glassware that has been dried in an electric oven (110°C, 2 h).

1. Weigh 0.105 g of compound **14b** in a 100 mL round-bottomed flask. Add 20 mL of ethanol and stir the mixture. Add to the stirring solution a solution of nitromalonodialdehyde sodium salt (0.55 g) in 1 mL of water followed by a solution of sodium hydroxide (0.037 g) in 0.5 mL of water.
2. Warm the reaction mixture to 46°C and stir for 16 h.
3. Allow the mixture to cool to room temperature and acidify with 2 M HCl. Remove the stirrer bar, pour the contents of the flask into a 100 mL separating funnel and extract three times with 10 mL of chloroform. Wash the combined organic layers with water.
4. Dry the organic solution with MgSO$_4$ and filter it through a filter funnel and a fluted filter paper into a 100 mL round-bottomed flask and remove the solvent on a rotary evaporator.
5. Recrystallize the residue from acetic acid to give hemispherand **15** (89% yield, m.p. 240–2°C).

The final strategy involves the synthesis of hemispherands containing a building block which allows further variation in the structure of the molecular cavity.[20] This strategy provides a general method for the modification of the central anisyl unit which is present in **2** for example. Such a synthetic building block must be readily available from simple starting compounds which are stable under the macrocyclization reaction conditions, and are easy to modify when incorporated in the molecular cavity. A 2,6-dianisyl-substituted 4*H*-pyran meets these requirements. In principle a 4*H*-pyran moiety can be converted into a pyrylium cation that has different reactive sites.

4*H*-Pyran hemispherands **16a,b** (prepared by macrocyclization of an appropriately substituted diaryl 4*H*-pyran) are oxidatively dehydrogenated to the pyrylium hemispherands **17a** (45%) and **17b** (98%), respectively (Scheme 8.5), upon reaction with triphenylcarbenium tetrafluoroborate in dimethoxyethane. Pyrylium salts generally react with nucleophiles at one of the 2,6-positions, due to high charge delocalization at these carbon atoms. The intermediate formed may cyclize to a new carbocyclic or heteroaromatic ring, the driving force being the aromatization energy from the isomerization. Reaction of **17b** with ammonium acetate in acetic acid gives hemispherand **18** in 89% yield. Starting from **17a** the corresponding hemispherand can only be obtained in very low yield. Introduction of further steric barriers can be accomplished by reacting **17b** with primary amines to yield the corresponding pyridinium hemispherands. For instance, reaction of **17b** with methylamine and aniline affords the corresponding *N*-methyl- and *N*-phenylpyridinium salts in 63% and 77% yield, respectively (Scheme 8.5).[20]

Protocol 9.
Synthesis of 25,26-dimethoxy-3,4,5,9,23-pentamethyl-13,16,19-trioxa-27-oxoniatetracyclo[19.3.1.1^{2,6}.1^{7,11}]heptacosa-1(25),2,4,6-(27), 7,9,11(26),21,23-nonene tetrafluoroborate (Structure 17b[20], Scheme 8.5)

Caution! Carry out all procedures in a well-ventilated hood, and wear disposable vinyl or latex gloves and chemical-resistant goggles.

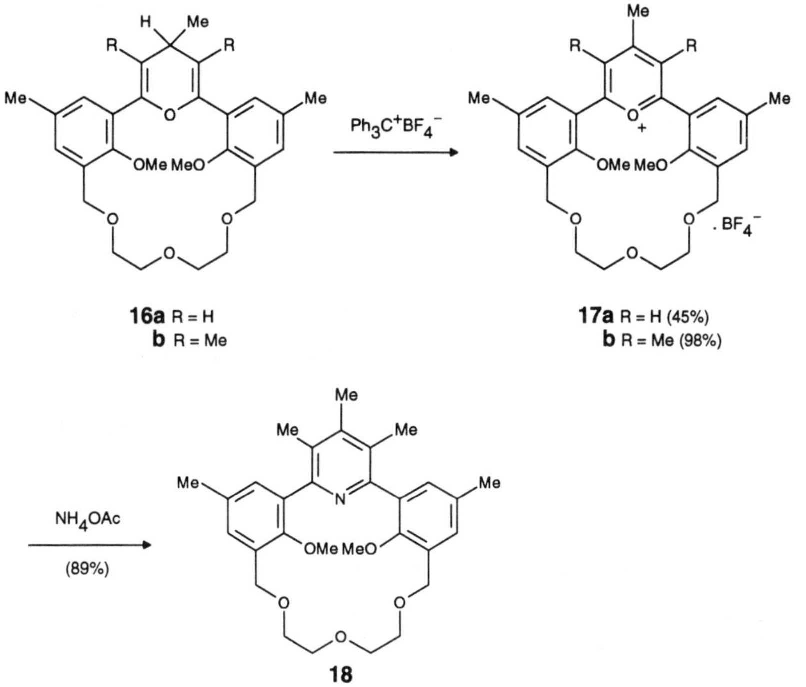

Scheme 8.5

Equipment

- Single-necked, round-bottomed flasks (2 × 50 mL)
- Buchner filter flask (100 mL)
- Neoprene adaptor
- Disposable argon balloon
- Filter paper (Whatman No. 1)
- Buchner funnel
- Gas line adaptor
- Magnetic stirrer bar
- Magnetic hot plate stirrer

Materials

- 25,26-Dimethoxy-3,4,5,9,23-pentamethyl-13,16,19,27-tetraoxatetracyclo-[19.3.1.1^{2,6}.1^{7,11}]heptacosa-1(25),2,5,7,9,11-(26),21,23-octaene **16b**,[20] 0.10 g, 0.21 mmol   **toxic**
- Triphenylcarbenium tetrafluoroborate (FW 330.1), 0.10 g, 0.30 mmol   **toxic**
- Ethylene glycol dimethyl ether (FW 90.1), 2 mL   **flammable, irritant**
- Diethyl ether, 100 mL   **flammable, irritant**

Protocol 9. *Continued*

1. Dry a 50 mL round-bottomed flask in an electric oven (110°C, 2 h). Allow the flask to cool under argon.
2. Weigh 0.10 g of 4*H*-pyran hemispherand **16b** in the flask. Subsequently add 2 mL of ethylene glycol dimethyl ether and 0.10 g of triphenylcarbenium tetrafluoroborate.
3. Stir the suspension for 16 h at room temperature under argon (argon balloon on the flask).
4. Remove the stirrer bar and filter the suspension using the Buchner flask and funnel with a filter paper. Wash the product with two 50 mL portions of diethyl ether.
5. Transfer the pale-yellow crystals of **17b** to a 50 mL round-bottomed flask and dry the product under vacuum (98%, m.p. 208–10°C).

Protocol 10.
Synthesis of 25,26-dimethoxy-3,4,5,9,23-pentamethyl-13,16,19-trioxa-27-azatetracyclo[19.3.1.1^{2,6}.1^{7,11}]heptacosa-1 (25),2,4,6(27),7,9,11(26), 21,23-nonaene (Structure 18[20], Scheme 8.5)

Caution! Carry out all procedures in a well-ventilated hood, and wear disposable vinyl or latex gloves and chemical-resistant safety goggles.

Equipment

- Single-necked, round-bottomed flasks (2 × 50 mL)
- Water-cooled condenser
- Gas line adaptor
- Magnetic stirrer bar
- Magnetic hot plate stirrer
- Oil bath
- Erlenmeyer flask (50 mL)
- Filter papers (Whatman No. 1)
- Separating funnel (100 mL)
- Filter funnel
- Source of dry argon
- Disposable argon balloon
- Rotary evporator

Materials

- 25,26-Dimethoxy-3,4,5,9,23-pentamethyl-13,16,19-trioxa-27-oxoniatetracyclo[19.3.1.1^{2,6}.1^{7,11}]heptacosa-1(25),2,4,6-(27),7,9,11(26),21,23-nonaene tetrafluoroborate **17b**, 0.10 g, 0.18 mmol   toxic
- Glacial acetic acid, 2 mL   corrosive, irritant
- Ammonium acetate (FW 77.1), 0.14 g, 1.8 mmol   irritant
- Chloroform, 30 mL   highly toxic, cancer-suspect agent
- Ethanol, for crystallization   flammable

1. Dry all glassware in an electric oven (110°C, 2 h). Assemble the flask, condenser, gas line adaptor and magnetic stirrer bar on the stirrer hot plate while warm and allow to cool under argon.

8: Spherands, hemispherands and calixspherands

2. Weigh 0.10 g of pyrylium hemispherand **17b** in the flask. Subsequently add 0.14 g of ammonium acetate and 2 mL of glacial acetic acid.
3. Boil the solution under reflux for 3 h.
4. Allow the reaction mixture to cool to room temperature and add 10 mL of water.
5. Transfer the mixture to a separating funnel and extract three times with 10 mL portions of chloroform.
6. Combine the organic layers and wash with water (25 mL) and dry over MgSO$_4$.
7. Filter the organic solution through a filter funnel and a fluted filter paper into a 50 mL round-bottomed flask.
8. Remove the solvent on a rotary evaporator and recrystallize the solid from ethanol to give **18** as colourless crystals (79 mg, 89% yield, m.p. 273–5 °C).

4. Calixspherands

Calix[4]arenes can be functionalized both at the phenolic OH groups (lower rim) and at the *para* positions of the phenol rings (upper rim).[21] A special class of selectively functionalized calix[4]arenes comprises bridged calix[4]arenes in which two phenol rings are connected by a cap.[22] An important subgroup of this class are the calixspherands in which a calix[4]arene is diametrically bridged with a rigid terphenyl moiety.

Initially, calixspherand **22a** was obtained by coupling of 26,28-dimethoxy-*p-t*-butylcalix[4]arene with *m*-teranisyl **20a** in yields of less than 30%.[8] A more efficient approach involves first coupling *p-t*-butylcalix[4]arene **19**[23], (Chapter 7, p. 147), with *m*-terphenyls **20a–c** giving the calixspherand diols **21a–c** and subsequent alkylation of the free hydroxy groups.[9] Therefore the polyanion of **19** is prepared by reaction with five equivalents of sodium hydride in THF in the presence of a catalytic amount of 18-crown-6, followed by addition of a solution of *m*-terphenyl **20** in THF to a suspension of the resulting calix[4]arene anion at reflux. In this way the calixspherand diols **21a–c** are obtained in 64–80% yield (Scheme 8.6). The sodium cation may act as a template around which the ligating sites of the *m*-terphenyl and the calix[4]arene fold, in favour of diametrical disubstitution. This argument is supported by the fact that, in the presence of the larger potassium cation, if potassium hydride is used instead of sodium hydride, only small amounts of calixspherand diols **21** are isolated, thus indicating that potassium is too large to act as a good template ion.

Subsequent alkylation of **21a–c** with iodomethane and potassium *t*-butoxide as a base in THF affords **22a–c** as the potassium complexes in 81–95% yield. Use of rubidium *t*-butoxide[24] under the same conditions gives the corresponding rubidium complexes. Alkylation of **21a** with iodoethane gives **22d** as the potassium complex in a yield of 83% (Scheme 8.6). However, attempted alkylations with 1- or 2-iodopropane, 1-iodobutane and 3-bromopropene are unsuccessful, indicating that there is insufficient space to accommodate larger alkyl groups. According to ^{1}H NMR spectroscopy and X-ray crystallography, all complexes of **22a–d** are in a partial cone conformation. In order to obtain the calixspherands **2** as free ligands, the potassium complexes are heated in a mixture of methanol/water (1:4) in a closed vessel at 120°C for 3 d.

The calixspherands form kinetically stable complexes with Na$^+$, K$^+$ and Rb$^+$. The kinetic stability of the different complexes, which can be determined by ^{1}H NMR spectroscopy or by a new method[25] based on the exchange of radioactive rubidium or sodium in the complexes for non-radioactive sodium in different solvents, is strongly increased when the size of the group on the central aromatic ring of the *m*-terphenyl is increased. Due to the highly shielded cavity, calixspherands **22b,c** also form kinetically stable complexes with Ag$^+$.[26]

Protocol 11.
Synthesis of calixspherand diols from *p-t*-butylcalix[4]arene and 3,3"-bis(bromomethyl)-1,1':3',1"-terphenyls[9] (Scheme 8.6)

Caution! Carry out all procedures in a well-ventilated hood, and wear disposable vinyl or latex gloves and chemical-resistant safety goggles.

The following procedure for the synthesis of 2,26,31,41-tetrakis(1,1-dimethylethyl)-44,45,46-trimethoxy-9,14,19-trimethylcalixspherand-35,38-diol **21a** is representative.

Equipment
- Magnetic stirrer and heater with contact thermometer
- Oil bath
- Triple-necked, round-bottomed flask (1 L)
- Teflon-coated magnetic stirrer bar
- 2 stoppers
- Water-cooled condenser with nitrogen inlet
- Pressure equalizing addition funnel (100 mL)
- Pear-shaped evaporation flask (1000 mL)
- Separating funnel (250 mL)
- Erlenmeyer flasks (2 × 150 mL)
- Column for flash chromatography
- Source of dry nitrogen

Materials
- Sodium hydride (FW 24.0, 80% suspension in oil), 0.45 g, 15 mmol — **moisture sensitive, flammable**
- *p-t*-butylcalix[4]arene toluene complex[23] (FW 741.1), 2.22 g, 3 mmol
- 3,3"-Bis(bromomethyl)-2,2',2"-trimethoxy-5,5',5"-trimethyl-1,1':3',1"-terphenyl[9] (FW 548.3), 1.65 g, 3 mmol
- 18-Crown-6 (FW 264.3), 0.04 g, 0.15 mmol — **toxic**
- Dry THF,^{a} 700 mL — **flammable**
- Dichloromethane for extraction, 100 mL — **toxic**
- 1 M aqueous hydrochloric acid — **corrosive, irritant**
- Silica gel for flash chromatography — **irritant dust**

21a R₁=Me
b R₁=Et
c R₁=*i*-Pr

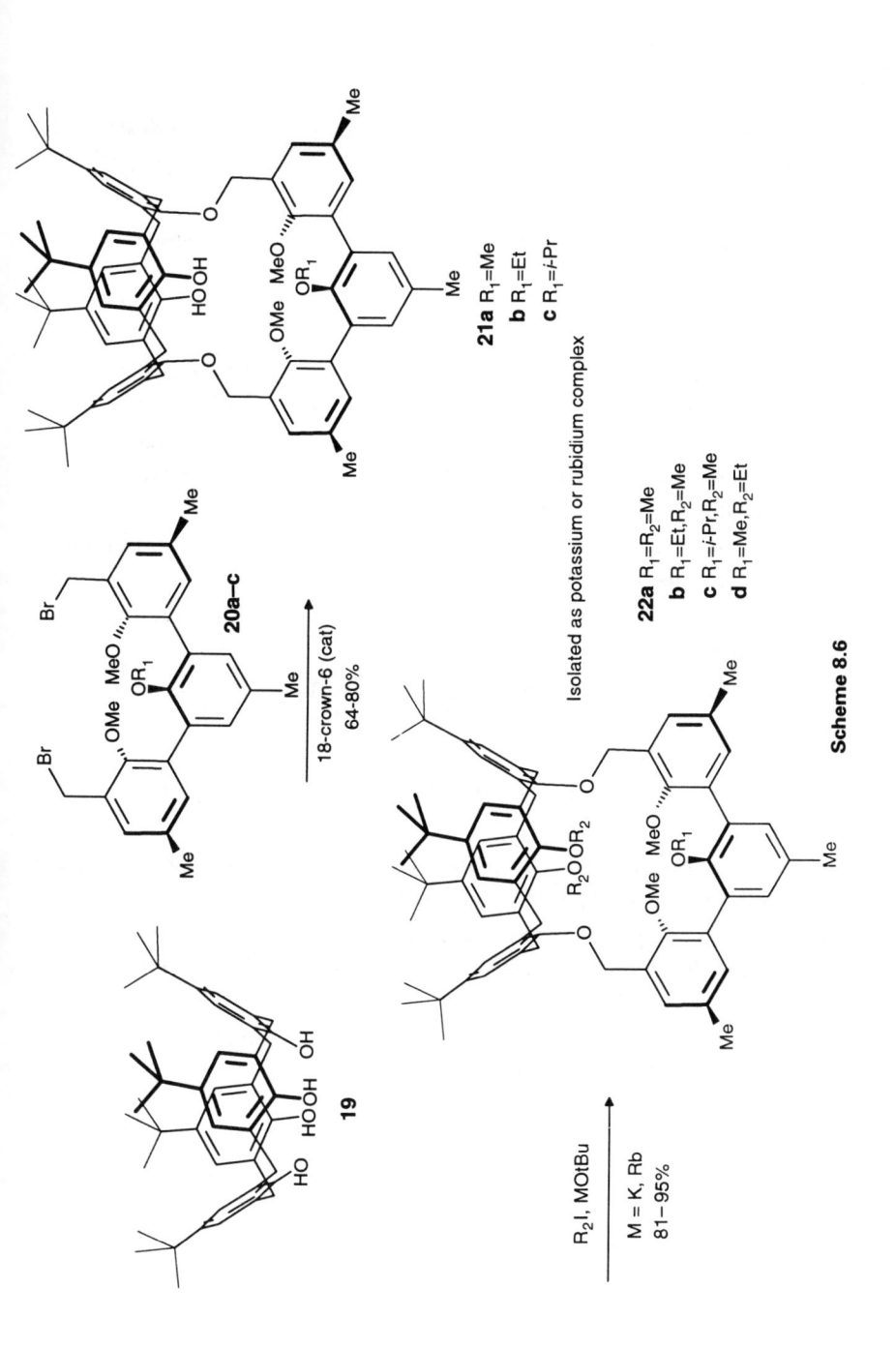

20a–c

18-crown-6 (cat)
64–80%

19

R₂I, MOtBu
─────────→
M = K, Rb
81–95%

Isolated as potassium or rubidium complex

22a R₁=R₂=Me
b R₁=Et, R₂=Me
c R₁=*i*-Pr, R₂=Me
d R₁=Me, R₂=Et

Scheme 8.6

Protocol 11. *Continued*

- Hexane for flash chromatography   flammable, irritant
- Ethyl acetate for flash chromatography   flammable
- Methanol for recrystallization, 50 mL   flammable, toxic

1. Make sure all glassware is clean and dry all glassware for at least 30 min in a 110°C electric oven before use.

2. Assemble the flask with the magnetic stirrer bar, the condenser with nitrogen inlet, the dropping funnel and two stoppers. Support the assembled flask, using a clamp and a stand with a heavy base, above the magnetic stirrer in such a way that it can be immersed in the oil bath.

3. Put the sodium hydride, the *p-t*-butylcalix[4]arene toluene complex and the 18-crown-6 in the flask and add 650 mL of dry THF and keep the flask under a slightly positive pressure of nitrogen.

4. Stir the suspension until gas evolution ceases (*c.* 0.5 h) and subsequently boil the suspension under reflux.

5. Meanwhile, add the terphenyl and 50 mL of dry THF to the dropping funnel and make sure the solution is homogeneous.

6. Add the solution of the terphenyl dropwise to the suspension of the calix[4]arene anion which is heated to reflux. After the addition is complete heat to reflux for another 3 h.

7. Cool the solution to room temperature and slowly add 5 mL of water dropwise.

8. Transfer the solution to a round-bottomed flask and remove the THF under reduced pressure using a rotary evaporator.

9. Dissolve the residue in dichloromethane (100 mL), transfer the suspension to a separating funnel and wash with 1 M aqueous HCl (50 mL) and brine (50 mL).

10. Dry the dichloromethane layers over $MgSO_4$ and filter through a filter paper. Concentrate the filtrate under reduced pressure using a rotary evaporator.

11. Apply the residue to a flash column packed with silica gel. Elute with a mixed solvent of hexane–ethyl acetate (4:1), collect the main fraction and remove the solvents under reduced pressure using a rotary evaporator.

12. Recrystallize the residue by dissolving it in a minimum amount of dichloromethane and carefully adding methanol until crystallization starts. Collect the solid and dry under vacuum (2.6 g, 84% yield; m.p. > 240°C dec).

[a] For the preparation of dry THF, see Protocol 1.

8: Spherands, hemispherands and calixspherands

Protocol 12.
Alkylation of calixspherand diols: formation of calixspherand potassium complexes (Scheme 8.6)

Caution! Carry out all procedures in a well-ventilated hood, and wear disposable vinyl or latex gloves and chemical-resistant safety goggles.

The following procedure for the synthesis of 2,26,31,41-tetrakis(1,1-dimethylethyl)-35,38,44,45,46-pentamethoxy-9,14,19-trimethylcalixspherand potassium complex **22a** is representative.

Equipment

- Magnetic stirrer and heater with contact thermometer
- Oil bath
- Triple-necked, round-bottomed flask (250 mL)
- Pressure-equalizing addition funnel (50 mL)
- Teflon-coated magnetic stirrer bar
- 2 stoppers
- Condenser with nitrogen inlet
- Pear-shaped evaporation flask (250 mL)
- Separating funnel (250 mL)
- Erlenmeyer flasks (2 × 150 mL)
- Single-necked round-bottomed flask (50 mL)
- Source of nitrogen

Materials

- Potassium t-butoxide (FW 112.2), 0.45 g, 4 mmol — **moisture sensitive, corrosive**
- 2,26,31,41-Tetrakis(1,1-dimethylethyl)-44,45,46-trimethoxy-9,14,19-trimethylcalixspherand-35,38-diol (FW 1035.4), 1.04 g, 1 mmol
- Iodomethane (FW 141.9, d 2.28), 0.62 mL, 10 mmol — **highly toxic, cancer-suspect agent**
- Dry THF,[a] 100 mL — **flammable**
- Dichloromethane for extraction and crystallization — **toxic**
- 1 M aqueous hydrochloric acid — **corrosive, irritant**
- Methanol, 50 mL — **flammable, toxic**
- Diisopropyl ether, c. 100 mL — **flammable, toxic**
- Potassium picrate, 2 g
- Chloroform, 25 mL — **highly toxic, cancer-suspect agent**

1. Make sure all glassware is clean and dry all glassware for at least 30 min in a 110°C electric oven before use.
2. Assemble the flask with the magnetic stirrer bar, the condenser with nitrogen inlet, the dropping funnel and two stoppers. Support the assembled flask, using a clamp and a stand with a heavy base, above the magnetic stirrer in such a way that it can be immersed in the oil bath.
3. Put the calixspherand diol and 0.25 mL of iodomethane in the flask, add 80 mL of dry THF and keep the flask under a slightly positive pressure of nitrogen.
4. Dissolve the potassium t-butoxide in 20 mL of dry THF in the dropping funnel and add this solution to the solution of the calixspherand diol at room temperature.
5. After addition of all the potassium t-butoxide, add the remaining iodomethane (0.37 mL). Heat the solution to 50°C and stir for 1 h.

203

Protocol 12. *Continued*

6. Cool the solution to room temperature and add 5 mL of 1 M aqueous HCl dropwise.
7. Transfer the solution to a round-bottomed flask and remove the THF under reduced pressure using a rotary evaporator.
8. Dissolve the residue in dichloromethane (100 mL), transfer the suspension to a separating funnel and wash with 1 M aqueous HCl (50 mL) and brine (50 mL).
9. Dry the dichloromethane layers over $MgSO_4$ and filter through a filter paper. Concentrate the filtrate under reduced pressure using a rotary evaporator.
10. Redissolve the residue in methanol (25 mL) and remove undissolved material by filtration. Remove the methanol of the filtrate under reduced pressure using a rotary evaporator.
11. Recrystallize the residue by dissolving it in a minimum amount of cold dichloromethane and carefully add diisopropyl ether until crystallization starts. Collect the solid and dry under vacuum to give the potassium iodide complex of the calixspherand as a white solid.
12. Dissolve the potassium iodide complex in methanol (25 mL) in a 50 mL round-bottomed flask, add excess potassium picrate, 2 g, and stir well for 15 min.
13. Remove the methanol under reduced pressure using a rotary evaporator.
14. Redissolve the residue in chloroform (25 mL) and remove undissolved material by filtration. Concentration of the filtrate under vacuum gives the potassium picrate complex of the calixspherand as a yellow solid (1.1 g, 81% yield).

[a] For the preparation of dry THF, see Protocol 1.

Protocol 13.
Decomplexation of calixspherand alkali metal complexes

Caution! This reaction is performed under pressure, so should be carried out behind a safety shield. Carry out all procedures in a well-ventilated hood, and wear disposable vinyl or latex gloves and chemical-resistant goggles.

Equipment
- Magnetic stirrer and heater with contact thermometer
- Oil bath
- Hydrogenation vessel with safety seal set to 6 atmospheres (e.g. Parr bench top hydrogenation apparatus)
- Small Teflon-coated magnetic stirrer bar
- Safety shield
- Separating funnel (100 mL)
- Erlenmeyer flasks (2 × 50 mL)
- Sintered glass filter (10 mL)

8: Spherands, hemispherands and calixspherands

Materials
- Methanol, 18 mL **flammable, toxic**
- Deionized water, 16 mL
- Calixspherand alkali metal complex, 200 mg
- Dichloromethane (for extraction) **toxic**

1. Dry all glassware in an electric oven at 110 °C for 2 h.
2. Support the reaction vessel using a clamp and stand with heavy base.
3. Put the calixpherand complex in the flask and add methanol (4 mL). Dissolve the calixspherand complex by stirring.
4. Add deionized water (16 mL) while stirring, and close the reaction vessel carefully but ensure it is tightly sealed.
5. Insert the vessel in the oil bath in such a way that the inside and outside levels are equal.
6. Set the contact thermometer to 120 °C.
7. Place the safety shield around the whole setup and heat the flask slowly to 120 °C.
8. Heat the reaction for 3 d.
9. Cool the solution to room temperature and open the reaction vessel carefully.
10. Transfer the suspension to a separating funnel and extract twice with 20 mL portions of dichloromethane.
11. Dry the dichloromethane layers over $MgSO_4$ and filter through a filter paper. Concentrate the filtrate under reduced pressure using a rotary evaporator.
12. Redissolve the residue in 10 mL of dichloromethane and add 10 mL of methanol.
13. Remove the solvents under reduced pressure until a white precipitate forms.
14. Collect the precipitate by filtration on a sintered glass filter and wash with methanol (2 × 2 mL).
15. Dry the solid under vacuum to give the pure free ligand.

References
1. Cram, D. J.; Kaneda, T.; Helgeson, R. C.; Lein, G. M. *J. Am. Chem. Soc.* **1979**, *101*, 6752–6754.
2. Cram, D. J. *Angew. Chem.* **1988**, *100*, 1041–1052.
3. de Jong, F.; Reinhoudt, D. N. *Adv. Phys. Org. Chem.* **1981**, *17*, 279–433.
4. Cram, D. J. *Angew. Chem.* **1986**, *98*, 1041–1060.

5. Judice, J. K.; Keipert, S. J.; Knobler, C. B.; Cram, D. J. *J. Chem. Soc., Chem. Commun.* **1993**, 1325–1327.
6. Koenig, K. E.; Lein, G. M.; Stuckler, P.; Kaneda, T.; Cram, D. J. *J. Am. Chem. Soc.* **1979**, *101*, 3553–3566.
7. Artz, S. P.; Cram, D. J. *J. Am. Chem. Soc.* **1984**, *106*, 2160–2171.
8. Dijkstra, P. J.; Brunink, J. A. J.; Bugge, K. E.; Reinhoudt, D. N.; Harkema, S.; Ungaro, R.; Ugozzoli, F.; Ghidini, E. *J. Am. Chem. Soc.* **1989**, *111*, 7567–7575.
9. Iwema Bakker, W. I.; Haas, M.; Khoo-Beattie, C.; Ostaszewski, R.; Franken, S. M.; den Hertog Jr., H. J.; Verboom, W.; de Zeeuw, D.; Harkema, S.; Reinhoudt, D. N. *J. Am. Chem. Soc.* **1994**, *116*, 123–133.
10. Iwema Bakker, W. I.; Haas, M.; den Hertog Jr., H. J.; Verboom, W.; de Zeeuw, D.; Bruins, A. P.; Reinhoudt, D. N. *J. Org. Chem.* **1994**, *59*, 972–976.
11. Cram, D. J.; Kaneda, T.; Helgeson, R. C.; Brown, S. B.; Knobler, C. B.; Maverick, E.; Trueblood, K. N. *J. Am. Chem. Soc.* **1985**, *107*, 3645–3657.
12. Dijkstra, P. J.; van Steen, B. J.; Reinhoudt, D. N. *J. Org. Chem.* **1986**, *51*, 5127–5133.
13. Danks, I. P.; Sutherland, I. O. *J. Inclusion Phenom.* **1992**, *12*, 223–236.
14. Dijkstra, P. J.; Skowronska-Ptasinska, M.; Reinhoudt, D. N.; den Hertog Jr., H. J.; van Eerden, J.; Harkema, S.; de Zeeuw, D. *J. Org. Chem.* **1987**, *52*, 4913–4921.
15. Dijkstra, P. J.; den Hertog Jr., H. J.; van Steen, B. J.; Zijlstra, S.; Skowronska-Ptasinska, M.; Reinhoudt, D. N.; van Eerden, J.; Harkema, S. *J. Org. Chem.* **1987**, *52*, 2433–2442.
16. Oude Wolbers, M. P.; Snellink-Ruël, B. H. M.; van Veggel, F. C. J. M.; Reinhoudt, D. N. unpublished results.
17. Ostaszewski, R.; Verboom, W.; Reinhoudt, D. N. *Synlett* **1992**, 354–356.
18. Miyaura, N.; Ishiyama, T.; Sasaki, H.; Ishikawa, M.; Satoh, M.; Suzuki, A. *J. Am. Chem. Soc.* **1989**, *111*, 314–321.
19. Keana, J. F. W.; Cuomo, J.; Lex, L.; Seyedrezai, S. E. *J. Org. Chem.* **1983**, *48*, 2647–2654.
20. Dijkstra, P. J.; den Hertog Jr., H. J.; van Eerden, J.; Harkema, S.; Reinhoudt, D. N. *J. Org. Chem.* **1988**, *53*, 374–382.
21. For a review see: Van Loon, J.-D.; Verboom, W.; Reinhoudt, D. N. *Org. Prep. Proc. Int.* **1992**, *24*, 437–462.
22. For a review see: Asfari, Z.; Weiss, J.; Vicens, J. *Synlett* **1993**, 719–725.
23. Gutsche, C. D.; Iqbal, M. *Org. Synth.* **1989**, *68*, 234–237.
24. Chisholm, M. H.; Drake, S. R.; Naüni, A. A.; Streib, W. E. *Polyhedron* **1991**, *10*, 337–345.
25. Iwema Bakker, W. I.; Haas, M.; den Hertog Jr., H. J.; Verboom, W.; de Zeeuw, D.; Reinhoudt, D. N. *J. Chem. Soc., Perkin Trans. 2* **1994**, 11–14.
26. Iwema Bakker, W. I.; Verboom, W.; Reinhoudt, D. N. *J. Chem. Soc., Chem. Commun.* **1994**, 71–72.

9

Transition metal-templated formation of [2]-catenanes and [2]-rotaxanes

J.-C. CHAMBRON, S. CHARDON-NOBLAT, C. O. DIETRICH-BUCHECKER, V. HEITZ and J.-P. SAUVAGE

1. Introduction

Catenanes and rotaxanes have been considered together for decades although the topologically different nature of each system has been clearly stated only in a limited number of discussions.[1] These molecules are made up of two components kept together by a mechanical link: in a catenane, two macrocycles are interlocked, whereas in a rotaxane a macrocycle is threaded on to a dumbbell-shaped molecular component whose stoppers are large enough so as to prevent unthreading of the ring (Fig. 9.1). Topologically speaking, catenanes are non-planar molecules, since they cannot be represented in a plane without two crossings at least. On the contrary, rotaxanes are planar molecules: it is easy to imagine the stretching of the chemical bonds of the ring so as to allow the blocking groups to pass through the ring and thus to obtain the cyclic and linear fragments separately.

A limited number of syntheses of catenanes and rotaxanes have been devised since the late 1960s.[2-8] They include totally directed[3,8] as well as statistical[2,7] syntheses, and, more recently, template syntheses.[4-6]

In the early 1980s, a transition-metal template strategy for the synthesis of catenanes was developed in our laboratory.[9] It is based on the ability of Cu(I) to entwine two diphenylphenanthroline molecules in such a way that the coordination geometry around the metal is tetrahedral. Thus catenanes

Fig. 9.1

synthesized in this manner are made up of two diphenylphenanthroline-containing macrocycles. This strategy was recently extended to rotaxanes:[10,11] in these molecules, the locked macrocycle is the same as the one used for making catenanes, and the dumbbell is a diphenylphenanthroline which bears two bulky porphyrinic moieties, whose size is easily controlled by the means of meso or β-substituents. Removal of the template transition metal cation (Cu$^+$) leads to the metal-free catenane and rotaxane. The principle of this generalized strategy is illustrated in Scheme 9.1.

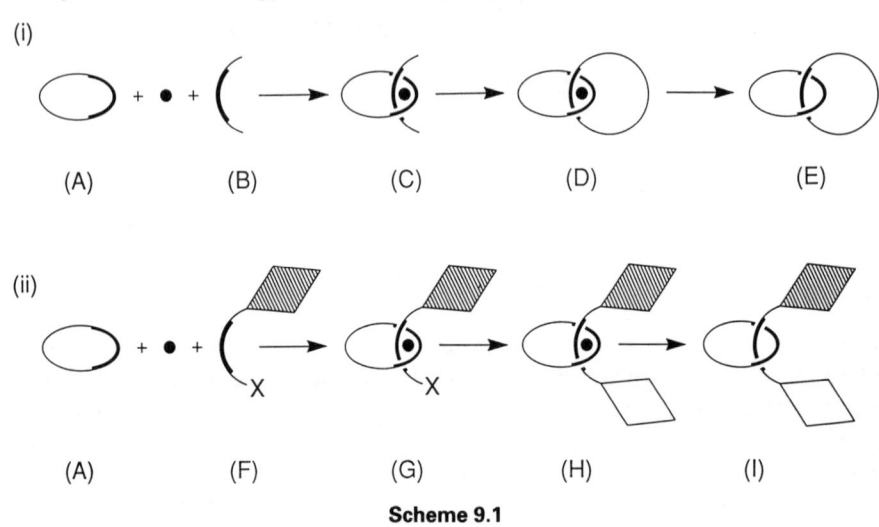

Scheme 9.1

2. Transition metal-templated synthesis of catenanes

To obtain a [2]-catenane following the strategy depicted in Scheme 9.1 requires the preliminary synthesis of a 2,9-diaryl-1,10-phenanthroline acyclic chelate and of a coordinating macrocycle. These chelating subunits are both obtained from a single precursor 2,9-dianisyl-1,10-phenanthroline, prepared in two steps from commercial 1,10-phenanthroline. The first step, preparation of a *p*-lithioanisole **2** solution, is achieved by the direct interaction of freshly cut lithium with *p*-bromoanisole **1** in ether under argon at room temperature (Scheme 9.2).[12] The resulting organolithium solution is titrated by the double titration method described by Gilman *et al.*[13] before subsequent use.

9: *Transition metal*

Protocol 1.
Preparation of *p*-anisyllithium (Structure 2, Scheme 9.2)

Caution! Carry out all procedures in a well-ventilated hood, and wear disposable vinyl or latex gloves and chemical-resistant safety goggles.

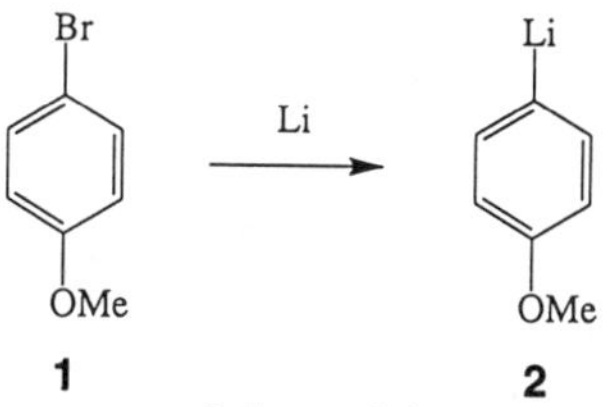

Scheme 9.2

The following procedure is representative of the preparation of almost all organolithium compounds with lithium metal and their titration.

Equipment

- Three-necked, round-bottomed flask (500 mL)
- Two-way Claisen adaptor
- Double surface water-cooled condenser
- Rubber septa
- Pasteur pipettes
- Oil bath
- Dual-bank vacuum manifold (vacuum–argon)
- Conical funnels
- Graduated cylinders
- Double-ended filter (frit porosity 3) with two male joints
- Two-necked, round-bottomed flask (250 mL, graduated every 20 mL) or 250 mL graduated storage flask with glass stopcock on side arm
- Dropping funnel with pressure equalization arm
- Conical flasks (4 × 50 mL)
- Graduated burette
- Thermometer (0–100 °C)
- Large Teflon-coated magnetic stirring bar
- 90°-angled adaptor with glass stopcock and hose barb
- Hot plate magnetic stirrer
- Flameless heat gun
- Large powder funnel
- 1 double-tipped steel cannula
- Hamilton syringe (2 × 1 mL)
- Small Teflon-coated magnetic stirring bars (4×)

Materials

- *p*-Bromoanisole, 56.1 g, 0.3 mol — irritant, toxic
- Lithium metal, 7 g, 1 mol — flammable, irritant, sensitive to moisture
- Anhydrous diethyl ether, 300 mL — extremely flammable
- Anhydrous toluene, 20 mL — harmful, highly flammable
- 1,2-dibromoethane, 20 mL — toxic
- HCl 0.1 M (Titrisol) — corrosive
- Phenolphthalein, 1% in ethanol
- Aluminium oxide standardized (activity II-III; 70–230 mesh ASTM)

1. Place the pieces of a broken Pasteur pipette with the large magnetic stirring bar in the three-necked, round-bottomed flask and dry it as well as all the other glassware in an electric oven at 110 °C overnight. Before use, allow to cool in a desiccator.
2. Weigh the lithium. For that, pour about 20 mL toluene into a small beaker and weigh it. Take a piece of lithium metal out of the oil, clean it with filter paper and rapidly cut large pieces with a sharp knife and put them into the

Protocol 1. Continued

beaker still on the balance until 7 g is weighed out. Cover the beaker with a watch glass.[a]

3. Fit the three-necked round-bottomed flask with a thermometer, the angled adaptor for argon inlet (side neck) and the solid funnel (large middle neck). Pour 300 mL ether into the cooled flask and cut the lithium into thin chips directly in the flask over the argon stream evolving from the large funnel.
4. Replace the funnel with the two-way Claisen adaptor fitted with reflux condenser and dropping funnel containing the p-bromoanisole (56.1 g). Place the argon inlet on top of the reflux condenser and close the side neck of the flask with a septum.
5. Start vigorous stirring and purge the whole reaction set-up by three argon–vacuum cycles.
6. To start the reaction, add rapidly a few drops of p-bromoanisole on to the stirred lithium suspension and heat the flask with a heat gun. Once the reaction has started (gas evolution) add the remaining p-bromoanisole dropwise at such a rate that the ether boils gently. At the end of the addition, place an oil bath under the flask and continue the stirring and heat under reflux for a further 15 min.
7. Equip the graduated storage flask with the double-ended filter closed on the top by a septum and purge this reaction set-up with argon.
8. Transfer the ethereal p-anisyllithium solution in the storage flask by means of a steel cannula under argon pressure. After transfer, remove the double-ended filter and close the storage flask with a septum.
9. Titration of the resulting organolithium solution: withdraw by means of a Hamilton syringe an aliquot (1 mL) of the organolithium solution and hydrolyse it in 20 mL of distilled water contained in a 50 mL conical flask. Titration with standard HCl using 3 drops of an ethanolic phenolphthalein solution as an indicator gives the total alkali concentration.
10. Pour 10 mL of dry, acid-free dibromoethane (obtained by filtration over a neat aluminium oxide column, collected and stored under argon) in a second 50 mL conical flask. Add another 1 mL aliquot of the organolithium solution into the 50 mL flask containing the dibromoethane, allow the mixture to stand for 1 min after the addition, then add 10 mL of distilled water, 3 drops of indicator and titrate with standard HCl under vigorous stirring. The difference between the two titrations gives the quantity of p-anisyllithium (200 mL of a 1.3 molar solution, 86% yield).[b] For greater accuracy the titration should be performed at least three times.

[a] All filter papers, knife, etc., which were in contact with lithium metal have to be treated with ethanol before cleaning with water or acetone. The best method of destroying excess lithium is to pour a large amount of technical grade ethanol in a large crystallizing dish and allow lithium-spoiled items to soak therein. It will also serve to destroy (under stirring) the lithium pieces left over after the reaction.
[b] If tightly closed, this solution may be stored in a refrigerator overnight but is best used shortly after its titration.

9: Transition metal

The reaction of 5 equivalents of *p*-lithioanisole **2** with 1,10-phenanthroline **3** suspended in toluene at room temperature under argon leads, after hydrolysis and oxidation by MnO$_2$, to the 2,9-dianisyl-1,10-phenanthroline **4** in 70% yield (Scheme 9.3).

Protocol 2.
Synthesis of 2,9-di(*p*-anisyl)-1,10-phenanthroline (Structure 4, Scheme 9.3)

Caution! Carry out all procedures in a well-ventilated hood, and wear disposable vinyl or latex gloves and chemical-resistant safety goggles.

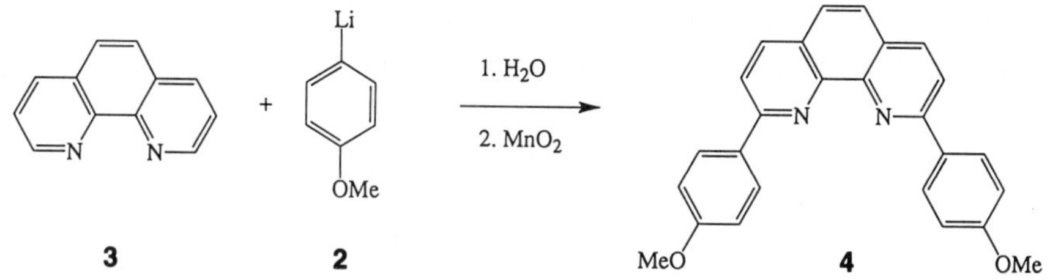

Scheme 9.3

This procedure can be used for the preparation of various alkyl or aryl 1,10-phenanthrolines disubstituted in their 2 and 9 positions.

Equipment

- Three-necked, round-bottomed flask (500 mL)
- Large Teflon-coated magnetic stirring bars (2×)
- Magnetic stirrer
- 90°-angled adaptor with glass stopcock and hose barb
- Graduated burette (100 mL)
- Water-cooled condenser
- Thermometer (0–100°C)
- Rubber septa
- Double-tipped steel cannula
- Conical funnels
- Powder funnels (2×)
- Filtering flask (500 mL)
- Small beaker
- Filter paper
- Dual-bank vacuum manifold (vacuum–argon)
- Separating funnel (500 mL)
- Conical flasks (750 and 200 mL)
- Sintered disc filter funnel (porosity 4)
- Column for chromatography

Materials

- Dry toluene, 50 mL — harmful, highly flammable
- 1,10-Phenanthroline monohydrate 4.95 g, 25 mmol — toxic
- Dichloromethane for extraction and chromatography, *c.* 1.5 L — harmful
- Methanol for chromatography, *c.* 50 mL — toxic, highly flammable
- Manganese dioxide Merck no. 805958 (~100 g) — harmful
- Magnesium sulfate for drying
- Toluene for recrystallization, 80 mL — harmful, highly flammable
- Silica gel (70–230 mesh ASTM)

1. Dry the three-necked, round-bottomed flask with magnetic stirring bar and double-tipped needle overnight in an electric oven at 110°C. Allow to cool in a desiccator before use.

Protocol 2. *Continued*

2. Weigh directly the phenanthroline (4.95 g) in the three-necked, round-bottomed flask, add 50 mL of dry toluene and equip the flask with a thermometer, septum and water-cooled condenser. Fix the angled adaptor on top of it.
3. Purge the reaction flask by three vacuum–argon cycles while the phenanthroline–toluene suspension is stirred.
4. By means of the double-tipped steel cannula add 100 mL of the *p*-anisyllithium solution to the suspension at room temperature. At the beginning of addition, the temperature rises to 40 °C and a bright-yellow colour appears in the flask. This yellow colour disappears progressively and at the end of the addition (100 mL, 130 mmol *p*-anisyllithium) a dark-red solution is obtained. Stir under argon at room temperature for a further 48 h.
5. Cool the round-bottomed flask with crushed ice and hydrolyse the dark-red solution at 0 °C under argon by injecting distilled water slowly and carefully through the septum. The solution turns bright yellow.
6. Decant the bright-yellow toluene layer and store it in the 750 mL conical flask.
7. Extract the remaining aqueous layer three times with 150 mL portions of CH_2Cl_2 and pour them also into the conical flask containing the toluene layer.
8. Rearomatize the combined organic layers by successive additions of MnO_2 under efficient magnetic stirring (~20 g for each batch).
9. Follow this reoxidation by TLC and disappearance of the yellow colour. TLC is carried out as follows: add a 20 g batch of MnO_2, allow the black suspension to stir for 15 min, remove with a Pasteur pipette a small aliquot of the solution (MnO_2 having first been allowed to settle down after stirring was stopped) and put directly on the TLC plate (silica gel plates eluted with CH_2Cl_2 and 4% MeOH; R_f of dianisylphenanthroline = 0.8; detecting agent: iodine vapour).
10. After reoxidation (after addition of *c*. 100 g MnO_2), dry the mixture by adding $MgSO_4$ (~100 g) to the conical flask. Stir for 30 min.
11. Filter the black slurry on a sintered disc filter funnel (of very fine porosity) and wash the filter cake thoroughly at least 5 times with CH_2Cl_2 (5 × 40 mL).
12. Evaporate the filtrate on a rotatory evaporator: crude compound **4** crystallizes spontaneously on cooling down.
13. Recrystallize crude **4** from hot toluene (*c*. 50 mL).
14. Filter the precipitated almost colourless crystals on a conical funnel fitted with filter paper or on a sintered disc filter funnel.
15. Wash rapidly the crystals with cold toluene; 4.20 g (10.7 mmol, 43% yield) of pure **4** are thus obtained.

9: Transition metal

16. A further 2.63 g (6.70 mmol) of **4** is obtained by column chromatography of the filtrate on silica gel (CH$_2$Cl$_2$/hexane, 50/50). Switch progressively to pure CH$_2$Cl$_2$; **4** is thereafter eluted with CH$_2$Cl$_2$/1% MeOH as a colourless solid.

The phenolic functions are deprotected under rather drastic conditions, by analogy with methods defined previously for the cleavage of aryl methyl ethers.[14] Treatment of dianisylphenanthroline **4** with pyridinium hydrochloride at 210°C affords 2,9-di(*p*-phenol)-1,10-phenanthroline **5** quantitatively (Scheme 9.4).

Protocol 3.
Preparation of 2,9-di(*p*-phenol)-1,10-phenanthroline (Structure 5, Scheme 9.4)

Caution! Carry out all procedures in a well-ventilated hood, and wear disposable vinyl or latex gloves and chemical-resistant safety goggles.

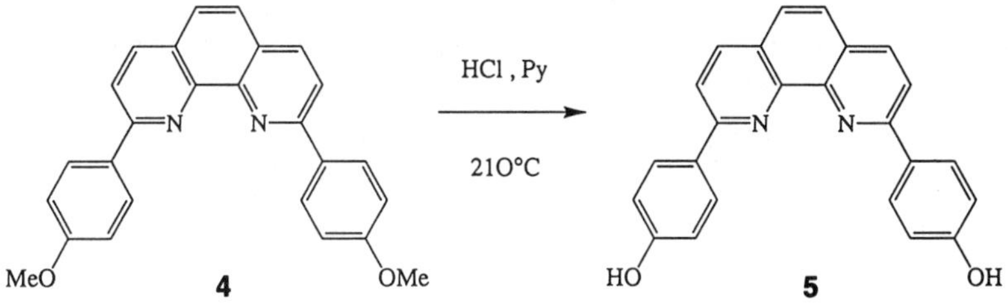

Scheme 9.4

Equipment

- Three-necked, round-bottomed flask (100 mL)
- Single-necked, round-bottomed flask (50 mL)
- Conical funnels (1 small, 1 large)
- Thermometer (250°C)
- Distillation head
- Graduated cylinders
- Powder funnel
- Teflon-coated magnetic stirring bars (1 small, 1 large)
- Heating mantle (100 mL)
- Pasteur pipettes
- Two-way Claisen adaptor
- Syringe (10 mL)

- Filter paper
- Vacuum desiccator
- Porous plate
- Source of argon
- Magnetic stirrer
- Water-cooled condenser
- Reversed-flow oil bubbler containing mineral oil
- Conical flasks (100 mL and 500 mL)
- pH meter
- Crystallizing dish
- Beaker (500 mL)
- Rubber septa

Materials

- Technical grade pyridine, 16 mL **highly flammable, harmful**
- Concentrated hydrochloric acid, 17.6 mL **corrosive**
- Buffer pH 7
- Sodium hydroxide pellets **corrosive**
- Phosphorus pentoxide powder **corrosive, irritant**

Protocol 3. *Continued*

1. Pour the pyridine (16 mL) into the three-necked, round-bottomed flask containing the small magnetic stirring bar.
2. While stirring, add progressively by means of a Pasteur pipette the hydrochloric acid (17.6 mL). *Caution!* Highly exothermic and fuming reaction: the three necks are best left open during the addition of HCl.
3. After HCl addition, fit the flask with a thermometer and distillation head.
4. Start to heat and distil the water from the mixture until its internal temperature rises to 210 °C (boiling point of the anhydrous pyridine hydrochloride).
5. Remove the distillation head and allow the mixture to cool to 140 °C.
6. Place the powder funnel on the large middle neck and pour, in a single batch, pure solid **4** (6.27 g, 16 mmol) into the flask.
7. Fit the reaction flask with the water-cooled condenser carrying on its top the two-way Claisen adaptor connected on one side to a source of argon and on the other to a mineral oil bubbler.
8. After establishment of an argon atmosphere in the flask, stir the contents and heat under reflux for 3 h (internal temperature 190–220 °C).
9. Dilute the hot yellow stirred mixture by slow careful injection (syringe) of 10 mL of hot water.
10. Pour the resulting bright yellow suspension into a conical flask containing 60 mL of hot water.
11. Allow the aqueous suspension to cool down to room temperature.
12. Filter the yellow precipitated solid on filter paper and wash it with cold water.
13. Transfer the crude acidic diphenol **5** from the filter paper to a 500 mL beaker and suspend it in an ethanol–water mixture (250/85 mL). It dissolves only partially.
14. Calibrate the pH meter to pH = 7.
15. Dissolve a few NaOH pellets in distilled water (~50 mL).
16. Under efficient magnetic stirring neutralize the acidic suspension by successive additions of the dilute NaOH solution. Monitor this neutralization with the pH-meter (end-point: pH 7.32).
17. After neutralization, dilute the suspension with hot water (300 mL).
18. Cool the suspension in a refrigerator: neutral **5** precipitates as a beige solid during cooling down.
19. Filter it on filter paper and wash it with cold water (3 × 30 mL).
20. Transfer the solid from the filter on a porous dish and dry it overnight in air to yield 5.85 g of an ochre solid.

9: Transition metal

21. Further drying is achieved in a high vacuum desiccator in the presence of P_2O_5 which is contained in a crystallizing dish.
22. During the latter drying procedure observe the change of colour: diphenol **5** turns from ochre to red. Overnight drying under high vacuum is usually enough to obtain the scarlet bright-red anhydrous diphenol **5** (5.31 g, 92% yield).[a] This can be used without further purification.

[a] It may be necessary to remove the hydrated P_2O_5 and replace it by a fresh amount and dry further under high vacuum until the diphenol becomes red.

The acyclic link required for the construction of a phenanthroline-containing macrocycle derives from pentaethyleneglycol di-*p*-toluenesulfonate **6** as shown in Scheme 9.5. By treating **6** in refluxing acetone with a large excess of sodium iodide, the 1,14-diiodo-3,6,9,10-tetraoxatetradecane **7** is obtained as a pale yellow liquid in 99% yield.

Protocol 4.
Preparation of the 1,14-diiodo-3,6,9,10-tetraoxatetradecane (Structure 7, Scheme 9.5)

Caution! Carry out all procedures in a well-ventilated hood, and wear disposable vinyl or latex gloves and chemical-resistant goggles.

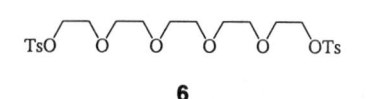

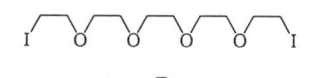

Scheme 9.5

Equipment
- Column for chromatography
- Powder funnel
- Water-cooled condenser
- Three-necked, round-bottomed flask (250 mL)
- Thermometer (100 °C)
- Teflon-coated magnetic stirring bar
- Magnetic stirrer

- Rubber septa
- Conical flask (500 mL)
- Separating funnel
- Oil bath
- Single-necked round-bottomed flask (500 mL)
- Suction flask
- Sintered disc filter funnel (medium porosity)

Materials
- Pentaethylene glycol di-*p*-toluenesulfonate (95%), 22.10 g, 40 mmol[a]
- Acetone, 150 mL (reagent grade)   **highly flammable**
- Sodium iodide
- Diethyl ether for extraction and chromatography   **extremely flammable**
- Sodium thiosulfate, 20 mL of a 10% aqueous solution
- Magnesium sulfate
- Silica gel (70–230 mesh ASTM)
- Ethanol for chromatography   **highly flammable**

Protocol 4. Continued

1. Purify the commercial (95% pure) ditoluenesulfonate by chromatography on silica gel (diethyl ether). The pentaethylene glycol ditoluenesulfonate is eluted as a colourless oil with $Et_2O/EtOH$ (15% EtOH).
2. Weigh 22.10 g of the ditoluenesulfonate into the three-necked, round-bottomed flask and add 150 mL acetone.
3. Under efficient stirring, at room temperature, add NaI to the resulting solution until saturation is reached.
4. Fit the three-necked, round-bottomed flask with a thermometer, water-cooled condenser and close the remaining opening with a septum. Place a drying tube containing $CaCl_2$ on top of the condenser.
5. Start to heat the stirred mixture and boil under reflux for 2 h. A white solid, sodium toluenesulfonate, begins to precipitate.
6. Turn off the heating, add more NaI to the mixture and allow it to stir overnight at room temperature
7. Filter the large amount of precipitated sodium toluenesulfonate on a sintered disc filter funnel and wash it with Et_2O (3 × 100 mL).
8. Transfer the organic filtrate to a round-bottomed flask and remove the solvents on a rotary evaporator.
9. The white slurry thus obtained is taken up in Et_2O/H_2O, 1:1 v/v, 300 mL.
10. Decant both phases and extract the aqueous layer with three 100 mL portions of Et_2O.
11. The combined pale-yellow Et_2O layers are transferred to a 500 mL conical flask; add ~20 mL of the 10% sodium thiosulfate solution and stir the mixture until complete discolouration occurs. Addition of more sodium thiosulfate solution may be necessary.
12. Decant the ether and aqueous layers in a separating funnel.
13. Wash the organic Et_2O layer with two 100 mL portions of distilled water.
14. Dry the organic colourless Et_2O layer over $MgSO_4$, filter it on a sintered disc filter funnel and evaporate the solvent on a rotary evaporator to yield 18.4 g of pure diiodo derivative **7** (pale-yellow liquid, 99% yield).

[a] Commercial pentaethylene glycol di-*p*-toluenesulfonate is quite expensive, so can be prepared from crude pentaethylene glycol obtained by distillation of PEG 200, followed by a tosylation reaction.[15,16]

Reaction of one equivalent of diiodo acyclic link molecule **7** with one equivalent of 2,9-di(*p*-phenol)-1,10-phenanthroline **5** under high-dilution conditions in the presence of a large excess of caesium carbonate affords the chelating macrocyle **8** in 45% yield (Scheme 9.6).

Protocol 5.
Preparation of macrocycle Structure 8 (Scheme 9.6)

Caution! Carry out all procedures in a well-ventilated hood, and wear disposable vinyl or latex gloves and chemical-resistant safety goggles.

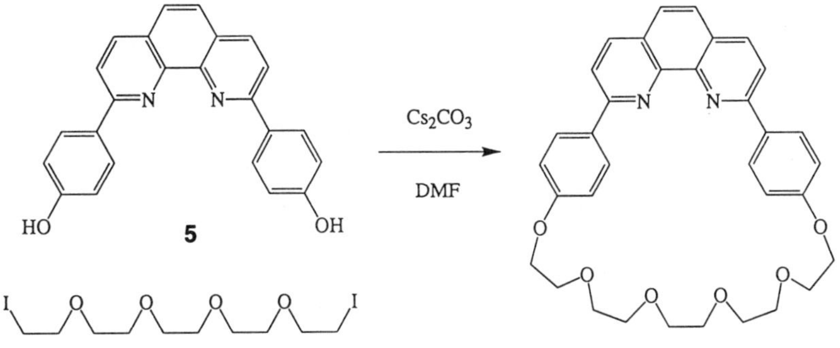

Scheme 9.6

Equipment

- Four-necked round-bottomed flask (1 L)
- Two-way Claisen adaptor
- Mechanical stirrer
- Rubber septa
- Mortar and pestle
- Pasteur pipettes
- PVC laboratory tubing
- Long syringe needle
- Conical funnels (1 large and 1 small)
- Conical flasks (100 and 500 mL)
- Suction flask (500 mL)
- Constant-pressure dropping funnel (200 mL) Normag, for high dilution
- Column for chromatography
- Thermometer (100°C)
- Oil bath
- Conical powder funnel
- Conical flask (100 mL)
- 90°-angled adaptor with glass stopcock
- Syringe and needles (5 mL)
- Single-necked round-bottomed flask (1 L)
- Source of argon or nitrogen
- Separating funnel (500 mL)
- Sintered disc filter funnel
- Graduated cylinders

Materials

- Diphenol 5, 1.045 g, 2.87 mmol
- Diiodo derivative 7, 1.445 g, 3.15 mmol
- Cs_2CO_3, 3 g, 9.2 mmol **harmful**
- Dry dimethylformamide (DMF), 500 mL **harmful**
- Dichloromethane, for chromatography **harmful**
- Methanol, for chromatography **highly flammable, toxic**
- Hexane, for chromatography **highly flammable**
- Magnesium sulfate

1. Dry the four-necked, round-bottomed flask, Claisen adaptor, conical funnels, mortar and pestle and graduated cylinders in an electric oven at 110°C overnight.

Protocol 5. *Continued*

2. Heat the oil bath to 50–60 °C and assemble the four-necked flask while hot under a stream of argon (provided by a long syringe needle plunging in the flask). Equip the flask with a thermometer and mechanical stirrer.
3. Remove the mortar and pestle from the oven and weigh 3 g of Cs_2CO_3 in it; crush this white very hygroscopic powder rapidly and pour it into the round-bottomed flask.
4. Add 400 mL of dry DMF in the flask, remove the conical funnel and close this opening with a septum. The DMF–Cs_2CO_3 suspension is allowed to degas (strong argon bubbling by means of the long needle now introduced through the septum) while the remaining side-arm is fitted with the two-way Claisen adaptor equipped with the high-dilution dropping funnel and the 90°-angled adaptor, respectively.
5. Connect both the dropping funnel and 4-necked flask to the argon source by means of a Y-shaped PVC tubing adaptor.
6. Weigh 1.045 g of diphenol **5**, 1.445 g of diiodo chain **7** and mix them together with 100 mL DMF in a conical flask.
7. Pour this solution into the Normag dropping funnel. Close it.
8. Remove the long argon needle from the septum.
9. Start efficient stirring and bring the internal temperature of the Cs_2CO_3–DMF suspension to 55–60 °C.
10. Start to add the diphenol–chain mixture to the argon-flushed Cs_2CO_3 suspension at such a rate that the addition will take between 24 and 32 h.
11. At the end of addition, stirring and heating is continued for 48 h.
12. Transfer the reaction mixture into a single-necked, round-bottomed, 1 L flask and evaporate all the DMF on a rotary evaporator connected to a high-vacuum pump.
13. Take up the dry yellow residue in CH_2Cl_2–H_2O (1:1, v/v, 250/250 mL).
14. Transfer both layers into a separating funnel, decant and extract the aqueous layer with CH_2Cl_2 (3 × 50 mL).
15. Collect the organic layers in a 500 mL conical flask and stir the solution for 30 min with $MgSO_4$.
16. Filter the $MgSO_4$ on a sintered disc filter funnel and evaporate the solvent on a rotary evaporator.
17. Purify the 2.00 g of crude yellow glassy product by column chromatography on silica gel (100 g in CH_2Cl_2:hexane, 50:50). Pure macrocycle **8** is eluted with CH_2Cl_2 containing 0.5 to 1% MeOH (0.731 g, 45% yield).

Copper(I)-directed threading of the acyclic diphenolic link into the coordinating macrocycle leads to a copper(I)-precatenate which is best pre-

9: Transition metal

pared the following way: macrocycle **8** is first complexed with Cu(I) by reaction with one equivalent of Cu(CH$_3$CN)$_4$·BF$_4$ in CH$_3$CN/CH$_2$Cl$_2$ solution.[17] Subsequent addition of one equivalent of diphenol **5** in DMF at room temperature affords, in quantitative yield, the complex **9**$^+$ which may be used in the next step without further purification (Scheme 9.7).

Protocol 6.
Preparation of precatenate Structure 9$^+$(BF$_4^-$) (Scheme 9.7)

Caution! Carry out all procedures in a well-ventilated hood, and wear disposable vinyl or latex gloves and chemical-resistant safety goggles.

Equipment

- Round-bottomed Schlenk flask (250 mL)
- Rubber septa
- Magnetic stirrer
- Three double-tipped steel cannulae
- Conical funnels
- Round-bottomed Schlenk flasks (3 × 100 mL)

- Three small Teflon-coated magnetic stirrer bars
- Dual-bank vacuum manifold (vacuum, argon)
- Small syringe needle
- Single-necked, round-bottomed flask (250 mL)

Materials

• Reagent grade acetonitrile, 50 mL	flammable, toxic
• Reagent grade dichloromethane, 50 mL	harmful
• Reagent grade DMF, 50 mL	harmful
• Cu(CH$_3$CN)$_4$BF$_4$, 0.383 g, 1.21 mmol	harmful
• Macrocycle **8**, 0.656 g, 1.16 mmol	
• Diphenol **5**, 0.422 g, 1.16 mmol	irritant

1. Connect a 100 mL Schlenk flask to the vacuum line and pour 50 mL of acetonitrile into it. Close the flask with a septum and degas this solvent by three vacuum–argon cycles.

2. Connect a second 100 mL Schlenk flask to the vacuum line and place a conical funnel on its main neck. Allow a light stream of argon to flow through the reaction set-up.

3. Weigh rapidly on a small sheet of paper 0.383 g of Cu(CH$_3$CN)$_4$BF$_4$ and pour it into the flask filled with argon. Remove the conical funnel and close the flask with a septum.

4. Dissolve Cu(CH$_3$CN)$_4$BF$_4$ in ~30 mL of previously degassed acetonitrile transferring the solvent by means of a double-tipped steel cannula. Stir until all the copper salt is dissolved.

5. Weigh in the third 100 mL Schlenk flask, 0.422 g of the diphenol **5** and dissolve it in 50 mL DMF. Close the flask with a septum, connect it to the vacuum line and degas the dark yellowish mixture by three vacuum–argon cycles.

Scheme 9.7

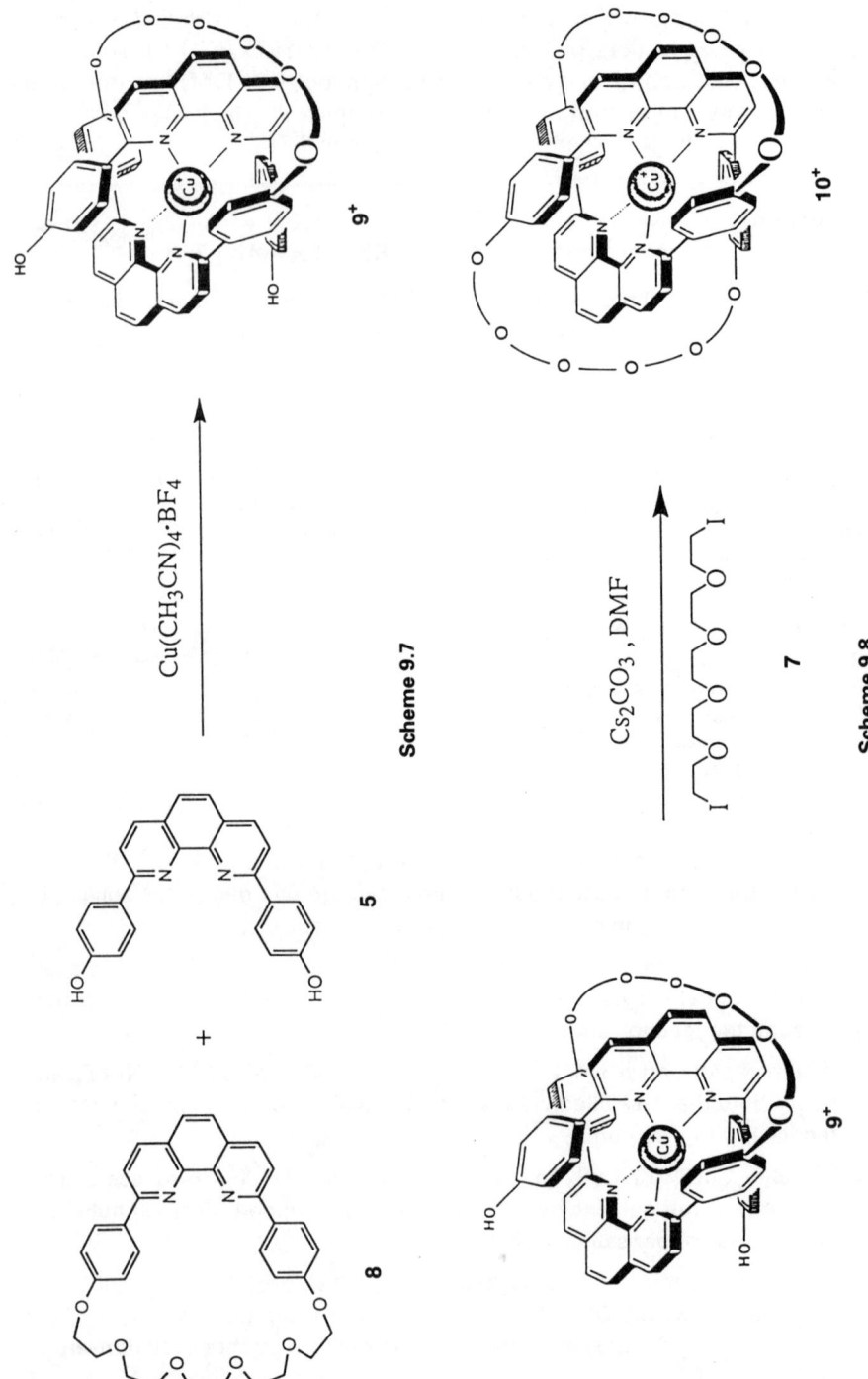

Scheme 9.8

Protocol 6. *Continued*

6. Weigh 0.656 g of macrocycle **8** in the 250 mL Schlenk flask, dissolve it in 50 mL of dichloromethane. Connect the flask to the vacuum line and degas the solution by three vacuum–argon cycles.

7. At room temperature transfer the Cu(CH$_3$CN)$_4$BF$_4$ solution to the stirred degassed solution of macrocycle **8** via a double-tipped steel cannula. A deep-orange colouration appears in the solution. Continue stirring for 0.5 h at room temperature.

8. Transfer the diphenol–DMF solution to the copper–macrocycle solution via a double-tipped steel cannula; the solution turns dark-red immediately.

9. Allow this mixture to stir under argon at room temperature for one additional hour.

10. Transfer the resulting solution to a 250 mL round-bottomed flask and evaporate to dryness on a rotary evaporator. A dark red solid of crude precatenate **9**$^+$(BF$_4^-$) is thus obtained in quantitative yield (1.252 g, 1.16 mmol).

Copper(I) catenate **10**$^+$, made of two interlocked 30-membered rings, is obtained in 42% yield by slow addition of an equimolar solution of precatenate **9**$^+$ and link **7** in DMF to a vigorously stirred suspension of Cs$_2$CO$_3$ in DMF under argon at 65 °C (high-dilution conditions) (Scheme 9.8).

Protocol 7.
Synthesis of catenate Structure 10$^+$(BF$_4^-$) (Scheme 9.8)

Caution! Carry out all procedures in a well-ventilated hood, and wear disposable vinyl or latex gloves and chemical-resistant safety goggles.

Equipment

- Four-necked, round-bottomed flask (1 L)
- Two-way Claisen adaptor
- Mechanical stirrer
- Conical powder funnel
- Mortar and pestle
- PVC laboratory tubing
- Long syringe needle
- Single-necked, round-bottomed flask (1 L)
- Separating funnel (500 mL)
- Sintered disc filter funnel
- Magnetic stirrer
- Constant-pressure dropping funnel, 200 mL, 'Normag' for high-dilution (BASF)
- Thermometer (100 °C)
- Oil bath
- Conical funnels
- 90°-angled adaptor with glass stopcock and hose barb
- 5 mL syringe and needles
- Rubber septa
- Conical flasks of various size (100 and 500 mL)
- Suction flask (500 mL)
- Teflon-coated magnetic stirring bars
- Graduated cylinders
- Column for chromatography
- Source of dry argon

Materials

- Precatenate **9**$^+$(BF$_4^-$), 1.252 g, 1.16 mmol
- Diiodo link **7**, 0.599 g, 1.31 mmol

Protocol 7. Continued

• Dry DMF, 300 mL	harmful
• Caesium carbonate, 1.300 g, 3.98 mmol	harmful
• Dichloromethane for extraction and chromatography	
• 34% aqueous HBF$_4$, 100 mL	corrosive
• NaBF$_4$, 15 to 20 g	harmful, corrosive
• MgSO$_4$	
• Silica gel (70–230 mesh)	
• Methanol for chromatography	toxic, highly flammable
• Hexane for chromatography	toxic, highly flammable

1. Dry the four-necked, round-bottomed flask, Claisen adaptor, conical funnels, mortar and pestle and graduated cylinders in an electric oven at 110 °C overnight.

2. Heat the oil bath to 50–60 °C and assemble the four-necked flask while hot under a stream of argon (provided by a long syringe needle plunging into the flask). Equip with a thermometer and a mechanical stirrer.

3. Remove the mortar and pestle from the oven and weigh directly 1.30 g of Cs$_2$CO$_3$ in it; crush it rapidly and pour it into the round-bottomed flask.

4. Add 200 mL of DMF to the flask, remove the conical powder funnel and close this opening with a septum. The DMF–Cs$_2$CO$_3$ suspension is allowed to degas (strong argon bubbling through the long needle now introduced via the septum) while the remaining side-arm is fitted with the two-way Claisen adaptor equipped with the high-dilution dropping funnel and the 90°-angled adaptor respectively.

5. Connect both the dropping funnel and the four-necked flask to the argon source with a Y-shaped PVC tubing adaptor.

6. Weigh 1.252 g of precatenate 9$^+$(BF$_4^-$), 0.599 g of diiodo link 7 and mix them together with 100 mL DMF in a 100 mL conical flask.

7. Pour the latter dark red solution into the 'Normag' dropping funnel; close it.

8. Remove the long argon needle from the septum.

9. Start an efficient stirring and bring the internal temperature of the DMF–Cs$_2$CO$_3$ suspension to 55–60 °C.

10. Start to add the precatenate–diiodo mixture to the argon-flushed Cs$_2$CO$_3$ suspension at such a rate that the whole addition will take about 30 h.

11. Observe that the mixture, first orange-yellow, turns progressively dark brown-red.

12. At the end of addition, continue stirring and heating for 40 h.

13. Transfer the reaction mixture into a single-necked, 1 L round-bottomed flask and evaporate the DMF on a rotary evaporator connected to a high-vacuum pump.

9: Transition metal

14. The dark-red dry residue is taken up in CH_2Cl_2–H_2O (250–250 mL).
15. Transfer the mixture into a separating funnel, decant and extract the aqueous layer with CH_2Cl_2 (3 × 100 mL).
16. Treat the organic layers collected in a 1 L conical flask with 100 mL 34% HBF_4 solution overnight (magnetic stirring, room temperature).[a]
17. Allow the decomplexed oligomers to settle down in the flask as a yellow glue: separate the supernatant by decanting or filtration on filter paper. These insoluble oligomers can be discarded.
18. After decanting, wash the remaining dark-red CH_2Cl_2 layer twice with water (in a separating funnel).
19. Pour the CH_2Cl_2 layers again into a 1 L conical flask and treat them for 4 h with a large excess of $NaBF_4$ (~15 g) dissolved in a minimum of water (magnetic stirring, room temperature).[b]
20. Transfer the mixture into a separating funnel, decant and wash the organic layer twice with water.
21. Dry the resulting organic layers collected in a conical flask over $MgSO_4$, filter on a sintered disc filter funnel and evaporate to dryness on a rotary evaporator to yield 977 mg of a dark-red solid.
22. Apply this crude mixture to a silica gel column (100 g in CH_2Cl_2:hexane, 50:50). Elute with increasing amounts of CH_2Cl_2 and finally with CH_2Cl_2 containing 0.5 to 5% MeOH. Macrocycle **8** is first eluted with CH_2Cl_2 containing 0.5 to 1% MeOH (0.289 g, 0.51 mmol, 45% yield) and thereafter pure catenate **10$^+$(BF$_4^-$)** with CH_2Cl_2 containing 1 to 5% MeOH (0.624 g, 0.49 mmol, 42% yield).

[a] This procedure allows the elimination of the open-chain copper(I) complexes (oligomers) also formed in the reaction.
[b] By the means of this exchange reaction, catenate **10$^+$**, originally formed as carbonate, iodide and tetrafluoroborate salts, is obtained exclusively as its BF_4^- salt.

The free catenand **11** is easily obtained in quantitative yield by treating the copper(I) catenate with a large excess of potassium cyanide (Scheme 9.9). This is the most convenient decomplexing agent, notwithstanding its toxicity.

Protocol 8.
Demetallation of copper(I) catenate Structure 10$^+$(BF$_4^-$) to afford the free catenand Stucture 11 (Scheme 9.9)

Caution! Carry out all procedures in a well-ventilated hood, and wear disposable vinyl or latex gloves and chemical-resistant goggles. Potassium cyanide is extremely toxic. Cyanide residues, as well as glassware which has been in contact with cyanide, have to be soaked 24 h in a concentrated sodium hypochlorite solution before usual cleaning.

Scheme 9.9

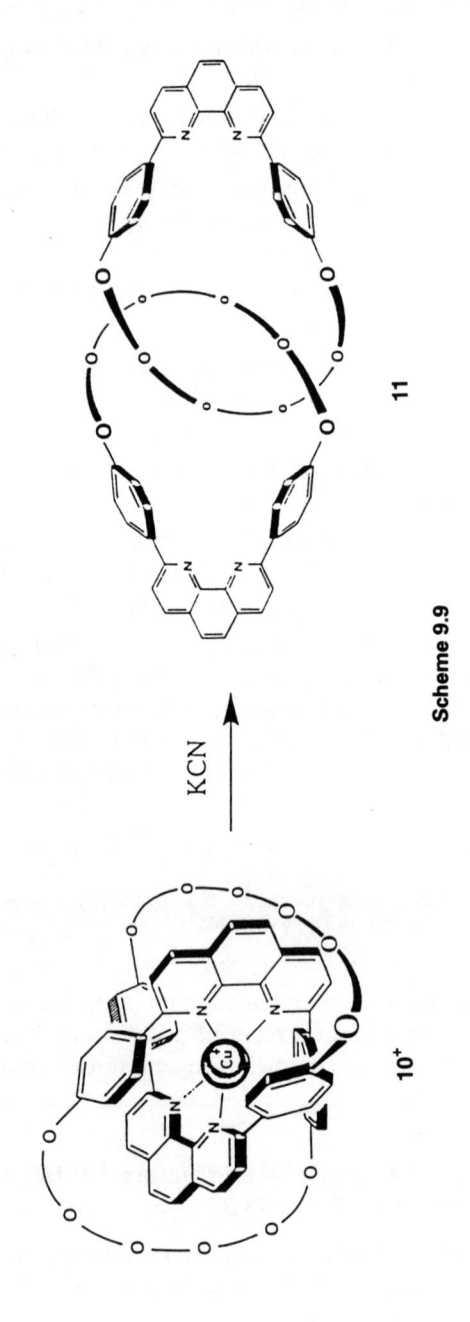

9: Transition metal

Protocol 8. *Continued*

Equipment

- Single-necked, round-bottomed flask (100 mL)
- Conical funnels
- Beaker (20 mL)
- Conical flask (100 mL)
- Single-necked, round-bottomed flask (250 mL)
- Beaker (500 mL)
- Teflon-coated magnetic stirring bars
- Magnetic stirrer
- Separating funnel (250 mL)
- Suction flask (500 mL)
- Sintered disc filter funnel

Materials

- Cu(I) catenate **10$^+$(BF$_4^-$)**, 0.956 g, 0.745 mmol
- Potassium cyanide, 1.00 g, 15.38 mmol **highly toxic**
- Reagent grade dichloromethane, 50 mL **harmful**
- Dichloromethane for extraction **harmful**
- MgSO$_4$
- Sodium hypochlorite **corrosive**

1. Weigh 0.956 g of catenate **10$^+$(BF$_4^-$)** in the 100 mL round-bottomed flask and dissolve it in 50 mL of dichloromethane.
2. Weigh out 1.00 g of KCN in a 20 mL beaker and dissolve it in 10 mL of distilled water.
3. Add the latter KCN solution to the stirred catenate solution.
4. Stir the resulting two-phase mixture at room temperature until the dark-red colour of the dichloromethane layer has totally disappeared.
5. When the reaction mixture is colourless, transfer it into a separating funnel and decant.
6. Extract the aqueous layer with dichloromethane (3 × 50 mL) and collect the organic layers in a conical flask.
7. Pour the aqueous KCN layer in the 500 mL beaker containing sodium hypochlorite.
8. Transfer the combined organic layers back into the separating funnel and wash them carefully three times with water. After each washing, discard the water by pouring it into the sodium hypochlorite solution.
9. Dry the organic layers over MgSO$_4$, filter on a sintered disc filter funnel and evaporate the solvent on a rotary evaporator. Pure catenand **11** is thus obtained as a colourless solid (0.843 g, 0.745 mmol, 100% yield).

---

3. Transition metal-templated synthesis of rotaxanes

Following the strategy represented in Scheme 9.1, the ideal precursor corresponding to (F) is a non-symmetrical diphenylphenanthroline (dpp) derivative attached to both a gold(III) porphyrin and an aromatic aldehyde (**22$^+$** of

J.-C. Chambron et al.

Scheme 9.17). The macrocycle (A) of Scheme 9.1 is the dpp-incorporating 30-membered ring **8** used earlier for making catenanes (see Protocol 5): it is sufficiently small to prevent release of the bis-porphyrin dumbbell. The preparation of **22⁺** takes place in seven steps.[18,19] One of the starting materials is 4-lithiotoluene **13**, which may be obtained by direct interaction of an excess of freshly cut lithium with 4-bromotoluene **12** in ether under argon at reflux (Scheme 9.10).[12] The resulting organolithium compound is titrated by the double titration method described by Gilman *et al.*[13] The experimental procedure is very similar to that described in Protocol 1.

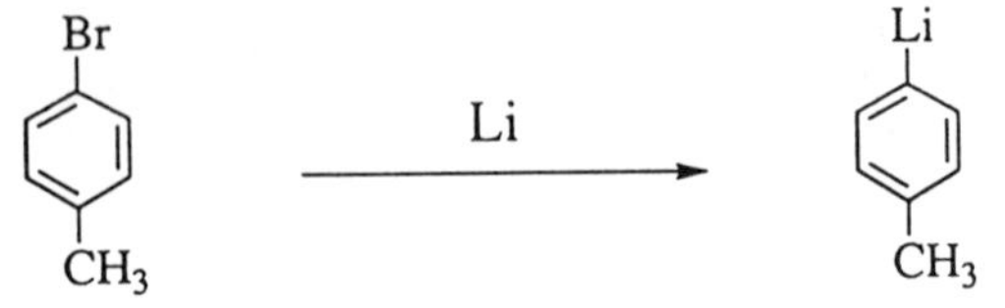

Scheme 9.10

2,9-Di(*p*-tolyl)-1,10-phenanthroline **14** is synthesized in 70% yield from 1,10-phenanthroline **3** (1 equiv) and 4-lithiotoluene **13** (5 equiv), after hydrolysis, oxidation with MnO_2 and recrystallization (Scheme 9.11). This reaction, detailed in Protocol 9, follows a literature procedure to make various polyimines substituted by alkyl or aryl groups α to the nitrogen atoms.[20]

Protocol 9.
Synthesis of 2,9-di(*p*-tolyl)-1,10-phenanthroline (Structure 14, Scheme 9.11)

Caution! Carry out all procedures in a well-ventilated hood, and wear disposable vinyl or latex gloves and chemical-resistant safety goggles.

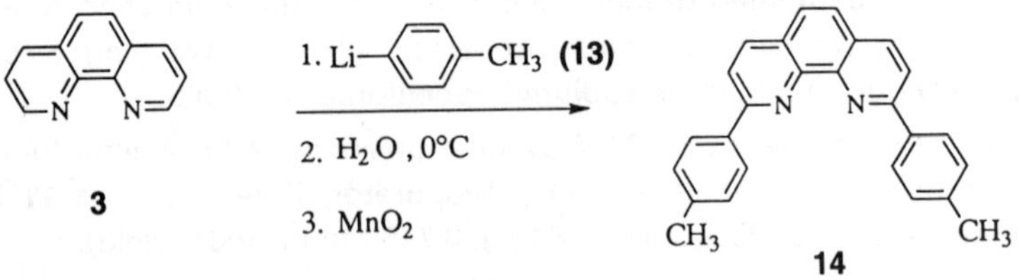

Scheme 9.11

Equipment
- Triple-necked, round-bottomed flask (250 mL)
- Water-cooled condenser
- Thermometer (–10 °C to 100 °C)
- Rubber septa
- 90°-angled adaptor with glass stopcock and hose barb
- Conical funnels
- Double-tipped steel cannula
- Syringe with needle (50 mL)
- Dual bank vacuum manifold (vacuum–argon)
- Column for chromatography
- Magnetic stirrer
- Ice bath
- Separating funnel

9: Transition metal

- Sintered glass funnels
- Suction filtering flasks
- Conical flasks
- Graduated cylinder
- Teflon-coated magnetic stirring bars
- Source of dry argon
- Powder funnels

Materials

• 1,10-Phenanthroline, monohydrate **3**, 2.62 g, 13.2 mmol	toxic
• 4-Lithiotoluene **13**, 66 mmol, 60 mL	flammable
• Dry toluene (30 mL)	harmful, highly flammable
• Manganese dioxide Merck no. 805958, 45 g	harmful
• Dichloromethane for extraction, 400 mL	irritant, harmful
• Silica gel (70–230 mesh)	
• Methanol for chromatography	toxic, highly flammable
• Toluene for chromatography and recrystallization	harmful, highly flammable
• Magnesium sulfate for drying	
• Distilled water, c. 10 mL.	

1. Dry the three-necked flask containing a stirring bar and the double-tipped steel cannula overnight in an electric oven at 110 °C.

2. Assemble apparatus used for the reaction consisting of the three-necked flask fitted with a condenser, a thermometer and a septum. Connect the top of the condenser to the dual bank via the angled adaptor.

3. To the flask, add 2.62 g of 1,10-phenanthroline, monohydrate[a] **3** and 30 mL of dry toluene.

4. Stir the suspension and degas it by three vacuum–argon cycles.

5. Cool the reaction flask in an ice bath. Add dropwise, via a double-tipped steel cannula and under argon, 60 mL of the 4-lithiotoluene solution. Adjust the rate of addition in order to maintain the reaction temperature below 30 °C. During the addition the colour of the mixture becomes first bright yellow, then dark red.

6. Stir the dark-red mixture under argon and at room temperature overnight.

7. Cool the reaction flask to 0 °C with an ice bath. Hydrolyse the mixture under argon by injecting dropwise, degassed water (c. 1 mL) with a syringe in order to keep the temperature below 15 °C. When the temperature remains stable, excess water can be added more rapidly.

8. Transfer the yellow mixture to a separating funnel and separate the two layers. Extract the aqueous layer three times with 100 mL of CH_2Cl_2. Pour the combined organic layers into a 1 L conical flask.

9. To rearomatize the resulting organic layer, add successively 15 g of MnO_2 while vigorously stirring the solution. Follow the evolution of the reaction by TLC (eluent: CH_2Cl_2:5% MeOH) since the final aromatic compound gives a more polar spot (R_f = 0.8). The reaction is completed after addition of 45 g of MnO_2.

10. Add 45 g of $MgSO_4$ to dry the mixture. Stir for 30 min, then filter on a

Protocol 9. Continued

sintered glass funnel to remove the black slurry. Wash thoroughly with CH$_2$Cl$_2$. Evaporate the filtrate to dryness to obtain crude compound **14**.

11. Recrystallize the crude product from hot toluene (c. 30 mL). Filter the pale-yellow precipitate on a sintered glass funnel and wash it with cold toluene.
12. To obtain further compound, apply the residue to a silica gel column. Elute with toluene and increase the polarity with MeOH. With toluene:5% MeOH, pure **14** is obtained. The overall yield is 70% (3.4 g).

[a] The amount of phenanthroline used in this step is calculated in order to use all the 4-lithiotoluene which has been prepared prior to this reaction. The proportion used is 5 equiv of 4-lithiotoluene per mole of phenanthroline.

Bromination of **14** is achieved by reaction with three equiv of *N*-bromosuccinimide in boiling benzene under light irradiation. This reaction gives a mixture of the desired tribromide derivative **15a** and also the dibromide and tetrabromide derivatives, **15b** and **15c** respectively (Scheme 9.12). The separation of these compounds is very difficult, and so the mixture is best used without purification for the next step. The experimental procedure is given in Protocol 10.

Protocol 10.
Synthesis of tribromide derivative (Structure 15a, Scheme 9.12)

Caution! Carry out all procedures in a well-ventilated hood, and wear disposable vinyl or latex gloves and chemical-resistant safety goggles. Benzene is a highly flammable and carcinogenic liquid.

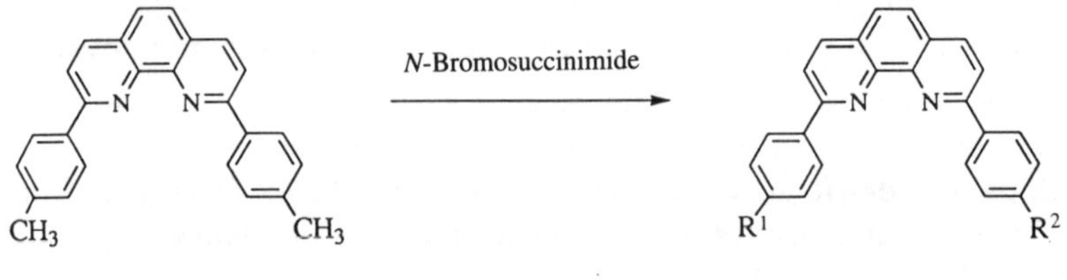

14

N-Bromosuccinimide

15a : R^1 = CHBr$_2$, R^2 = CH$_2$Br

15b : R^1 = R^2 = CH$_2$Br

15c : R^1 = R^2 = CHBr$_2$

Scheme 9.12

Equipment
- Single-necked, round-bottomed flask (50 mL)
- Water-cooled condenser
- Sintered glass funnels (porosity 4)
- Drying tube
- Hot plate magnetic stirrer
- Water bath

9: Transition metal

- Slide projector (150 W lamp, λ > 320 nm)
- Graduated cylinder (50 mL)
- Suction filtering flasks
- Conical funnel
- Teflon-coated magnetic stirring bars
- Separating funnel (250 mL)
- Conical flasks (100 mL)

Materials

- Dry benzene, 20 mL   **carcinogenic, highly flammable**
- 2,9-Di(p-tolyl)-1,10-phenanthroline **14**, 408 mg, 1.1 mmol
- N-Bromosuccinimide, 621 mg, 3.5 mmol   **irritant**
- Dichloromethane for extraction, 50 mL   **irritant, harmful**
- Silica gel, granulated, self-indicating desiccant
- Magnesium sulfate for drying

1. Put the round-bottomed flask, condenser, stirring bar and conical funnel in an electric oven overnight, at 110°C.
2. Place in the flask, 408 mg of **14**, 621 mg of N-bromosuccinimide,[a] 20 mL of benzene and a stirring bar. Place the flask in a water bath.
3. Attach a water-cooled condenser and fix on the top of the condenser a drying tube fitted with silica gel. Place the slide projector close to the flask in order to have the maximum intensity of light passing through the solution.
4. Heat the stirred yellow solution under reflux.
5. After 30 min start to irradiate the solution with the projector (λ > 320 nm). During the irradiation a white fine precipitate of succinimide appears.
6. After 20 min stop the irradiation, and allow the orange mixture to cool down.
7. Filter the mixture on a sintered glass funnel and evaporate the filtrate to dryness.
8. Dissolve the residue in CH_2Cl_2 (50 mL), pour the solution in a separating funnel and wash the organic layer three times with water (3 × 30 mL).
9. Dry the organic layer on $MgSO_4$, filter it through a sintered glass, evaporate the solvent on a rotary evaporator.
10. The crude product (550 mg) contains **15a** and two other bromide derivatives **15b** and **15c** which are very difficult to separate. It is used without further purification for the next reaction (Protocol 11).

[a] N-Bromosuccinimide should be recrystallized before use. Dissolve 20 g of N-bromosuccinimide in 1000 mL of hot distillated water (80°C). Work under the hood because of release of gaseous bromine. Rapidly cool the orange solution obtained in an ice bath to precipitate pure N-bromosuccinimide. Filter on a sintered glass funnel, wash the precipitate with iced water. Dry the compound in a desiccator under vacuum overnight with P_2O_5 as drying agent.

The crude mixture of compounds **15a**, **15b** and **15c** is treated by sodium propionate in refluxing propionic acid for 2 h leading to a mixture of the desired compound **16a** besides compounds **16b** and **16c** (Scheme 9.13). After

J.-C. Chambron et al.

chromatographic separation, 37% of the pure aldehyde-protected alcohol **16a** is obtained.

Protocol 11.
Synthesis of aldehyde-protected alcohol compound Structure 16a (Scheme 9.13)

Caution! Carry out all procedures in a well-ventilated hood, and wear disposable vinyl or latex gloves and chemical-resistant safety goggles.

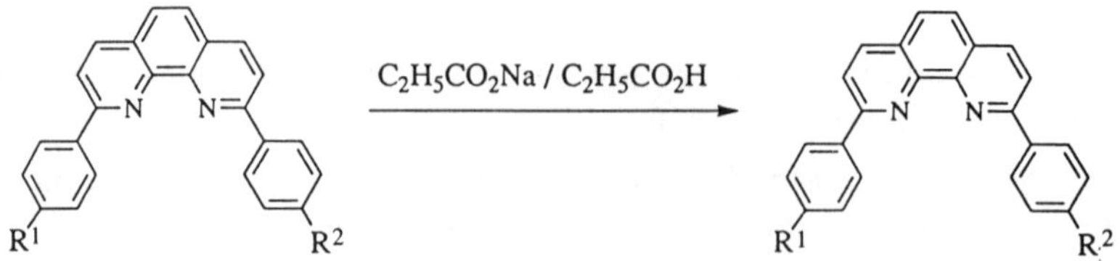

15a: $R^1 = CHBr_2$, $R^2 = CH_2Br$

15b: $R^1 = R^2 = CH_2Br$

15c: $R^1 = R^2 = CHBr_2$

16a: $R^1 = CHO$, $R^2 = CH_2OCOC_2H_5$

16b: $R^1 = R^2 = CH_2OCOC_2H_5$

16c: $R^1 = R^2 = CHO$

Scheme 9.13

Equipment

- Single-necked, round bottomed flasks (250 mL and 500 mL)
- Water-cooled condenser
- Teflon-coated magnetic stirring bars
- Conical flasks
- Suction filtering flasks
- Graduated cylinder
- Pestle and mortar
- Separating funnel (500 mL)
- Powder funnel
- Sintered glass funnels (porosity 4)
- Ice bath
- Oil bath
- Column for chromatography
- Hot plate magnetic stirrer
- Conical funnel

Materials

- Crude product containing **15a**, 15.28 g
- Sodium hydroxide, 1.46 g, 0.37 mmol   **corrosive**
- Propionic acid, 110 mL   **corrosive**
- Saturated sodium hydrogencarbonate solution, 450 mL
- Magnesium sulfate for drying
- Silica gel (70–230 mesh)
- Ethyl acetate for chromatography, *c.* 1 L   **flammable**
- Dichloromethane for extraction and chromatography, *c.* 2 L   **irritant, harmful**
- Hexane for chromatography, *c.* 1 L   **highly flammable, harmful**

1. Measure 110 mL of propionic acid in a graduated cylinder and pour the solvent into the 250 mL round-bottomed flask.

2. Cool the flask in an ice bath. Crush 1.46 g of NaOH in a mortar and add it slowly to the stirred solution.
3. When all the NaOH is dissolved, add the mixture of bromide compounds, attach the condenser and replace the ice bath by an oil bath.
4. Heat the reaction mixture under reflux for 2 h (128 °C). A white precipitate appears in the brown solution.
5. After cooling to 20 °C, filter off the NaBr on a sintered glass funnel and wash it with CH_2Cl_2. Pour the filtrate into the 500 mL round-bottomed flask, and evaporate the solvent on a rotary evaporator and the propionic acid under reduced pressure.
6. Dissolve the residue in CH_2Cl_2 (200 mL) and stir the solution vigorously. Add slowly (effervescence) a saturated solution of $NaHCO_3$ (150 mL) to neutralize the remaining acid.
7. Decant the organic layer in a separating funnel and repeat the neutralization process twice.
8. Wash the organic layer three times with water (3 × 100 mL) and dry it over $MgSO_4$. Filter the solution through a sintered glass funnel and evaporate the solvent.
9. Apply the residue on a chromatography column[a] (silica gel 300 g, CH_2Cl_2:hexane, 1:1 as eluent). Di-protected alcohol **16b** is obtained with CH_2Cl_2:hexane 75:25 as eluent. Increase the polarity to CH_2Cl_2:0–5% ethyl acetate to obtain pure compound **16a** (37% yield). CH_2Cl_2:10% ethyl acetate gives the dialdehyde compound **16c**.

[a] The crude compound contains a mixture of the desired compound **16a**, the dialdehyde **16c** and the di-protected alcohol **16b**. Since these compounds display similar polarities, the purification must be done with care.

The synthesis of the monoporphyrin **19** is achieved using Adler's reaction conditions.[21] A mixture of **16a** (1 equiv), 3,5-di-*t*-butylbenzaldehyde[18] **18** (22 equiv) and pyrrole **17** (24 equiv) is stirred and heated to reflux in propionic acid for 16 h. After work-up and chromatographic separation, two compounds are isolated: the desired monoporphyrin **19** in 20% yield (Scheme 9.14) and the major compound mesotetrakis-5,10,15,20 (3,5-di-*t*-butylphenyl)porphyrin in 24% yield, based on pyrrole in excess.

Protocol 12.
Synthesis of monoporphyrin (Structure 19, Scheme 9.14)

Caution! Carry out all procedures in a well-ventilated hood, and wear disposable vinyl or latex gloves and chemical-resistant safety goggles.

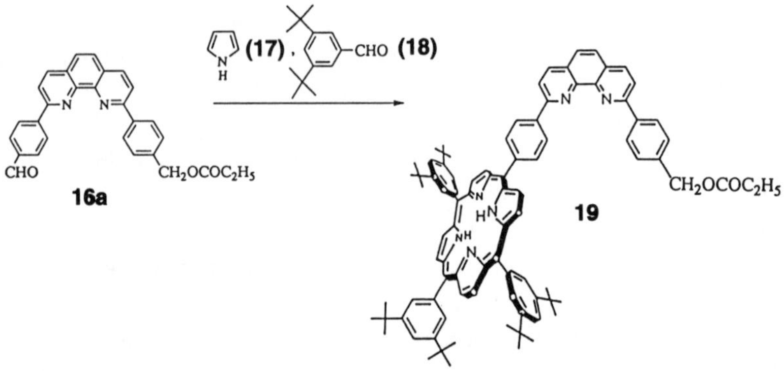

Scheme 9.14

Equipment

- Two-necked, round-bottomed flask (250 mL)
- Single-necked, round-bottomed flask (500 mL)
- Water-cooled condenser
- Teflon-coated magnetic stirring bars
- Thermometer (0 to +200 °C)
- Graduated cylinders
- Syringe with needle (2 mL)
- Conical funnel
- Separating funnel
- Suction filtering flasks
- Sintered glass funnels
- Heating mantle
- Column for chromatography
- Magnetic stirrer
- Conical flasks

Materials

- Compound **16a**, 526 mg, 1.18 mmol
- 3,5-di-*t*-butylbenzaldehyde **18**, 5.79 g, 26.51 mmol
- Pyrrole **17**, 2 mL, 28.89 mmol — **harmful**
- Propionic acid, 60 mL — **corrosive**
- 10% sodium carbonate solution, 300 mL — **irritant**
- Silica gel (70–230 mesh)
- Magnesium sulfate for drying
- Toluene, for chromatography — **harmful, highly flammable**
- Dichloromethane, for chromatography — **irritant, harmful**
- Ethyl acetate, for chromatography — **flammable**

1. In the double-necked, round-bottomed flask, place 526 mg of compound **16a**, 5.79 g of 3,5-di-*t*-butylbenzaldehyde **18** and 60 mL of propionic acid. Add 2 mL of pyrrole[a] **17** with a syringe.
2. Attach a water-cooled condenser and place a thermometer in the mixture.

3. Stir the mixture and heat under reflux for 16 h.
4. Allow the black mixture to cool down. Pour it into a 500 mL round-bottomed flask.
5. Remove as much as possible of the propionic acid under reduced pressure. To remove the remaining acid, add 100 mL of toluene (toluene:propionic acid forms an azeotrope, b.p. 80 °C) and evaporate to dryness. Repeat this process twice.
6. Dissolve the residue in CH_2Cl_2 (100 mL), pour it into a 500 mL conical flask, stir it vigorously and neutralize by adding slowly a 10% Na_2CO_3 solution (100 mL). Decant and repeat the process twice. Wash the resulting organic layer three times with water (3 × 100 mL).
7. Add 10 g of silica gel for chromatography to the organic layer to adsorb the black polymers and stir the mixture for 30 min.
8. Filter on a sintered glass funnel and wash the black solid several times with CH_2Cl_2 (5 × 50 mL), until the solvent coming through the funnel is almost colourless.
9. Dry the organic layer over $MgSO_4$, filter on a sintered glass funnel and evaporate to dryness.
10. Column chromatography on silica gel of the crude product (7 g) gives first the monomer, mesotetrakis-5,10,15,20 (3,5-di-*t*-butylphenyl)porphyrin with toluene as eluant (yield: 24% based on pyrrole in excess). The monoporphyrin **19** is eluted with toluene: 6% EtOAc. The pure compound **19** (313 mg) is obtained as a brown powder after a second chromatography on silica gel eluted with toluene. Yield: 20% from **16a**.

[a] If the pyrrole is a brown liquid instead of a pale-yellow one, pass it through an alumina column under argon to remove the polymers. Keep it under argon and protect it from light.

Hydrolysis of the protected alcohol group in **19** is realized by reaction of aqueous sodium hydroxide in *N,N*-dimethylformamide at room temperature and under argon. Chromatographic purification affords the benzylic alcohol derivative **20** in 65% yield (Scheme 9.15).

Protocol 13.
Preparation of the benzylic alcohol derivative Structure 20 (Scheme 9.15)

Caution! Carry out all procedures in a well-ventilated hood, and wear disposable vinyl or latex gloves and chemical-resistant safety goggles.

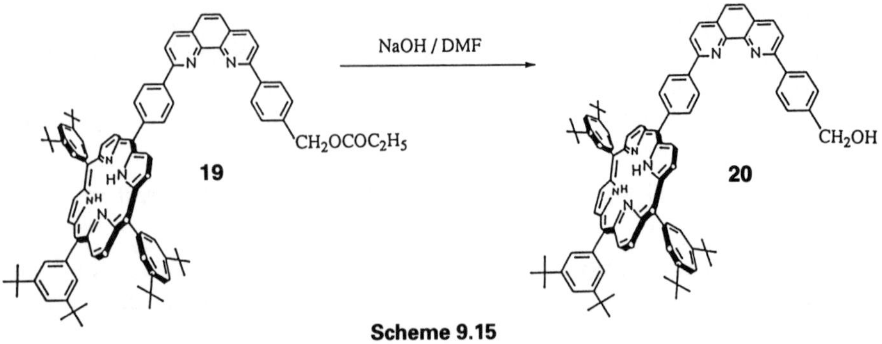

Scheme 9.15

Equipment
- Round-bottomed Schlenk flask (100 mL)
- Round-bottomed flask (500 mL)
- Teflon-coated magnetic stirring bars
- Rubber septum
- Conical funnel
- Hot plate magnetic stirrer
- Syringe with needle lock Luer (5 mL)
- Dual bank vacuum manifold (vacuum–argon)

- Graduated cylinder
- Suction filtering flask
- Sintered glass funnel
- Separating funnel (250 mL)
- Conical flasks
- Oil bath
- Column for chromatography
- Source of dry argon

Materials
- Monoporphyrin **19**, 281 mg, 0.22 mmol
- Sodium hydroxide 1.8 M, 2.8 mL, 5.04 mmol — **corrosive**
- Dry N,N-dimethylformamide, 15 mL — **harmful**
- 10% sodium carbonate aqueous solution, 200 mL — **irritant**
- Magnesium sulfate for drying
- Aluminium oxide standardized (activity II-III; 70–230 mesh)
- Silica gel (70–230 mesh)
- Toluene for chromatography — **harmful, highly flammable**
- Ethyl acetate for chromatography — **flammable**
- Dichloromethane for extraction, 300 mL — **irritant, harmful**

1. In the Schlenk flask, place 281 mg of compound **19** and 15 mL of DMF. Close the opening with a septum.
2. Degas the suspension while stirring, by three vacuum–argon cycles.
3. Heat the mixture at 60 °C until dissolution of the solid.
4. Allow the solution to cool down, then add dropwise under argon and with a syringe 2.8 mL of the sodium hydroxide solution.
5. Continue to stir for 1 h at room temperature.

9: Transition metal

6. Add 20 mL of water and transfer the mixture into a separating funnel. Extract the aqueous layer three times with 100 mL CH$_2$Cl$_2$.
7. Combine the different organic layers and wash them twice with 100 mL of a 10% Na$_2$CO$_3$ aqueous solution.
8. Dry the resulting organic layer over MgSO$_4$, filter and evaporate the maximum of solvent on a rotary evaporator. Remove *N,N*-dimethylformamide under reduced pressure.
9. Adsorb the crude product on alumina and apply it on a silica gel column chromatography. The column is eluted with toluene under argon pressure. Pure alcohol compound **20** (purple powder) is obtained with toluene containing 0 to 0.5% ethyl acetate as eluant. Yield: 65%.

To oxidize the benzylic alcohol function of compound **20**, a mixture of this compound and a large excess of manganese dioxide is stirred in dichloromethane at room temperature under argon for 1.5 h. After a short work-up, compound **21** may be obtained quantitatively (Scheme 9.16).

Protocol 14.
Preparation of the aldehyde derivative Stucture 21 (Scheme 9.16)

Caution! Carry out all procedures in a well-ventilated hood, and wear disposable vinyl or latex gloves and chemical-resistant safety goggles.

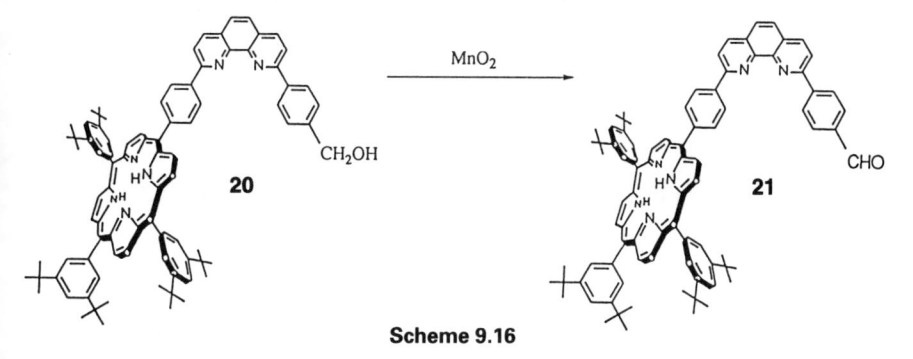

Scheme 9.16

Equipment

- Round-bottomed Schlenk flask (250 mL)
- Teflon-coated magnetic stirring bars
- Sintered glass funnel (porosity 4)
- Conical funnel
- Dual bank vacuum manifold (vacuum–argon)
- Suction filtering flask
- Conical flasks
- Graduated cylinder
- Rubber septum
- Source of dry argon
- Magnetic stirrer
- Powder funnel

Materials

- Compound **20**, 107 mg, 0.09 mmol
- Manganese dioxide, 3.9 g, 22.4 mmol **harmful**

Protocol 14. Continued

• Dichloromethane, reagent grade, 120 mL   irritant, harmful
• Magnesium sulfate for drying
• Toluene for TLC   harmful, highly flammable
• Ethyl acetate for TLC   flammable

1. Place 107 mg of compound **20**, 120 mL of dichloromethane and a stirring bar in the Schlenk flask. Close the opening with a septum.
2. Stir the mixture to dissolve the solid.
3. Degas the purple solution by three vacuum–argon cycles.
4. Under argon, add 3.9 g of manganese dioxide.
5. Stir the mixture under argon for 1.5 h. The end of the reaction can be checked by TLC (silica gel, toluene:10% EtOAc). The aldehyde gives a less polar spot than the alcohol derivative.
6. Add MgSO$_4$ to dry the mixture and to aid the removal of MnO$_2$.
7. After 10 min, filter on a sintered glass funnel and evaporate the filtrate to dryness.
8. The pure aldehyde derivative **21** is obtained as a purple powder in quantitative yield.

The free-base porphyrin **21** is metallated with KAuCl$_4$ (2.5 equiv) in refluxing acetic acid with 4 equiv of sodium acetate, affording the gold(III) porphyrin **22$^+$** in 95% yield after column chromatography (Scheme 9.17).

Protocol 15.
Synthesis of gold porphyrin (Structure 22$^+$(PF$_6^-$) Scheme 9.17)

Caution! Carry out all procedures in a well-ventilated hood, and wear disposable vinyl or latex gloves and chemical-resistant safety goggles.

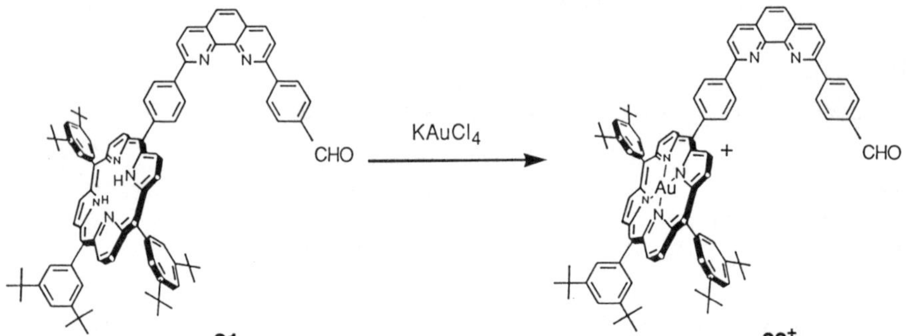

Scheme 9.17

9: Transition metal

Equipment

- Double-necked round-bottomed flask (250 mL)
- Water-cooled condenser
- Rubber septum
- Teflon-coated magnetic stirring bar
- Column for chromatography
- Heating jacket
- Magnetic stirrer
- Two 250 mL separating funnels (2 × 250 mL)
- Syringe with needle lock Luer and needle (1 mL)

Materials

- Acetic acid, 100 mL corrosive
- Porphyrin **21**, 0.307 g, 0.25 mmol
- Potassium tetrachloroaurate, 0.236 g, 0.62 mmol
- Sodium acetate, 0.079 g, 0.99 mmol
- Potassium hexafluorophosphate, saturated solution corrosive
- Sodium carbonate, 10% aqueous solution irritant
- Dichloromethane, for chromatography harmful
- Methanol, for chromatography highly flammable, toxic
- Silica gel (70–230 mesh)
- Silica gel plates for TLC

1. Assemble the two-necked, round-bottomed 250 mL flask, the heating jacket and the stirrer. Dissolve the porphyrin (0.31 g) in 100 mL of acetic acid. Add potassium tetrachloroaurate (0.236 g) and sodium acetate (0.079 g). Introduce the stirring bar, assemble the condenser and stopper the remaining opening of the flask with a septum.
2. Heat the reaction mixture to reflux for 2 d. Monitor the reaction by TLC (silica, CH_2Cl_2:CH_3OH, 100:6), taking aliquots with a syringe.
3. Remove the solvents on a rotary evaporator.
4. Add 100 mL of dichloromethane. Neutralize the mixture with 100 mL of 10% aqueous sodium carbonate.
5. Pour the mixture into a 250 mL separating funnel. Collect the organic layer in a 500 mL round-bottomed flask. Add 100 mL of a saturated potassium hexafluorophosphate solution and stir overnight with a magnetic stirrer.
6. Pour the mixture into a 250 mL separating funnel and wash the organic layer with 3 × 50 mL of water. Separate the organic layer carefully.
7. Transfer the organic solution to a 250 mL round-bottomed flask and remove the solvent on a rotary evaporator.
8. Apply the residue to a silica gel column (60 g of silica in CH_2Cl_2) and elute with CH_2Cl_2:MeOH, 100:1, to yield a red-orange solid (0.360 g, 95% yield).

The rotaxane synthesis itself starts with copper(I)-directed threading of gold(III) porphyrin **22$^+$** into macrocycle **8**.[10a] Macrocycle **8** is first complexed with Cu(I) by reaction with 1 equiv of Cu(CH$_3$CN)$_4$PF$_6$ in CH$_3$CN/CH$_2$Cl$_2$ solution.[22] Then, the gold(III) porphyrin **22$^+$**(PF$_6^-$) is added in stoichiometric amount, leading to the quantitative formation of complex **23^{2+}**(PF$_6^-$)$_2$ (Scheme 9.18), which is used in the next step without purification.

Protocol 16.
Synthesis of prerotaxane (Structure 23^{2+}(PF$_6^-$)$_2$, Scheme 9.18)

Caution! Carry out all procedures in a well-ventilated hood, and wear disposable vinyl or latex gloves and chemical-resistant safety goggles.

Equipment
- Single-necked, round-bottomed Schlenk flasks (2 × 25 mL, 3 × 10 mL)
- Teflon-coated magnetic stirring bar
- Magnetic stirrer
- Dual bank vacuum manifold (vacuum–argon)
- Syringe and needle (5 mL)
- Syringe and needle (2 mL)
- Two double-tipped steel cannulae
- Rubber septa

Materials
- Gold porphyrin 22$^+$(PF$_6^-$), 0.157 g, 0.10 mmol
- Macrocycle 8, 0.057 g, 0.10 mmol
- Cu(CH$_3$CN)$_4$PF$_6$, 0.0373 g, 0.10 mmol
- Dichloromethane, 10 mL   **harmful**
- Acetonitrile, 3 mL   **flammable, toxic**

1. Connect a 25 mL round-bottomed Schlenk flask and a 10 mL round-bottomed Schlenk flask to the vacuum line and fill them with dichloromethane and acetonitrile, respectively. Stopper the flasks with septa. Evacuate the solvents by vacuum/argon cycles (3 times).

2. Weigh 0.057 g of macrocycle, put it into a 25 mL Schlenk flask and stopper it with a septum. Connect the flask to the vacuum line. Apply three vacuum–argon cycles to the apparatus.

3. Dissolve the macrocycle in 5 mL of dichloromethane (added via syringe).

4. Connect a 10 mL Schlenk flask to the vacuum line, stopper it with a septum and subject it to three vacuum/argon cycles.

5. Weigh rapidly 0.0373 g of Cu(CH$_3$CN)$_4$PF$_6$. Transfer the powder into the flask filled with argon. Apply one more vacuum–argon cycle and dissolve Cu(CH$_3$CN)$_4$PF$_6$ in 2 mL of acetonitrile (added via syringe).

6. Transfer the solution of Cu(CH$_3$CN)$_4$PF$_6$ into the solution of macrocycle via a double-tipped steel cannula. The solution turns bright orange. Rinse the flask containing Cu(CH$_3$CN)$_4$PF$_6$ with 1 mL of CH$_3$CN. Stir the reaction mixture for 30 min.

7. Weigh 0.157 g of porphyrin. Put it into a 10 mL Schlenk flask and stopper it with a septum. Connect the flask to the vacuum line. Subject the apparatus to three vacuum/argon cycles.

8. Dissolve the porphyrin in 2 mL of dichloromethane (added via syringe). Transfer the resulting solution into the reaction mixture via a double-tipped

9: Transition metal

steel cannula. Rinse the flask containing the porphyrin with 2 × 1 mL of dichloromethane. Stir for at least 1 h.

9. Evaporate the solvents directly on the vacuum line. The product is obtained quantitatively and may be used in the next step without further purification.

The next step is shown in Scheme 9.19: prerotaxane $23^{2+}(PF_6^-)_2$, 3,5-di-*t*-butylbenzaldehyde 18^{18} and (diethyl-3,3'-dimethyl-4,4'-dipyrryl-2,2')methane 24^{23} (molar ratio 1:4:10) are mixed and stirred in dichloromethane, in the presence of trifluoroacetic acid (3:1) for 17 h. Subsequently, a large excess of *p*-chloranil (30:1) is added in order to oxidize the intermediate porphyrinogens and the reaction mixture is heated under reflux for 1.5 h. After work-up and chromatographic separation, three porphyrins are isolated: the etio-porphyrin **24**, the desired Cu(I) [2]-rotaxane $25^{2+}(PF_6^-)_2$ and the compartmental bis-copper(I) [3]-rotaxane $26^{4+}(PF_6^-)_4$. Copper(I) [2]-rotaxane $25^{2+}(PF_6^-)_2$ is isolated in 25% yield (Scheme 9.19).

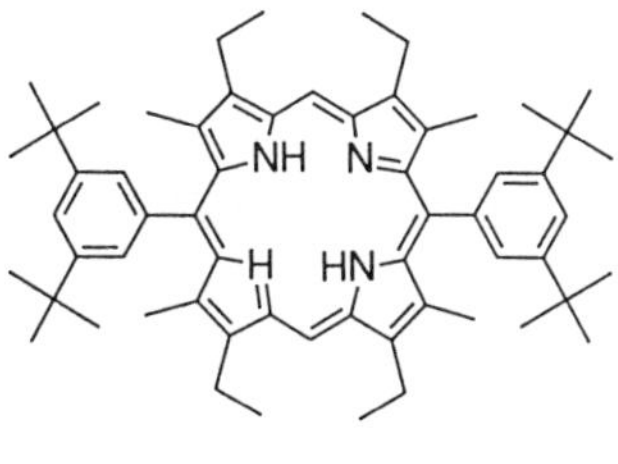

24

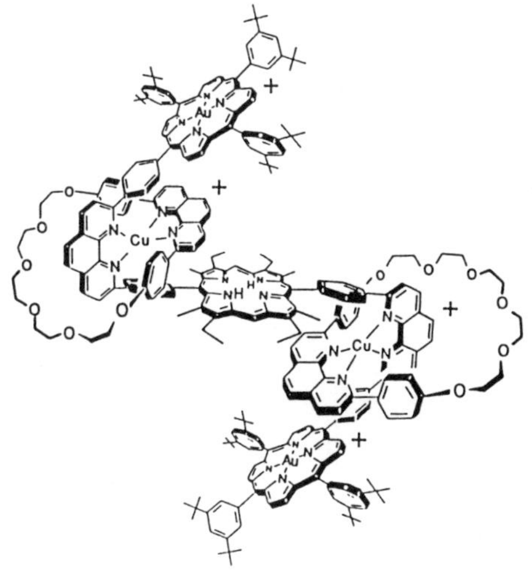

26^{4+}

Protocol 17.
Synthesis of Cu(I)-complexed rotaxanes Structure $25^{2+}(PF_6^-)_2$ and Structure $26^{4+}(PF_6^-)_4$ (Scheme 9.19)

Caution! Carry out all procedures in a well-ventilated hood, and wear disposable vinyl or latex gloves and chemical-resistant safety goggles.

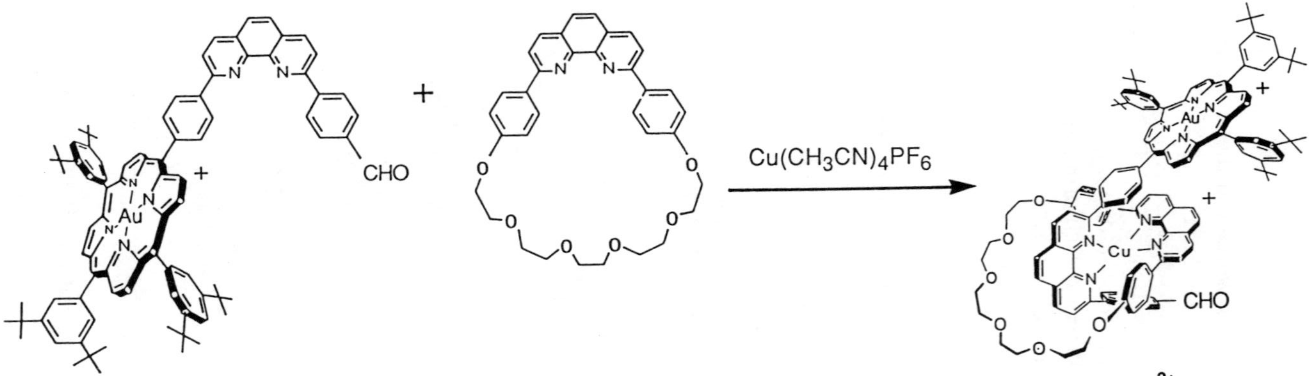

Scheme 9.18

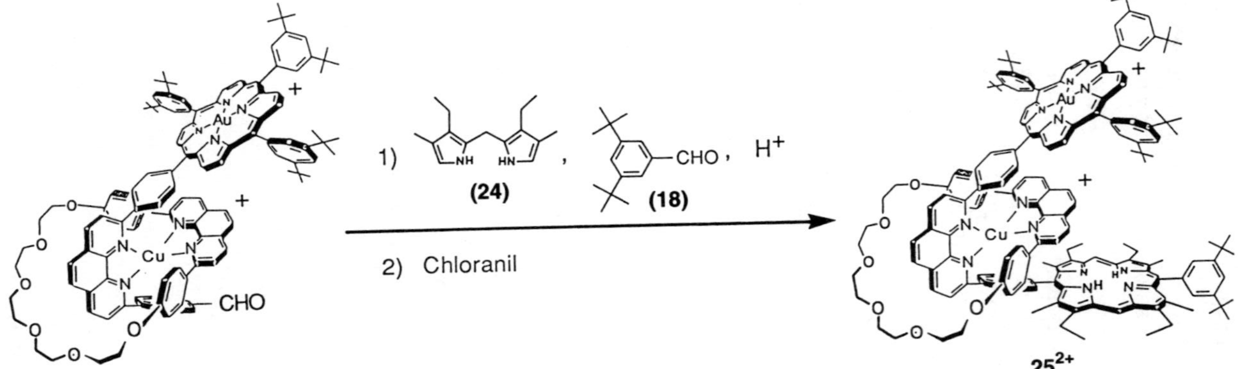

Scheme 9.19

9: Transition metal

Protocol 17. Continued

Equipment

- Double-necked, round-bottomed flask (100 mL)
- Water-cooled condenser
- Rubber septum
- 90°-angled adaptor with glass stopcock and hose barb
- Two Teflon-coated magnetic stirring bars
- Single-necked, round-bottomed Schlenk flask (250 mL)
- Single-necked, round-bottomed flask (250 mL)
- Dual bank vacuum manifold (vacuum/argon)
- Hot plate magnetic stirrer
- Separating funnel (100 mL)
- Graduated pipette (1 mL)
- Syringe with needle-lock Luer (50 mL)
- Column for chromatography
- Graduated cylinders
- Source of dry argon
- Oil bath

Materials

- Prerotaxane $23^{2+}(PF_6^-)_2$, 0.194 g, 0.10 mmol
- 3,5-di-t-butylbenzaldehyde 18, 0.091 g, 0.42 mmol
- (Diethyl-3,3'-dimethyl-4,4'-dipyrryl-2,2')methane 24, 0.250 g, 1.08 mmol
- Trifluoroacetic acid, 0.25 mL, 0.3 mmol — **harmful by inhalation, corrosive**
- p-Chloranil, 0.744 g, 2.97 mmol — **irritant**
- Dichloromethane, distilled over P_2O_5, 150 mL — **harmful**
- 10% sodium carbonate aqueous solution — **irritant**
- Saturated potassium hexafluorophosphate aqueous solution — **corrosive**
- Dichloromethane, for chromatography, c. 1700 mL — **harmful**
- Hexane, for chromatography, c. 700 mL — **highly flammable harmful**
- Methanol, for chromatography, c. 50 mL — **highly flammable, toxic**
- Silica gel (70–230 mesh)

1. Dry the double-necked 100 mL round-bottomed flask containing a stirring bar, the condenser, the adaptor, and the single-necked Schlenk flask in an electric oven at 110 °C for 2 h.

2. Assemble the oil bath and the magnetic stirrer, the double-necked 100 mL round-bottomed flask, the condenser and the adaptor. Connect the apparatus to the vacuum line and allow it to cool down under an argon flow.

3. Introduce 0.091 g of 3,5-di-t-butylbenzaldehyde 18, 0.194 g of prerotaxane $23^{2+}(PF_6^-)_2$ and 0.25 g of (diethyl-3,3'-dimethyl-4,4'-dipyrryl-2,2')methane. Stopper the flask with a rubber septum.

4. Connect the single-necked 250 mL Schlenk flask to the vacuum line and allow it to cool down under an argon flow. Pour c. 100 mL of dichloromethane into it and stopper it with a rubber septum. Evacuate the solvent by vacuum/argon cycles (3 times).

5. Transfer, via syringe, 45 mL of dichloromethane into the reaction flask.

6. Remove the septum and introduce rapidly 0.35 mL of trifluoroacetic acid with a graduated pipette.

7. Allow the reaction mixture to stir overnight at room temperature.

8. Add p-chloranil (0.744 g) to the reaction mixture and heat under reflux for 1 h.

Protocol 17. *Continued*

9. Transfer the reaction mixture to a 100 mL separating funnel and shake it with 20 mL of 10% sodium carbonate aqueous solution.
10. Collect the organic layer, pour it again into a 100 mL separating funnel, wash it with water (2 × 50 mL), and transfer it into a 250 mL round-bottomed flask. Add 50 mL of saturated aqueous KPF_6 solution and stir the mixture overnight.
11. Transfer the mixture to a 100 mL separating funnel, and wash it with water (3 × 50 mL). Collect the organic solution into a 250 mL round-bottomed flask and evaporate the solvent to dryness on a rotary evaporator.
12. Apply the residue to a silica gel column (170 g of silica in CH_2Cl_2:hexane, 50:50). Elution with 0.5% MeOH in CH_2Cl_2 affords etioporphyrin 5,15-bis-(3,5-di-*t*-butylphenyl)-3,7,13,17-tetraethyl-2,8,12,18-tetramethylporphyrin (0.213 g), which is discarded. Elution with 1.2% MeOH in CH_2Cl_2 affords crude $25^{2+}(PF_6^-)_2$ and elution with 2–3.5% MeOH in CH_2Cl_2 affords crude $26^{4+}(PF_6^-)_4$. $25^{2+}(PF_6^-)_2$ is further purified by column chromatography (42 g of silica in CH_2Cl_2:hexane, 50:50). Elution with 0.13-0.15% MeOH in CH_2Cl_2 affords $25^{2+}(PF_6^-)_2$ (0.073 g, 25% yield). $26^{4+}(PF_6^-)_4$ is further purified by column chromatography (42 g of silica in CH_2Cl_2:hexane, 50:50). Elution with 0.55% of MeOH in CH_2Cl_2 affords $26^{4+}(PF_6^-)_4$ (0.081 g, 32% yield).

Free-base porphyrin in copper(I) [2]-rotaxane $25^{2+}(PF_6^-)_2$ is metallated with $Zn(OAc)_2 \cdot 2H_2O$ in a boiling mixture of CH_2Cl_2 and CH_3OH, to afford the Cu(I) [2]-rotaxane $27^{2+}(PF_6^-)_2$ in 66% yield (Scheme 9.20).

Protocol 18.
Synthesis of Cu(I) complexed rotaxane Structure $27^{2+}(PF_6^-)_2$ (Scheme 9.20)

Caution! Carry out all procedures in a well-ventilated hood, and wear disposable vinyl or latex gloves and chemical-resistant safety goggles.

Equipment

- Two-necked, round-bottomed flask (50 mL)
- Single-necked, round-bottomed Schlenk flasks (10 mL, 20 mL, 50 mL)
- Water-cooled condenser
- 90°-angled adaptor with glass stopcock and hose barb
- Double-tipped steel cannula
- Dual bank vacuum manifold (vacuum–argon)
- Syringes with needle lock Luer (5 mL, 20 mL)
- Teflon-coated magnetic stirring bar
- Hot plate magnetic stirrer
- Oil bath
- Rubber septum
- Separating funnel (100 mL)
- Source of dry argon
- Single-necked, round-bottomed flasks (50 mL, 100 mL)
- Column, for chromatography

Materials

- Rotaxane $25^{2+}(PF_6^-)_2$, 0.066 g, 22 µmol
- $Zn(OAc)_2 \cdot 2H_2O$, 0.010 g, 45 µmol   **very hygroscopic**

9: Transition metal

- Dichloromethane, c. 85 mL — **harmful**
- Methanol, c. 15 mL — **highly flammable, toxic**
- Sodium carbonate aqueous solution (5%) — **irritant**
- Saturated potassium hexafluorophosphate aqueous solution — **corrosive**
- Hexane, for chromatography — **highly flammable, harmful**
- Dichloromethane, for chromatography — **harmful**
- Methanol, for chromatography — **highly flammable, toxic**
- Silica gel (70–230 mesh)
- Aluminium oxide standardized (activity II-III, 70–230 mesh)

1. Connect the two single-necked 50 and 20 mL Schlenk flasks to the vacuum line. Fill them with 20 mL of dichloromethane and 10 mL of methanol, respectively. Evacuate the solvents by three vacuum–argon cycles.

2. Assemble the oil bath and the magnetic stirrer, the double-necked 50 mL round-bottomed flask with the stirring bar, the condenser and the adaptor.

3. Weigh 0.066 g of rotaxane $25^{2+}(PF_6^-)_2$ and place the compound into the reaction flask. Stopper the apparatus with a rubber septum and submit it to three vacuum–argon cycles. Dissolve the rotaxane in 15 mL of degassed CH_2Cl_2 (use a syringe).

4. Put 0.010 g of $Zn(OAc)_2 \cdot 2H_2O$ into the single-necked 10 mL Schlenk flask. Stopper the apparatus with a rubber septum and submit it to three vacuum/argon cycles.

5. Dissolve the zinc acetate in 5 mL of degassed methanol (use a syringe).

6. Heat the reaction mixture under reflux. Transfer the zinc acetate solution into the reaction mixture via a double-tipped steel cannula. Continue heating for 1.5 h.

7. Transfer the reaction mixture into the 50 mL round-bottomed flask. Evaporate the solvents to dryness with a rotary evaporator.

8. Take up the residue into 50 mL of dichloromethane. Transfer the resulting solution into a separating funnel and shake it with 30 mL of 5% sodium carbonate aqueous solution. Wash the organic layer with water (30 mL). Collect the organic phase into a 100 mL round-bottomed flask and stir overnight with 25 mL of saturated KPF_6 aqueous solution.

9. Transfer the mixture into a separating funnel. Wash it with 3 × 50 mL of water. Collect the organic phase into a 100 mL round-bottomed flask and evaporate the solvent with a rotary evaporator.

10. Apply the residue to an alumina column (50 g of Al_2O_3 in CH_2Cl_2:hexane, 50:50) and elute with 0.16–0.27% MeOH in CH_2Cl_2. This yields 0.0444 g (66%) of the desired product.

Finally, the rotaxane $28^+(PF_6^-)$ is obtained in 79% yield by treatment of $27^{2+}(PF_6^-)_2$ with an excess of KCN in a ternary mixture of water, CH_3CN and CH_2Cl_2 (Scheme 9.21).

Scheme 9.20

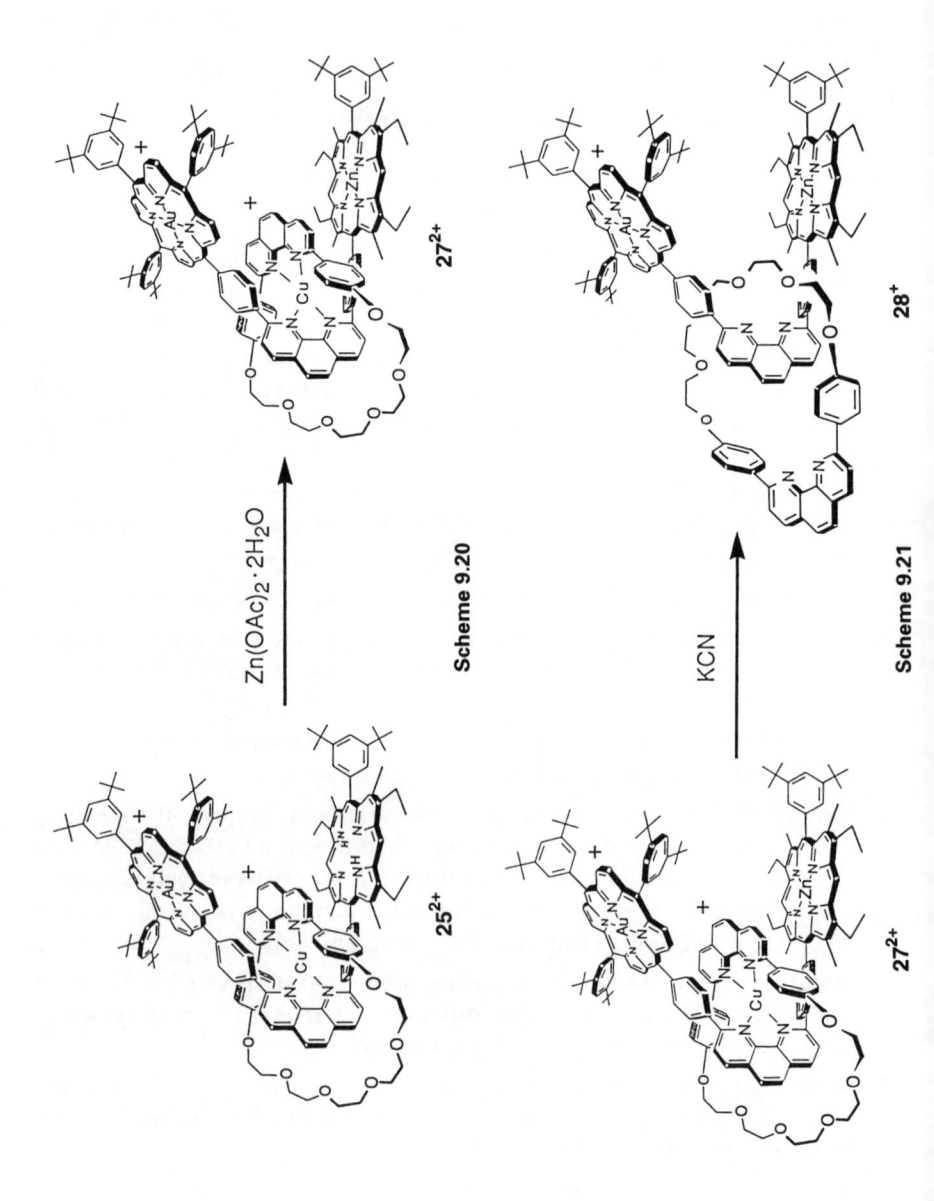

Scheme 9.21

25²⁺ →[Zn(OAc)₂·2H₂O] 27²⁺

27²⁺ →[KCN] 28⁺

Protocol 19.
Synthesis of metal-free rotaxane Structure 28⁺(PF₆⁻) (Scheme 9.21)

Caution! Carry out all procedures in a well-ventilated hood, and wear disposable vinyl or latex gloves and chemical-resistant safety goggles. Potassium cyanide is extremely dangerous. Cyanide residues, as well as glassware which has been in contact with cyanide, should be washed with a sodium hypochlorite solution.

Equipment

- Single-necked round-bottomed flask (20 mL)
- Teflon-coated magnetic stirring bar
- Magnetic stirrer
- Pasteur pipette
- 3 ml vials
- Separating funnel (100 mL)
- Single-necked round-bottom (50 mL)
- Graduated cylinder (10 mL)
- Column for chromatography

Materials

- Cu(I)-complexed rotaxane $27^{2+}(PF_6^-)_2$, 0.011 g, 3.61 µmol
- Potassium cyanide, 0.0081 g, 124 µmol — highly toxic
- Acetonitrile, 10 mL — flammable, toxic
- Dichloromethane, 25 mL — harmful
- Saturated potassium hexafluorophosphate aqueous solution — corrosive
- Aluminium oxide standardized (activity II-III, 70–230 mesh)
- Hexane, for chromatography — highly flammable, toxic
- Dichloromethane, for chromatography — harmful

1. Assemble the single-necked 20 mL round-bottomed flask with stirring bar, and the magnetic stirrer.

2. Weigh 0.011 g of rotaxane $27^{2+}(PF_6^-)_2$ and carefully transfer the compound into the reaction flask. Dissolve it in 7 mL of CH_3CN and 1.2 mL of CH_2Cl_2. Add 0.0055 g of KCN dissolved in 1 mL of water, and stir the reaction mixture for 2 h.

3. Add 0.0026 g of KCN dissolved in 0.8 mL of water. Continue stirring for 2 h.

4. Transfer the reaction mixture into the separating funnel. Dilute with CH_2Cl_2 (20 mL) and wash the organic phase with water (3 × 20 mL).

5. Collect the organic phase into a 50 mL round-bottomed flask and stir overnight with a saturated potassium hexafluorophosphate aqueous solution.

6. Transfer the mixture into the separating funnel. Discard the aqueous phase and wash the organic layer with water (3 × 20 mL). Collect the organic layer into a 50 mL round-bottomed flask and evaporate it to dryness on a rotary evaporator.

7. Apply the residue to an alumina column (7 g in CH_2Cl_2:hexane, 50:50) and elute with CH_2Cl_2. This yields 0.0081 g (79%) of the desired metal-free rotaxane.

References
1. Walba, D. *Tetrahedron* **1985**, *41*, 3161–3212 and references therein.
2. Harrison, I. T.; Harrison, S. *J. Am. Chem. Soc.* **1967**, *89*, 5723–5724.
3. Schill, G. *Chem. Ber.* **1967**, 2021–2037.
4. Hunter, C. A. *J. Am. Chem. Soc.* **1992**, *114*, 5303–5311.
5. Anelli, P. L.; Ashton, P. R.; Ballardini, R.; Balzani, V.; Delgado, M.; Gandolfi, M. T.; Goodnow, T. T.; Kaifer, A. E.; Philp, D.; Pietraszkiewicz, M.; Prodi, L.; Reddington, M. V.; Slawin, A. M. Z.; Spencer, N.; Stoddart, J. F.; Vicent, C.; Williams, D. J. *J. Am. Chem. Soc.* **1992**, *114*, 193–218.
6. Ogino, H.; Ohata, K. *Inorg. Chem.* **1984**, *23*, 3312–3316.
7. (a) Agam, G.; Graiver, D.; Zilkha, A. *J. Am. Chem. Soc.* **1976**, *98*, 5206–5214. (b) Agam, G.; Zilkha, A. *J. Am. Chem. Soc.* **1976**, *98*, 5214–5216.
8. Schill, G.; Zollenkopf, E. *Liebigs Ann. Chem.* **1969**, *721*, 53–74.
9. (a) Dietrich-Buchecker, C. O.; Sauvage J.-P.; Kern, J.-M. *J. Am. Chem. Soc.* **1984**, *106*, 3043–3045. (b) Dietrich-Buchecker, C. O.; Sauvage, J.-P. *Chem. Rev.* **1987**, *87*, 795–810.
10. (a) Chambron, J.-C.; Heitz, V.; Sauvage, J.-P. *J. Am. Chem. Soc.* **1993**, *115*, 12378–12384. (b) Ashton, P. R.; Johnston, M. R.; Stoddart, J. F.; Tolley, M. S.; Wheeler, J. W. *J. Chem. Soc., Chem. Commun.* **1992**, 1128–1131.
11. Wu, C.; Lecavalier, P. R.; Shen, Y. X.; Gibson, H. W. *Chem Mater.* **1991**, *3*, 569–572.
12. Gilman, H.; Zoellner, E. A.; Selby, W. M. *J. Am. Chem. Soc.* **1932**, *54*, 1957–1962.
13. Gilman, H.; Haubein, A. H. *J. Am. Chem. Soc.* **1944**, *66*, 1515–1516.
14. Curphey, T. J.; Hoffman, E. J.; McDonald, C. *Chem. Ind.* **1967**, 1138.
15. Rasshofer, W.; Wehner, W.; Vögtle, F. *Liebigs Ann. Chem.* **1976**, 916–923.
16. Fenton, D. E.; Parkin, D.; Newton, R. F. *J. Chem. Soc., Perkin Trans. I* **1981**, 449–454.
17. Meerwein, H.; Hederich, V.; Wunderlich, K. *Ber. Dtsch. Pharm. Ges.* **1958**, *63*, 548. Alternatively, Cu(CH$_3$CN)$_4$BF$_4$ was prepared by reduction of Cu(BF$_4$)$_2$ by copper powder (excess) in CH$_3$CN under argon at room temperature, the mixture being stirred until complete bleaching of the solution was effected. See also Ref. 22.
18. Chardon-Noblat, S.; Sauvage, J.-P. *Tetrahedron* **1991**, *47*, 5123–5132.
19. Heitz, V.; Chardon-Noblat, S.; Sauvage, J.-P. *Tetrahedron Lett.* **1991**, *32*, 197–198.
20. Dietrich-Buchecker, C. O.; Marnot, P. A.; Sauvage, J.-P. *Tetrahedron Lett.* **1982**, *23*, 5291–5294.
21. Adler, A. D.; Longo, F. R.; Finarelli, J. D.; Goldmacher, J.; Assour, J.; Korsakoff, L. *J. Org. Chem.* **1967**, *32*, 476.
22. Kubas, G. J. *Inorg. Synth.* **1990**, *28*, 90–92.
23. (a) Young, R.; Chang, C. K. *J. Am. Chem. Soc.* **1985**, *107*, 898–909. (b) Bullock, E.; Johnson, A. W.; Markham, E.; Shaw, K. B. *J. Chem. Soc.* **1958**, 1430–1440.

A1

List of suppliers

Acros Chimica
see **Janssen Chimica**

Aldrich Chemical Co. Ltd
France: 80 Rue de Lozais, BP 701, 38070, St Quentin, Fallavier Cedex, LYON. Tel. 74822800
Germany: SAF, Messerschmitt Strasse 17, D-7910 Neu-Ulm. Tel. 0731-9733640
Japan: Aldrich Chemical Co., Inc. (Japan), Kyodo-building Shinkanda, 10 Kandamikura-chou, Chiyoda-ku, Tokyo 101. Tel. 03-54344712
UK: The Old Brickyard, New Road, Gillingham, Dorset SP8 4JL. Tel. 0800-717181
USA: PO Box 355, Milwaukee, WI 53201. Tel. 0414-2733850

Alfa
France: Johnson Matthey SA, BP 50240, Rue de la Perdix, Z1 Paris Nord LL, 95956 Roissy, Charles De Gaulle Cedex. Tel. 1-48632299
Germany: Johnson-Matthey GmbH, Zeppelinstrasse 7, D-7500 Karlsruhe-1. Tel. 0721-840070
UK: Catalogue Sales, Materials Technology Division, Orchard Road, Royston, Herts. SG8 5HE. Tel. 0763-253715
USA: Alfa/Johnson Matthey, PO Box 8247, Ward Hill, MA 01835-0747. Tel. 0508-5216300

BDH
UK (Head Office and International Sales): Merck Ltd, Merck House, Poole, Dorset BH15 1TD. Tel. 0202-665599

Boulder Scientific Co
USA: 598 3rd Street, PO Box 548, Mead, CO 80542. Tel. 03035354494

Fluka Chemika-BioChemika
France: Fluka S.a.r.l., F-38297 St Quentin, Fallavier Cedex, Lyon. Tel. 74822800
Germany: Fluka Feinchemikalien GmbH, D-7910 Neu-Ulm. Tel. 0731–729670
Japan: Fluka Fine Chemical, Chiyoda-Ku, Tokyo. Tel. 03-32554787
UK: Fluka Chemicals Ltd, Gillingham, Dorset SP8 4JL. Tel. 0747-823097

List of suppliers

USA: Fluka Chemical Corp., Ronkonkoma, NY 11779-7238. Tel. 0516-4670980

FMC Corporation, Lithium Division
Japan: Asia Lithium Corporation (ALCO), Shin-Osaka Daiichi-Seimei Building 11F, 5-24, Miyahara 3-Chome, Yodogawa-Ku, Osaka. Tel. 06-3992331
UK and Mainland Europe: Commercial Road, Bromborough, Merseyside L62 3NL. Tel. 051-3348085
USA: 449 North Cox Road, Gastonia, NC 28054. Tel. 0704-8685300

Heraeus
Germany: Alter Weinberg, D-7500, Karlsruhe 41-Ho. Tel. 0721-4716769

ICN Biomedicals/K and K Rare and Fine Chemicals
France: ICN Biomedicals France, Parc Club Orsay, 4 rue Jean Rostand, 91893 Orsay Cedex. Tel. 1-60193460
Germany: ICN Biomedicals, GmbH, Mühlgrabenstrasse 10, Postfach 1249, D-5309, Meckenheim. Tel. 02225-88050
Japan: ICN Biomedicals Japan Co., Ltd, 8th Floor, Iidabashi Central Building, 4-7-10 Iidabashi, Chiyoda-ku, Tokyo 102. *Tel. 03-32370938*
UK: ICN Biomedicals, Ltd, Eagles House, Peregrine Business Park, Gomm Road, High Wycombe, Bucks. HP13 7DL. Tel. 04940443826
USA: ICN Biomedicals, Inc., 3300 Hyland Avenue, Costa Mesa, CA 92626. Tel. 0800-8540530

Janssen Chimica (Acros Chimica)
Mainland Europe, Central Offices, Belgium: Janssen Pharmaceuticalaan 3, 2440 Geel. Tel. 014-604200
UK: Hyde Park House, Cartwright Street, Newton, Hyde, Cheshire SK14 4EH. Tel. 0613-244161
USA: Spectrum Chemical Mgf. Corp., 14422 South San Pedro Street, Gardena, CA 90248. Tel. 0800-7728786

Johnson Matthey Chemical Products
France: Johnson Matthey SA, BP 50240, Rue de la Perdix, Z1 Paris Nord LL, 95956 Roissy, Charles De Gaulle Cedex. Tel. 48632299
Germany: Johnson-Matthey GmbH, Zeppelinstrasse 7, D-7500 Karlsruhe-1. Tel. 0721-840070
UK: Catalogue Sales, Materials Technology Division, Orchard Road, Royston, Herts. SG8 5HE. Tel. 0763-253715
USA: Alfa/Johnson Matthey, PO Box 8247, Ward Hill MA 01835-0747. Tel. 0508-5216300

List of suppliers

Kanto Chemical Co., Inc.
Japan: 2-8, Nihonbashi-honcho-3-chome, Chuo-ku, Tokyo 103. Tel. 03-2791751

Lancaster Synthesis
France: Lancaster Synthesis Ltd, 15 Rue de l'Atome, Zone Industrielle, 67800 Bischheim, Strasbourg. Tel. 05035147
Germany: Lancaster Synthesis GmbH, Postfach 15 18, D-63155 Mülheim am Main. Tel. 0130-6562
Japan: Hydrus Chemical Inc., Tomitaka Building, 8-1, Uchikanda 2-chome, Chiyoda-ku, Tokyo 101. Tel. 03-32585031
UK: Lancaster Synthesis, Eastgate, White Lund, Morecambe, Lancashire LA3 3DY. Tel. 0800-262336
USA: Lancaster Synthesis Inc., PO Box 1000, Windham, NH 03087-9977. Tel. 0800-2382324

Merck
Germany: Promochem, PO Box 101340, Mercatorstrasse 51, D-46469 Wesel. Tel. 0281-530081
Japan: Merck Japan Ltd, ARCO Tower, SF, 8-1, Shimomeguro-1-chome, Merguro-ku, Tokyo 153. Tel. 03-54344712
UK: Merck Ltd, Merck House, Poole, Dorset BH15 1TD. Tel. 0202-669700
USA: Gallard Schlesinger Companies, 584 Mineola Avenue, Carle Place, NY. Tel. 0516-3335600

Mitsuwa Scientific Corp.
Japan: 11-1, Tenma-1-Chome, Kita-ku, Osaka 530. Tel. 06-3519631

Nacalai Tesque, Inc.
Japan: Karasuma-nishi-iru, Nijo-dori, Nakagyo-Ku, Kyoto 604. Tel. 075-2315301

'Normag' Labor und Verfahrenstechnik GmbH
Germany: Postfach 1269, Feldstrasse 1, D-6238 Hofheim a. Ts.

Prolabo
UK etc.: See **Rhône-Poulenc**
France: 12, Rue Pelee, BP 369, 75526, Paris Cedex 11. Tel. 1-49231500

Rhône-Poulenc
France: Rhône-Poulenc SA, 25 Quai Paul Donmer, F-92408 Courbezoie Cedex. Tel. 1-47681234

List of suppliers

Germany: Rhône-Poulenc GmbH, Staedelstrasse 10, Postfach 700862, Frankfurt Am Main 70. Tel. 069-60930
Japan: Rhône-Poulenc Japan Ltd, 15 Kowa Building Annexe, Central PO Box 1649, Tokyo 107. Tel. 03-35854691
UK: Rhône-Poulenc Chemicals Ltd, Laboratory Products, Liverpool Road, Barton Moss, Eccles, Manchester M30 7RT. Tel. 061-7895878
USA: Rhône-Poulenc Basic Chemicals Co., 1 Corporate Dr., Shelton, CT 06484. Tel. 0203-9253300 *or* Rhône-Poulenc Inc., Fine Organics, CN 7500, Cranbury, NJ 08512-7500. Tel. 0609-860400

Riedel de Haen
Germany: Wunstorfer Strasse 40, Postfach, D-3016 Seelze 1. Tel. 05137-7070

Scientific and Medical Products Ltd
UK: Shirley Institute, 856 Wilmslow Road, Didsbury, Manchester M20 2SH.

Strem
France: Strem Chemicals, Inc., 15 Rue de l'Atome, Zone Industrielle, 67800 Bischheim. Tel. 88625260
Germany: Strem Chemicals GmbH, Querstrasse 2, D-7640 Karlsruhe. Tel. 0721-75879
Japan: Hydrus Chemical Co., Tomikata Building, 8-1, Uchkanda, 2-Chome, Chiyoda-ku, Tokyo 101. Tel. 03-2585031
UK: Fluorochem Limited, Wesley Street, Old Glossop, Derbyshire SK13 9RY. Tel. 0457-868921
USA: Strem Chemicals, Inc., Dexter Industrial Park, 7 Mulliken Way, Newburyport, MA 01950-4098. Tel. 0508-4623191

Tokyo Kasei Kogyo Co., Ltd
Japan: 3-1-13, Nihonbashi-Honcho, Chuo-ku, Tokyo 103. Tel. 03-38082821
UK: Fluorochem Limited, Wesley Street, Old Glossop, Derbyshire SK13 9RY. Tel. 0457-868921
USA: PCI America, 9211 North Harbourgate Street, Portland, OR 97203. Tel. 503-2831681

Whatman Scientific Ltd
UK: 20 St. Leonard's Road, Maidstone, Kent ME16 0LS

Index

*Page entries in **bold** refer to protocols on those pages*

acidity 2, 8, 25, 28, 152
alkylation
  nitrogen 4–**6**, 10, 20
  oxygen **155**, **159**, **164**, 199, **203**
aza crowns 1–23
aza-oxa crowns 25–47
aza-thia crowns 59–70

N-bromosuccinimide 228

calixarenes 145–73, 177, 199–204
  lower rim functionalisation **155**, **157**, **159**, **161**, **163**
  upper rim functionalisation 164, **167**, **169**, **171**
calixspherands 199–205
catenands 207
chiral macrocycles
  aza crowns 13, 14
  crown ethers 71, **78**, **80**
complex formation
  aza crowns 1
  calixspherands 199, 200, **203**, **204**
  catenands 207, 208, 218–19
  crown ethers 71, 80, 82
  cryptands 93, 104
  porphyrins 236, 242
  rotaxanes 236
  spherands 175, 180
  thia crowns 52, 53
  torands 119, 140
condensation reactions
  acetone 18
  arylaldehydes **124**, **136**, 232
  Bredereck's reagent 136, **139**
  diethyl malonate 11, 13–14, **15**, 25
  formaldehyde 5, 20, 67, **68**
  glyoxal 18
conformation 53, 67, 158–9, 167, 192
conjugate addition 4, 18, 130–**2**, 140
crown ethers 71–91
cryptands 93–118

deprotection
  aza crowns 2, **3**
  aza-oxa crowns **29**

aza-thia crowns 61, **62**
calixarenes **165**
catenands 213, 234
crown ethers 83, **87**
cryptands **112**, **116**
hemispherands 192, **193**

ether bond formation **76**, **78**, **80**, **85**, **88**,

hemispherands 181–99
high dilution
  apparatus 58, 95
  reactions 25, **54**, **65**, **88**, 95, **97**, 100, 181, 216

iodination 169, 215
ion exchange 30, **69**, 104, 107, 117

kinetic control 146–7, 152
kinetic stability 1, 176, 200

lithiation 177, **179**, 181–2, 188–**90**, **209**, 226

medium dilution reactions
  aza crowns 2
  aza-oxa crowns 25, **29**, 31
  crown ethers **80**
  thia crowns 49–**52**, **56**
metal decomplexation 18, 44, **204**, 223–4, 243–**5**

nitration 170–1

oxidation
  alcohols 128, 211, **226**, 235
  alkenes 133, **134**, **137**
  pyridines **121**

porphyrins 231, **232**
preorganisation 175
protecting groups
  aza crowns 5–10

Index

protecting groups (*cont.*)
  aza-oxa crowns 28, 31, **35**, 39, 42
  crown ethers **73**, **76**
  cryptands 106–7, **108**, **110**
  porphyrins 236, 242
  spherands 177
  torands 123, 136
pyridine
  ring containing 39, **44**, 55
  ring formation 123, 130, **132**, 140
pyrilium salts 140, 196

rearrangement 123, 196
reduction
  diborane 25, **26**, **65**, **102**
  LiAlH$_4$ 25, **99**
  Zn–HCl **68**
Richman Atkins 28–37, 61
ring formation
  aza crowns 2, 12, 14–15, 18–22, **97**
  aza-oxa crowns **26**, **29**, **32**, **35**, **40**, **44**
  calixarenes 147, 149
  catenands **217**

crown ethers **78**, **80**, **88**
cryptands **94**, **115**
hemispherands **186**, **192**, 196
porphyrins **232**
spherands **178**
thia crowns 50–1, **56**–9
torands **140**
rotaxanes 207, 225–45

Schiff base 44
sulfonation 167
Suzuki coupling 188

templated cyclisations 17, 20, 39, **68**, 80, **147**, **149**, **186**, 188–9, 199, 207, 221, **240**
thermodynamic control 146
thia-crowns 49–59
Thorpe–Ingold effect 53
topology 207
torands 119–43
tosylation **2**, **9**, **84**, **109**, 110–**11**